D1557504

Cambridge Imperial and Post-Colonial Studies Series

General Editors: **Megan Vaughan**, Kings' College, Cambridge and **Richard Drayton**, Corpus Christi College, Cambridge

This informative series covers the broad span of modern imperial history while also exploring the recent developments in former colonial states where residues of empire can still be found. The books provide in-depth examinations of empires as competing and complementary power structures encouraging the reader to reconsider their understanding of international and world history during recent centuries.

Titles include:

Sunil S. Amrith
DECOLONIZING INTERNATIONAL HEALTH
India and Southeast Asia, 1930–65

Tony Ballantyne
ORIENTALISM AND RACE
Aryanism in the British Empire

Robert J. Blyth
THE EMPIRE OF THE RAJ
Eastern Africa and the Middle East, 1858–1947

Roy Bridges *(editor)*
IMPERIALISM, DECOLONIZATION AND AFRICA
Studies Presented to John Hargreaves

L.J. Butler
COPPER EMPIRE
Mining and the Colonial State in Northern Rhodesia, c.1930–64

Hilary M. Carey *(editor)*
EMPIRES OF RELIGION

T.J. Cribb *(editor)*
IMAGINED COMMONWEALTH
Cambridge Essays on Commonwealth and International Literature in English

Michael S. Dodson
ORIENTALISM, EMPIRE AND NATIONAL CULTURE
India, 1770–1880

B.D. Hopkins
THE MAKING OF MODERN AFGHANISTAN

Ronald Hyam
BRITAIN'S IMPERIAL CENTURY, 1815–1914
A Study of Empire and Expansion
Third Edition

Robin Jeffrey
POLITICS, WOMEN AND WELL-BEING
How Kerala became a 'Model'

Gerold Krozewski
MONEY AND THE END OF EMPIRE
British International Economic Policy and the Colonies, 1947–58

Sloan Mahone and Megan Vaughan (*editors*)
PSYCHIATRY AND EMPIRE

Javed Majeed
AUTOBIOGRAPHY, TRAVEL AND POST-NATIONAL IDENTITY

Francine McKenzie
REDEFINING THE BONDS OF COMMONWEALTH 1939–1948
The Politics of Preference

Gabriel Paquette
ENLIGHTENMENT, GOVERNANCE AND REFORM IN SPAIN
AND ITS EMPIRE 1759–1808

John Singleton and Paul Robertson
ECONOMIC RELATIONS BETWEEN BRITAIN AND
AUSTRALASIA 1945–1970

Kim A. Wagner (*editor*)
THUGGEE
Banditry and the British in Early Nineteenth-Century India

Jon E. Wilson
THE DOMINATION OF STRANGERS
Modern Governance in Eastern India, 1780–1835

Cambridge Imperial and Post-Colonial Studies Series
Series Standing Order ISBN 0–333–91908–4 (Hardback) 0–333–91909–2 (Paperback)
(*outside North America only*)

You can receive future titles in this series as they are published by placing a standing order. Please contact your bookseller or, in case of difficulty, write to us at the address below with your name and address, the title of the series and the ISBN quoted above.

Customer Services Department, Macmillan Distribution Ltd, Houndmills, Basingstoke, Hampshire RG21 6XS, England

---

Empires of Religion

Edited By

Hilary M. Carey
Professor of History, University of Newcastle, NSW

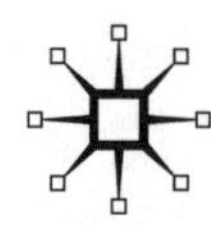

First published 2008 by
PALGRAVE MACMILLAN

Palgrave Macmillan in the UK is an imprint of Macmillan Publishers Limited, registered in England, company number 785998, of Houndmills, Basingstoke, Hampshire RG21 6XS.

Palgrave Macmillan in the US is a division of St Martin's Press LLC, 175 Fifth Avenue, New York, NY 10010.

Palgrave Macmillan is the global academic imprint of the above companies and has companies and representatives throughout the world.

Palgrave® and Macmillan® are registered trademarks in the United States, the United Kingdom, Europe and other countries.

ISBN-13: 978–0–230–20880–3 hardback
ISBN-10: 0–230–20880–0 hardback

This book is printed on paper suitable for recycling and made from fully managed and sustained forest sources. Logging, pulping and manufacturing processes are expected to conform to the environmental regulations of the country of origin.

A catalogue record for this book is available from the British Library.

A catalog record for this book is available from the Library of Congress.

10 9 8 7 6 5 4 3 2 1
17 16 15 14 13 12 11 10 09 08

Printed and bound in Great Britain by
CPI Antony Rowe, Chippenham and Eastbourne

Contents

Abbreviations

BBC	British broadcasting Corporation
BCC	British Council of Churches
BWM	Board of World Mission
CBMS	Conference of British Missionary Societies
CMAI	Christian Medical Association of India
CMC	Christian Medical College
CRO	Commonwealth Relations Office
CUSO	Canadian University Service Overseas
DWO	Division of World Outreach
EMMS	Edinburgh Medical Missionary Society
FMCNA	Foreign Missions Conference of North America
GAA	Gaelic Athletic Association
GEB	General Education Board
GMS	Glasgow Missionary Society
IAI	International African Institute
ICCLA	International Committee on Christian Literature for Africa
IMC	International Missionary Council
IPF	Imperial Protestant Federation
LMS	London Missionary Society
MU	Mothers' Union
NCC	National Christian Council
SCM	Students Christian Movement
SMS	Scottish Missionary Society
SPG	Society for the Propagation of the Gospel
UCD	University College Dublin
UN	United Nations
UPC	United Presbyterian Church
WCC	World Council of Churches
WFMS	Woman's Foreign Missionary Society
WSCF	World Student Christian Federation

Preface

When I visited Dublin for the first time at the end of 2004, it was at the beginning of a two-year term as Keith Cameron Professor of Australian History at University College Dublin. As the Aircoach made its slow progress across the city from north to south towards UCD, I made note of the landmarks as they materialized through the window. The tourist sites were familiar enough from picture books and my hasty and imperfect historical reading: there was O'Connell Street, named after the nationalist leader; now we were passing the GPO, the main scene of the Easter Rising of 1916; the River Liffey looked much smaller than I was expecting, and Trinity College rather larger; why, I wondered, was the Bank of Ireland housed in such a grand building and why did it appear to be doing service as a traffic island? Later I would head to the countryside and tour monasteries, castles and round towers: Glendalough, Trim and Clonmacnoise. I viewed all these with great interest as one does any new place, but they felt no more or no less intriguing than, say, Westminster Abbey, Stonehenge or Edinburgh Castle: they were all bits of the picturesque British Isles, that in Ireland I was to learn to call 'these islands'. But passing through St Stephen's Green, I felt an unexpected pang of recognition as we passed the familiar crest and nameplate of the religious order whose schools I had attended in the Australian cities of Perth, Melbourne and Sydney during a peripatetic childhood. A little later, I had another pang as we passed the Mater Hospital – that must be run by the Sisters of Mercy who had commissioned me to write the history of their hospital at North Sydney; St Vincent's must be another hospital – that would be run by the Sisters of Charity. With what was probably an over-confident sense of familiarity, I felt sure that I could anticipate what the function, class and social context of these institutions and others which I assumed lay scattered through the suburbs. This stemmed in no way from expatriate Irish internationalism, which for Australians tends to beat with an irregular and unsentimental pulse: I was a foreigner on my first visit to Ireland; my Irish surname came courtesy of my husband, the son of Irish emigrants; my mother's family, all the Irish blood I could claim, had come to the colony of New South Wales via India 150 years ago courtesy of the British army; my father was a New Zealander of Scottish descent. What I was identifying were not therefore Irish roots but religious ones: Dublin was the template for the urban and cultural geography of my colonial subculture, a religious rather than a political or national metropole. What an excellent place, I felt myself decide before we reached UCD in Dublin's leafy southern suburbs, to have a conference on religion and imperialism.

More concrete plans for this conference developed in 2005 when Hugh McLeod visited Sydney to attend the 20th International Congress of the Historical Sciences during which we participated in a number of sessions which the Ecclesiastical History Society shared with the Australia-based Religious History Society. I wanted to ask Anne O'Brien and Catherine Hall to keynote the conference; Hugh thought this was a good idea and agreed to invite John MacKenzie and John Wolffe. At one point we were joined by Andrew Porter, also in Sydney attending the Congress, in discussing the conference theme, which had begun to solidify around the idea of 'Empires of religion'. This sounded impressive, he agreed, 'but what exactly does it mean?' This was sobering and both Hugh and I felt the need to ensure that the conference and this set of proceedings did not dissolve into the metaphorical mists. We therefore decided that the theme of the conference would focus on the forms of religious imperialism in Britain and the colonies of Greater Britain, paying less attention to the foreign missionary movement which has its own vast literature and which there seems no particular need to extend. Missions have found their way into this volume nonetheless and indeed it has proven impossible to do imperial religious history without them.

The conference was held at UCD in the Global Irish Institute from 20–21 June 2006. It was made possible with financial support from the UCD Centre for Australian Studies, the UCD School of History and Archives, the Micheál Ó Cléirigh Institute and a Discovery grant from the Australian Research Council. For their contributions to the conference and/or this volume I thank Hugh McLeod, Kate Breslin, Bernard Carey, Beatrice Carey, Howard Clarke, Mary Daly, Judith Devlin, Sarah Feehon, Deana Heath, Brian Jackson, Edward James, Jane Koustas, Michael Laffan, Emma Lyons, Peter Martin, Susan O'Reilly and the Australian Ambassador to Dublin, Anne Plunkett. Michael Strang from Palgrave Macmillan has been supportive since visiting me in Dublin and expressing interest in the volume. The Palgrave reader made many useful suggestions, including a plan for the overall arrangement of the chapters, which I have silently incorporated.

Hilary M. Carey
University of Newcastle, NSW

Notes on Contributors

Fiona Bateman completed her PhD in 2003 at University College Galway on the subject of Ireland's Spiritual Empire. She currently works in NUI (National University of Ireland) Galway, where she has recently been appointed as postdoctoral researcher on the 'Globalisation, Empire and Culture' strand of the Texts, Contexts, Cultures Project at the Moore Institute.

Esther Breitenbach is a Research Fellow in the School of History, Classics and Archaeology at the University of Edinburgh. She has written widely on women in Scotland, including on women's history in nineteenth- and twentieth-century Scotland and on women in contemporary Scottish politics.

Ruth Compton Brouwer teaches history at King's University College at The University of Western Ontario, London, Canada. She has published extensively on women and missions and is now focusing on transitions from missions to secular development work.

Hilary M. Carey is a Professor of History at the University of Newcastle, NSW. From 2004–2006 she was Keith Cameron Professor History at University College Dublin. She is currently preparing a history of colonial missions to British settlers.

Peter Clayworth is a freelance historian based in Wellington, New Zealand. He has previously worked as the Historian for New Zealand's Department of Conservation. and as a Research Officer for the Waitangi Tribunal. He has researched extensively on Maori–Pakeha relationships and intellectual history in New Zealand.

Peter Cunich is Associate Professor in the Department of History at the University of Hong Kong and is currently researching a biography of Archbishop Roger Bede Vaughan.

Catherine Hall is Professor of Modern British Social and Cultural History at University College London. Since the 1990s she has been working on questions of race, nation, empire and identity and has published extensively including *Civilising Subjects: Metropole and Colony in the English Imagination 1830–1867* (2002) and a collection edited with

Sonya Rose, *At Home with the Empire: Metropolitan Culture and the Imperial World* (2006). Her current research focuses on Macaulay and the writing of history.

John MacKenzie is Honorary Research Professor of the Research Institute of Irish and Scottish Studies at the University of Aberdeen and Honorary Professor at the Centre for Environmental History and Policy at the University of Stirling. He currently holds an AHRB grant to conduct research on the Scots in South Africa.

John McAleer is the Curator of Eighteenth-Century Imperial and Maritime History at the National Maritime Museum, Greenwich. He is currently involved in developing the Museum's exhibition and research programmes on various aspects of imperial and maritime history, with a particular focus on the British engagement with Africa and the Caribbean in the late-eighteenth and early-nineteenth centuries.

Anne O'Brien teaches history at the University of New South Wales. Her most recent major publication is *God's Willing Workers: Women and Religion in Australia* (UNSW Press, 2005). She is currently writing a history of religion, welfare and 'problem populations'.

Tadhg Ó hAnnracháin is a Senior Lecturer in the School of History and Archives at University College Dublin where his current research is on the European Counter–Reformation.

Elizabeth E. Prevost is Assistant Professor of History at Grinnell College, Iowa. She is currently revising a book project on women's missionary encounters in colonial Africa and their impact on British feminism.

John Stuart teaches history at Kings College London and Kingston University. He is writing a book on British Protestant missionaries and the 'end of empire' in Africa.

Shurlee Swain is a Professor of the Australian Catholic University and a Senior Research Fellow in the School of Historical Studies at the University of Melbourne. Her chapter has its origins in a project funded by the Australian Research Council entitled Child-Race-Nation-Empire.

John Wolffe is Professor of Religious History and Associate Dean for Research and Postgraduate Policy in the Faculty of Arts at The Open University. He has published extensively on anti-Catholicism, evangelicalism and religious influences in British national identities.

1
Introduction: Empires of Religion

Hilary M. Carey

> Religion is the only bond, the only pledge of perpetuity, the one only source of unity among mankind.[1]
>
> (Archdeacon Henry Manning, 17 March 1846)

Histories

At the height of the imperial age church people liked to argue that religion and the British empire were inseparable – that the visible, commercial and political empire was woven into the fabric of another, invisible, country – a spiritual empire. But, then as now, historians and politicians have not been convinced that the empire had significant religious roots. Although religious interpretations of imperial history have undergone a revival in recent years, it is probably going too far to argue, with Tony Ballantyne, that religion can now be found at 'the very centre of understandings of British colonialism', and that this has come about because of the recent wave of post-colonial and ethnographic studies.[2] On the contrary, religion is still only a minority concern within the broader field of imperial history if it is measured by the space afforded to it in the five-volume *Oxford History of the British Empire*, where religion is largely confined to the contributions of Andrew Porter.[3] There is rather more religion in the companion volumes to the same series, including those on missions, Ireland and Australia.[4] But if a new imperial religious history is emerging, it remains both patchy and lacking in consensus – and it is not easy to summarize neatly. Debate continues to rage on at least four fronts, with historians from the older and the newer imperial history significantly divided on many points.[5]

The first of these debates concerns the relationship between missions and imperialism and colonialism.[6] Disagreement has focused on whether missions facilitated the advance of European empires or were essentially religious and apolitical in nature; others have emphasized that the issues are complex and that while some missionaries supported the imperial project, others can be found among its fiercest critics.[7] It seems reasonable to give some weight to the views of missionaries themselves, who repeatedly prioritized the religious over the political and 'civilizing' aspects of their work. But such protestations grow less convincing in the light of the ongoing historical research.[8] In this volume, for example, John MacKenzie gives an account of Scottish missionary work in the Eastern Cape arguing that missionaries were deeply engaged at every stage of the opening of the frontier and the subsequent negotiations with African Christians. The imperial significance of widely divergent missionary ideologies is also emphasized in the papers by Esther Breitenbach for Scottish missionary literature, Shurlee Swain in relation to child rescue and Fiona Bateman for Irish missions in the twentieth century.

The currently contested status of missionary history flows in part from its origins in the moral critique of imperialism, which began in the late nineteenth century but erupted following the break up of the empire after the Second World War. Never entirely without critics, the old enthusiasm for empire, amounting in some cases to a kind of civil religion with its own rituals, priestly castes and sacred sites, began to seem dubious to the point of immorality.[9] The cultural critic Edward Said was particularly influential for his denunciation of imperial ideologies which were seen to infantilize, demonize and feminize non-Western cultures and peoples as 'other'. 'Orientalism', the word he coined for this process, entered the lexicon of post-colonial studies of empire, much of it written by de-colonized and post-colonial intellectuals in Western universities.[10] In their most celebrated book, Jean and John Comaroff extended the Orientalist paradigm to the analysis of the London Missionary Society's mission to the Tswana of southern Africa, arguing that missions were essential to the hegemonic establishment of imperial rule in Africa because of their control of education and the representation of newly colonized peoples.[11] European critics of both Said and the Comaroffs have tended to argue with earlier historians that religion is simply too marginal to the major imperial action (which was handled by guns and trading blocs rather than bibles, schools and missions) to bear the weight of condemnation placed upon it. Despite what Susan Thorne has recently called the 'missionary betrayal' of Christianity,[12] missionaries have also continued to find defenders, especially in the post-colonial churches, where nostalgia and filial piety generates ongoing respect for founding clergy and there is an understandable tendency to downplay imperial origins and celebrate national beginnings.[13]

A second, more fruitful, discussion concerns the nature of the relationship between religion and British national identity, a topic which lies a little outside the main field of imperial history but is highly relevant to it. Linda Colley has argued that the British were not a natural people but were rather yoked together because of the 1707 Act of Union which bound Scotland to England and Wales under 'one Protestant ruler, one legislature and one system of free trade'.[14] Two powerful forces worked to weld the union throughout the eighteenth century and into the nineteenth century, the age of empires: the forces were religion and war. For Colley, the British were able to set aside their differences and focus on enemies abroad because they had a shared identity as Protestants with a destiny to overcome Catholic forces in Spain, Portugal and France. Stewart J. Brown has also suggested that religious arguments – or at least the idea of Providence – continued to play a vital role in the promotion of the imperial ideal throughout the nineteenth century.[15] While historians have argued for the continuing significance of religion in bolstering national identity in nineteenth-century Britain,[16] there has been less interest in how religion worked to hold together the imagined British World, especially its settler colonies.

A third strand of inquiry flows from the growing appreciation that the British Empire was as much a cultural as an economic and political construction. While commercial understanding of the growth of empire has remained paramount, this has led to a general withdrawal from the earlier view that the empire happened by chance, as a by-blow of trade, that it was in some respects 'accidental', as Sir John Seeley suggested.[17] Historians such as Cannadine, though criticized for trivializing the oppressive and racist reality of British imperial rule, have drawn appropriate attention to the way in which the empire was performed as much as imposed. Jane Samson has argued that Cannadine might have strengthened his thesis if he had looked at the important ways in which religion created and sustained the empire.[18] To the extent that there is a new imperial religious history, historians such as Susan Thorne and Jeffrey Cox have been leading the way in providing nuanced interpretations of the complex ways in which religion was, and to some extent still is, entangled with other imperial networks and relationships.[19]

Finally, there are a range of new interpretations of the imperial past influenced by theories of gender, race and class, and the post-colonial inquiry into the 'mutual constitution' of colonizers and the colonized. There is increasing recognition that the imperial relationship, in religious matters as much as any others, did not flow in one direction from London to the peripheries of empire. The key work here is Catherine Hall's *Civilizing Subjects*, a study which focuses on the work of Baptist missionaries in Jamaica, and which uncovers the ways in which Jamaican and British Baptists were mutually transformed in the course of the long process of conversion and the political and religious campaigns to end the slave trade.[20] Such studies recognize that women played a key role in the cultural constitution of empire, particularly through their influence and numerical strength in metropolitan religious organizations. It is now argued that the empire was a gendered domain that was constituted just as much 'at home' as it was abroad, and that the masculine work of soldiers, traders and colonial administrators was sustained by the feminine work of missionary auxiliaries, readers, letter writers and tenders of the fireside.[21] In Chapter 10, in this volume, John McAleer examines a transforming relationship of this kind between colonizer and colonized in his reading of Alison Blyth's missionary journal. Mrs Blyth was shocked by not only the debauched standards of British settlers in Jamaica but also by their Anglican clergy.[22] Other theory-rich studies of church institutions, making use of the tools of gender and class analysis, include Susan Thorne's work on home missions and Congregational foreign missions.[23] More directly influenced by Said and the central notion of the mutual entanglement of colonizer and colonized, Peter van den Veer has attempted a double history of modern, secularizing Britain and its relationship with India, conceived as a religious and medieval society open to civilization and modernization.[24] Linguistic theory is also beginning to be applied to the study of the language of missionary writing, for example by Anna Johnston on the Australian missionary, Lancelot Threlkeld.[25]

But despite the emergence of a rich, contested historiography, there is still not much common ground between traditional and post-colonial approaches to imperial religious history.[26] Few scholars on either side of the theoretical divide appear to be reading each other's books: the 'tensions of empire', to quote the title of an important collection edited by Frederick Cooper and Ann Laura Stoler, include ongoing tensions in the historiography of religion and empire.[27] There is also room for new research. One significant omission, particularly because of the contrast it makes to the wealth of studies of missionaries and the churches that rose up in their wake in the foreign mission fields of Africa, India, the West Indies and the Pacific, are studies of the settler churches which give due weight to the significance of religion in the overall constitution of Britain's settler colonies. The essays in this collection make a contribution to this project, organized around three themes: 'religious metropoles', 'colonies and mission fields', and 'post-colonial transformations'; we begin in the sixteenth century and continue to the second half of the twentieth when the empire has faded but the colonial churches continued a vigorous, post-imperial existence.

Creating spiritual metropoles

The first British Empire was established at a time when states across Europe were seeking to impose religious uniformity on the general principle of *cuius regio eius religio*[28] (whose realm – his religion). Following the Peace of Augsburg (1555), the German Empire and later much of Europe was divided into mutually hostile Protestant and Catholic camps according to the religious persuasion of ruling princes. Yet, as Tadhg Ó hAnnracháin suggests in this volume, in Chapter 2, questions remain as to why, in states as disparate as Ireland and Bohemia, there were successful attempts to resist the imposition of uniform state religion, Protestantism in the case of Ireland and Catholicism in the case of Bohemia. In the British Isles, it can be argued that the ideal of a unified Protestant nation was only ever partially realized: multiple ethnic and religious tensions were persistently expressed throughout the British realms, and colonies and missions were ever spaces for free spirits. As John MacKenzie comments, it is not so much that these divergent religious modalities re-emerge in the midst of the trials and opportunities of the imperial frontier and its missions, colonies and dominions, they were never actually submerged. As John Wolffe shows, the empire and its white settler colonies were a significant forum for both Protestant loyalism and anti-Catholic rhetoric. His study focuses on the Imperial Protestant Federation in the 1890s and 1900s which he sees as the most successful of a range of attempts to win the British Empire for militant Protestantism. While he concludes that ultimately the Victorian Empire was the vehicle for the deconstruction of the British Protestant nation, the strength of residual Protestant anti-Catholicism is testimony to the vigour of an earlier national ideal whose seed was planted at the height of the Counter Reformation.

The establishment of the cultural and political ascendancy of the metropole – whether London, Edinburgh or Dublin – was a labour of the clerical and intellectual classes. As Catherine Hall outlines, the project of establishing a more moral British Empire was a task taken up by men and women such as Zachary Macaulay and his son Thomas Babington Macaulay whose lives were played out in the company of those conservative, religious and politically active families into which they were married. Such families set the standard for the empire, ensuring that activity in the colonial peripheries, including spiritual ones, was measured by the home gauge.

There is no space to do more than sketch a few elements of the frame in which the national and colonial churches developed in the Victorian age.[29] There were, of course, two established churches in Britain, the episcopal Church of England and Ireland and the presbyterian Church of Scotland;[30] there was a majority Catholic Church in Ireland and vigorous dissenting churches throughout the British islands. After the disruption of the Church of Scotland in 1843 and the denominational separation of the Wesleyan Methodists from the Church of England, dissenting Protestants and free Presbyterians not only formed a majority in some parts of Britain, but they were also active in missionary societies and in articulating the notion of Protestant identity which was a significant component of imperial ideology.[31] Until the 1830s, the Church of England and Ireland was established officially in the colonies and recognized as part of the civil establishment by the endowment of clergy and church lands and the occasional appointment of colonial bishops. From the 1840s, however, changes in government policy towards the churches overseas as well as the looming reality of large-scale emigration led to a transformation in the rhetoric and action relating to colonial and foreign missions.[32]

As representative of the changing spirit of these times, we might take note of a meeting which was held in London in the Mansion House on 17 March 1846, in which Charles Blomfield, bishop of London, and the Lord Mayor of London hosted a gathering of city notables and senior clerics. The speakers' objective was to establish a fund in aid of emigrants and settlers, a rather modest proposal but one that was assaulted by volley after volley of grand speeches urging the need to extend the religious basis of the empire.[33] The longest speech was delivered by the then Archdeacon, later Cardinal, Henry Manning (1807–1892) who also got to deliver some of the most quotable lines: 'Empires are not things of chance: they are expressions of Divine will, of that continuous law of God's Providence by which the kingdoms of the world are ruled.'[34] From India, from the Cape of Good Hope, from Australia and Tasmania, Manning provided testimonies to the lack of clergy and inadequate spiritual services which, he suggested, were nothing less than a challenge to Providence. Empires, he went on, are not bound together by language, legislation or armies and would inevitably fail unless they were provided with some higher purpose:

> [U]nless our Commerce and our Empire be based upon somewhat that has perpetuity in itself, the day will come when the smoke of our colonial cities shall go up to heaven, and the wreck of our merchandize shall strew the strand of every shore. And what then, is that one only basis of perpetuity? It is the Church, and the Gospel of our Redeemer. I will not dwell upon it. It is too obvious. Religion is the only bond, – the only pledge of perpetuity, – the one only source of unity among mankind.

It is an extract from this quotation which heads this chapter and it encapsulates the Anglican imperialism of the high church party in the 1840s. There is no need to expand on the other speeches but it is salutary to find that the collection which followed them is unlikely to have assured Manning that the fate of the Empire was secure: it came to a little over one hundred pounds, and for the fund itself, only ten individuals came forward to subscribe as much as £100. The message from the commercial part of the city would seem to be that they had rather more confidence than the Archdeacon that the Empire would continue to cohere without – or despite – the binding power of religion. It isn't really relevant, but possibly somewhat ironic that five years later, on 6 April 1851, Manning himself had left the established Church of England for the one true Catholic and Apostolic church in communion with Rome.[35]

Whatever their denominational allegiance, the Erastian notion that religion, particularly religion prescribed by the established Church of England, was actually necessary for the maintenance of the empire was not something that would have been met with universal agreement anywhere in Britain or the colonies in the mid-nineteenth century. While, as Brian Stanley has demonstrated, dissenters and evangelicals warmed to the idea that the extension of Protestantism to the Americas was ordained by Providence under a special dispensation to the British nation,[36] the heightened rhetoric favoured by Manning at the Mansion House meeting would have appealed to a relatively small number such as the supporters of the colonial missionary ideals of the Society for the Propagation of the Gospel (SPG).[37] Outside that tradition, Anglican claims for imperial dominion were viewed with suspicion or downright hostility: distaste for the established church and its imperial claims was one factor that prompted the rebellion of the American colonists; to the north, the Church of Scotland was confident in the superiority of the presbyterian mode of church governance and was active in the conduct of its own missionary work; elsewhere dissenting sects had been successful in creating circuits and chaplaincies to convicts, emigrants and heathen. The empire had provided unprecedented opportunities for Catholic soldiers, clergy, emigrants and settlers away from the penal conditions and rural poverty that distressed Catholics in Britain. Established pretensions generally received short shrift throughout the colonies even where endowments and other privileges for the established clergy had been set in place. Religious pluralism rather than Anglican precedence triumphed in the settler colonies of Canada, the Cape, New Zealand and Australia, especially in sites such as systematically colonized South Australia where dissenters were warmly received.[38] Governments too were reluctant to appoint Anglican bishops because of the enormous cost of maintaining the clerical and physical establishments to which they claimed title. Nevertheless, as Rowan Strong has argued, Anglican aspirations for an imperial church underwent a remarkable transformation in the early decades of the nineteenth century.[39] The speeches that Manning and Blomfield made on behalf of the Emigrants' Spiritual Aid Fund were something of a re-run of the even grander addresses delivered on the occasion of the launch of Blomfield's much more successful venture into privatizing the funding of the colonial church, the Colonial Bishoprics Fund, established in April 1841.[40] The period from 1840–1850 is described by Cnattingius as an 'epoch-making' one which set the stage for the emergence of Anglicanism as a world church: 'The heads of the Church of England had shown that, as builders of a spiritual empire, they were worthy to take their place besides the builders of the political empire.'[41] Yet the arrival of colonial bishops was not generally welcomed by the missionary societies who were already active and successful in missionary fields such as New Zealand and India and had no reason to imagine that bishops would do a better job.[42]

While the Anglican Communion was successfully seeking private funding for the expansion of its churches and episcopal hierarchy in the empire, the other denominations did not stand idle. Although the British Parliament had been centralized in London after the Act of Union in Westminster, in many significant ways the established and dissenting churches remained rooted to the four nations from where – as the papers by Esther Breitenbach, John MacKenzie and Ruth Compton Brouwer discuss in relation to Scotland and missionary and settler churches in South Africa and Canada, and Fiona Bateman discusses in relation to Ireland – they were to be highly successful in launching imperial religious ventures of their own. Scottish foreign missionary societies were formed as early as those in southern Britain. After 1843, there were three major Scottish churches supporting missions and they all operated to advance British imperial and Scottish national pride. Breitenbach shows that missionary literature represented Scots as 'moral' empire builders led by figures such as David Livingstone and Mary Slessor who were icons for the entire international missionary movement. Alison Bythe is a striking example of a Scottish missionary wife who records her disappointment at the moral degeneration of the sons of Scotland from whom more was to be expected but who were corrupted by the inherent immorality of the slave economy. Her previously unrecorded journal was acquired in 2005 by the British Empire & Commonwealth Museum in Bristol and John McAleer has transcribed and analysed it for the first time in his chapter (Chapter 10) for this volume.

Of the other Protestant denominations, Susan Thorne has examined the extent to which Congregationalist missions acted as vehicles for Liberal British sentiment and aspirations.[43] There were significant differences between the forms of imperialism espoused by different political parties and denominations. The Methodists were consistently pro-imperial and anti-Catholic and their foreign and colonial missionary societies were the most internationally active of the dissenting congregations. When John Wesley declared that the world was his parish, he was speaking of an obligation which he felt arose directly from scripture, 'to declare unto all that are willing to hear the glad tidings of salvation'.[44] However, that religious impulse could only be translated physically and politically by preaching wherever English was spoken, including across the Irish Sea which Wesley crossed 40 times. Methodists spread across the Atlantic and the Pacific, down the imperial trading and communication networks with great rapidity.

Perhaps the case of Irish Catholicism is even more remarkable than the Wesleyans given their poverty and the social and humanitarian catastrophe of the Irish famine of 1845–1849, and the depopulation and mass emigration which followed it. Even before the passage of the Catholic Emancipation Act (1829) which allowed Catholics to hold public office, Catholic chaplains had been appointed to imperial military and penal establishments, such as British Army bases in India and the convict settlements in Australia, where there were significant numbers of Roman Catholics. However, this did not occur without some

Protestant protest at a perceived papist takeover of the empire as John Wolffe demonstrates. While training for colonial and missionary clergy was provided by most denominations either at home or abroad, of all the missionary training institutions in the Empire none was as active and successful as All Hallows' College in Drumcondra, Dublin. Founded in 1842, its first priests were sent out at the outset of the famine; eventually it was to train over 4000 fiercely Irish priests for South America, South Africa, India, Canada, Australia, the West Indies, New Zealand, the United States, England, Scotland and Wales.[45] The anti-Catholics were probably right to fear the strength of these Irish-trained clergy in the colonies. Irish religious imperialism, as Peter Cunich shows, was asserted not only in relation to Protestants and other heretics and heathen, but also against the English Benedictines who were bold enough to attempt to establish a colonial province in New South Wales in the teeth of the majority Irish Catholic population. The Irish were slower to send out foreign missionaries than the Scots mostly because this was the responsibility of other non-Irish religious orders including the English Mill Hill Missionaries. However, Fiona Bateman's chapter (Chapter 13) considers the long-delayed creation of an Irish Catholic missionary movement in the twentieth century.

Funding for religious work overseas came partly from government in the case of the established Churches of England and Scotland and colonial chaplains of all denominations, but by the second half of the century, it came mostly from private fund-raising either by missionary auxiliaries at home, or newly enriched philanthropists in the colonies. All the major denominations supplied clergy to the colonies through the agency of colonial and foreign mission societies, with the SPG in England and the missionary committees of the Church of Scotland aided in this work by government grants. Where they had the numbers, colonial Presbyterians asked and received state subvention for schools, churches and chaplains which was their right as members of Britain's second established church.[46] In Canada, the larger denominations also benefited from the abolition of the Canadian clergy reserves in 1843 (there were also clergy reserves in New South Wales and Tasmania) when funds were redistributed to other denominations including Catholics – once again to the annoyance of the anti-Catholic forces.[47] The Church of Ireland was to be disestablished in 1869 and the Church of Wales in 1920, but ideas of disestablishment had long been anticipated in the colonies which in this respect acted as laboratories for religious experiments back home. It was the decision by the Privy Council in the Colenso case (1866) which established that there could be no legal connection between Canterbury and her colonial churches once the colonies had achieved responsible government: the churches, and the colonial bishops, were on their own. The first Lambeth Conference was held the following year to assert the doctrinal supremacy of Canterbury but it was too late; the bishops had, as it were, already bolted and it was too late to lock the church door.[48]

Nevertheless, let us return to Manning's point: is religion essential to empire and does it provide a force for cultural cohesion which is stronger than commercial advantage or military duress? Is religion, in his words, the 'one only source of unity'? To many of his contemporaries, this might well have sounded like a rather dangerous and theologically suspect proposition.[49] Scripture is generally unsympathetic to empires of all kinds and especially to the emperor worship which was the norm in Hellenistic and near-Eastern empires of the ancient world and against which the scriptural people of Israel were forced to struggle. In the Reformation, anti-Catholic polemic borrowed freely from texts such as 'Revelation 13: 1–8' to denounce the papal monarchy and its claims for imperial dominance throughout Christendom. In the 'age of empires' which followed, religion was associated most closely with the first European powers to establish themselves in the New World, namely Catholic Spain and Portugal.[50] Prior to the French Revolution and the Napoleonic wars, the French Empire was also associated overtly with the determination to spread the religion of the imperial power, again Catholicism. Overall, the imposition of religious uniformity by military means was associated with the most reviled aspects of militant Catholicism and was not something that would naturally endear itself to freedom-loving Protestants.

These reservations about empire and about linking the church too closely with secular power are evident in both the contemporary and the current debates about the morality of missions and the role of the church in the world, and they have a very ancient pedigree, as Adrian Hastings has shown.[51] It is also important to stress that although the churches and their clergy could supply propaganda in favour of empire, they might just as easily be found subverting colonial power structures for Christian purposes.[52] Apart from the SPG and Society for the Promotion of Christian Knowledge (SPCK), Protestant missionary societies only really began their work at the end of the eighteenth century when they were swept up in the twin forces of the Enlightenment and the evangelical revival. Missions could be represented as part of a new kind of imperial expansion, that of an enlightened Protestant power which sought to bring order, rationality and the benefits of European civilization to the word.[53] The Scottish missionary societies founded at the end of the eighteenth century, many run by local committees of devout lay evangelicals, were contemporary with and a direct product of the Scottish enlightenment with all its attendant enthusiasm for education, reform and discovery.[54] It was this kind of enlightened religion, increasingly maintained by private fund-raising rather than direct government endowment that characterized the colonial churches of Greater Britain. Establishment in the form of endowed bishoprics, clergy reserves and places in the legislature was a feature of some colonies but did not survive the 1840s. The effective separation of church and state from the mid-nineteenth century in the colonies meant that churches, like most of the missionary societies, were voluntary associations; competition between religious communities for the financial and emotional support of their adherents was the norm. Like other empires, the British Empire became characterized by the diversity of its religious practice, albeit with established churches at home continuing to extend support to the daughter churches overseas.

If no one (other than Manning) would wish to argue that religion was necessary for empires, can we ask the question in a modified way: is religion *useful* for imperial expansion? Here, there is much more historical support with many scholars willing to accept that religion has generally provided essential cultural justification for imperial expansion. On the other hand, since the time of Gibbon, some historians at least have seen religion of all kinds, but especially Christianity, as more of a hindrance than a help to imperial culture and government. Pocock has shown that Gibbon's anti-clericalism arises from currents in Enlightenment thought which saw religion as inherently irrational and effeminate. Gibbon was himself rather in favour of imperialism, or at least the British version of it with which his work is contemporary.[55] He published the first volume of the *Decline and Fall of the Roman Empire* in 1776, the year of the American Declaration of Independence, six years after Captain James Cook claimed the east coast of Australia for the British crown on his first voyage of exploration, and he published the final volume in 1788, the year Captain Arthur Philip settled in Port Jackson with his load of convicts and free men including one chaplain,[56] to establish the first colony on the last continent to be incorporated into a European empire. Cook's voyages, Gibbon tells us, unlike other imperial ventures: 'were inspired by the pure and generous love of science and of mankind'.[57] Of course, religion also had its Enlightened manifestations in the settlement of Australia, as John Gascoigne has argued very well.[58] But despite respectable arrangements for colonial chaplains, at first exclusively of members of the established church and later Presbyterians, Catholics and Methodists as well, there was a notable lack of enthusiasm for the institutional forms of religion in the first Australian colonies which was the despair of those bishops who were eventually enthroned there, courtesy of letters patent (some of them) and the Colonial Bishoprics Fund.

Piety was linked to commerce in the establishment of British colonies in the Americas, but this was essentially a rather low key business as Louis B. Wright outlined in a small work first published in 1943: chaplains travelled with the first Elizabethan adventurers and the clergy consistently argued that God had set aside parts of the Western hemisphere just for the English.[59] The idea that commercial success was a mark of God's favour was a notion that initially had the support of some, but by no means all, British missionaries, but this support did not survive the nineteenth century.[60] It is probably significant that in both 1841 and 1846, the strongest arguments in favour of religion expanding with the British Empire were made to allow this expansion to occur independently of the government by providing an independent source of funding. Whatever a saintly man might choose to say about money and commerce and the need to avoid too much intermingling of religious and commercial purposes, money was as necessary to the regimes of religion as it was to other functions of empire.[61]

Mission fields

Beyond the metropole, as far as the churches were concerned, the empire was made up of colonial and mission fields from which a rich harvest was always to be anticipated. Large-scale emigration to new centres of settlement in British North America, Australia, New Zealand and the Cape Colony created new challenges and opportunities for religious institutions, much of it little studied these days. Historians have been energetic in charting the establishment of missions in places where Europeans remained in the minority of the population, but in the case of the relationship between settler churches and the metropole there is still all the work to do. Surveys such as the newly emerging volumes of the *Cambridge History of Christianity* make comparative history much more enticing,[62] and there are dense survey histories of most of the national churches, but, as we have already noted, imperial history still tends to prioritize secular issues and the wealth of studies of missions and world Christianity reflect Victorian prejudices which saw colonial work and home missions as less prestigious than foreign missions.

Anxieties about the lesser standing of religious work in the colonies, or indeed most other colonial activity, readily translated into heightened enthusiasm for the imperial ideal. Elizabeth Prevost provides a rich study of the dizzy rhetorical reaches achieved by the Anglican Mother's Union in Madagascar, not on the face of it a promising site for a vision of imperial maternal sisterhood. Her chapter reviews current work on gender, mission and empire as she investigates the empire's largest Anglican women's organization, indeed possibly the largest women's organization of any kind with the possible exception of the much more extensively studied Women's Christian Temperance Union.[63] Yet despite the unadulterated enthusiasm for their maternal subjection to Britain, she argues that in the particular context of Madagascar, the extravagant language of imperial and maternal affiliation included a genuine vision of international female solidarity across empire, class and race.

Australia was one of the widest domains of Greater Britain, and the one with the smallest indigenous population; its frontier was still moving until at least 1928, the date of the last well-attested massacre by government police of indigenous Australian people in the Northern Territory. Anne O'Brien provides an analysis of how the different perspectives of two churchmen from these northern regions, Gilbert White (1859–1932), the founding bishop of the Anglican Diocese of Carpentaria, and James Noble (1876?–1941), an Aboriginal deacon and missionary, shed light on the vexed relationship between inclusion and hierarchy at the heart of missionary endeavour. Australia's northern frontier was opened after the establishment of responsible government in Queensland and the seeding of its missions and dioceses, aided by the SPG, staffed in rural areas by the bush brothers, is a good example of the way in which imperial religious expansion rolled on well after the direct connection with the British metropole had been broken. White's life reveals the complex ways in which churchmen could simultaneously resist and aid the ends of Empire. An outspoken critic of the brutality of northern expansion, he nevertheless found it impossible to publicly denounce the failure of the rule of law to protect Aboriginal people. For Noble, the Anglican Church provided him with an arena in which he could carry out an effective ministry to his own people but he was never ordained to the priesthood and, though he felt a strong attachment to white missionaries, his conception of the ideal mission was quite different from theirs.

The life and religious experiences of White and Noble provide an antidote to the over-optimistic depictions of the carefree life which was the mainstay of promotional literature directed at metropolitan audiences of potential emigrants, investors and donors. The existence of such sunny southern domains was a necessary backdrop to other kinds of missionary work at home as well as abroad. Shurlee Swain has immersed herself in the literature of the international child rescue movement in which images of the distant, depraved heathen were first likened to the white heathen at home and then contrasted to the colonies of settlement, the 'brighter Britain' which was seen as a kind of birthright which might be inherited, through emigration and rescue work, by denizens of darkest England.[64]

New Zealand was generally seen by British humanitarians as an example to the rest of the empire for its success in pacifying, Christianizing and uplifting the Maori from warrior enemies to British subjects through the vehicle of the Treaty of Waitangi (1840). Because of its perceived similarity in terms of climate, geography and dimensions to the British Isles, as well as its record of free settlement on Wakefield's moral principles, it was sometimes represented as a true and perfect Britain, a 'Better Britain'.[65] Peter Clayworth considers the theories of Richard Taylor (1805–1873), an Anglican clergyman who was a witness to the Treaty of Waitangi and a subscriber to the general principles of the Select Committee on Aborigines whose reports were released in 1836 and 1837 and which condemned the systematic destruction of Aboriginal people through the immoral actions of white settlers. As Elizabeth Elbourne argues, this was the high water mark for the influence of humanitarians working from the British metropole to influence and contain the action of colonists on the reaches of imperial settlement.[66] Taylor's contribution is a striking example of what might be called scientific, ethnographic, biblical literalism by a man fully linked to imperial knowledge networks. Seeking the ultimate sanction for the incorporation of the Maori into the British race, Taylor sought to establish that the Maori were descendants of one of the ten lost tribes of Israel; though, for reasons that are unclear, he eventually modified this view, writing it out of the 1870 edition of *Te Ika a Maui*. While ultimately acting, as Clayworth argues, as an apologist for British conquest and occupation of New Zealand, Taylor's views also indicate the necessary adaptations and reformulations that occurred between colony and metropole at the edges of empire. Indeed the three chapters by McAleer, Clayworth and Prevost indicate not only the potential but also the limitations of humanitarian and Christian discourses, very active in the necessary encounters between religious people and foreign, 'different' native peoples in both mission territories and Greater Britain.

After empire

The final chapters in this volume take us beyond empire, or rather into the period in which the political bonds between Britain and her colonies were being renounced. Canada had flourished as an independent Dominion since 1867 though, as Ruth Compton Brouwer notes, with increasingly close interaction with the United States; in the case of Ireland, considered by Fiona Bateman, the link had been severed by revolution and war; the tiny Ngwato kingdom, considered by John Stuart, was part of the British Protectorate of Bechuanaland in what became independent Botswana in 1966. British missionaries in decolonizing post-war Africa were confronted with innumerable difficulties as they were challenged by emerging nationalist and separatist churches, as well as financial and theological difficulties compounded by the dislocation of war and its aftermath. Into this cauldron, the heir to the Bangwato throne introduced a white English wife. Stuart analyses the controversy about the standing of the royal couple and the role of the interdenominational London Missionary Society (LMS), which was effectively the Bangwato state church. The LMS failed to support the incoming king unequivocally when his legitimacy was challenged by outside forces; these included resistance by the neighbouring states of Southern Rhodesia and South Africa, where the formal institution of apartheid had begun. Both states were appalled by the royal miscegenation. The infinite complexities of the case suggest the fissuring of colonial rule as missions became detached from home churches, and as national governments emerged to challenge what had formerly been asserted as imperial and universal religious rights and duties. In the Bangwato Reserve, Stuart argues that there were many losers from the collapse of empire, above all the LMS.

There are more continuities with former imperial arrangements in Canada, despite or perhaps because of its early political detachment from Britain.[67] However, Brouwer is able to chart the increasing impact of United States and ecumenical patterns on the missionary activities of the Canadian Protestant churches. And, by the 1960s, as Canadian society moved irrevocably beyond its colonial past, a new perception of Canadians' obligations to the developing world emerged. As a result, the churches' historic participation in overseas missions was increasingly deprecated as a racist legacy best overlooked. Shame at past imperial enthusiasms is characteristic of settler churches, as indeed of all parts of 'Greater Britain', a term which now tends to evoke cringing rather than pride. The contrast with the awakening of imperial religious ambitions for the overseas missionary field by the Irish Catholic Church could not be more striking. As Peter Cunich shows in the case of the colonial church in New South Wales, both Irish and English members of the Catholic hierarchy in Australia were happy to adopt imperial language to speak about the extension of the Catholic Church throughout British territory. However, Fiona Bateman, using the missionary propaganda written in the main for a young audience, sees the outbreak of Irish foreign missions as tantamount to a second scramble for Africa.[68] If the Scots had failed to create Black Scotsmen and women, Irish designs for a shamrock-shaped church and a round tower in Sierra Leone continued the tradition of assembling a European religious bricolage and calling it the Kingdom of God.

In conclusion, we might return to Archdeacon Manning and the enthusiasm with which he preached a doctrine of religious imperialism as a solution to the otherwise inevitable collapse of empires. It may not be entirely flippant to suggest that he might, after all, have been right. The high summer of imperial idealism, for the churches was after all, as John Stuart notes, synchronous with the meridian of imperialism itself. The commitment by missionary and colonial agencies to establish English, Scottish or Irish churches which acted, almost inevitably as conduits for imperial ideology is an excellent marker of the cultural expansion of the empire. Churches, like missions, are cultural constructions nurtured and negotiated within the haven of the imperial world. Yet of all imperial institutions, those of the churches would appear to have been the most long-lived and they bear the marks of their imperial origins. Historians will continue to debate whether the churches are implicated or honoured by the connection with empire; churchmen and women never entirely set aside the realization that the empires of religion must ultimately be sought in another world.

Notes

1. *Proceedings at a Meeting Held the Egyptian Hall, Mansion House, on Tuesday, March 17, 1846, to Increase the Means of Religious Instruction for the Emigrants and Settlers in the British Colonies through the Society for the Propagation of the Gospel* (London: Richard Clay, 1846), p. 27.
2. T. Ballantyne, 'Religion, Difference, and the Limits of British Imperial History', *Victorian Studies,* LXVII, no. 3 (2005) 397–426; for the increasing

importance of cultural history in imperial studies, see J. Gascoigne, 'The Expanding Historiography of British Imperialism', *The Historical Journal*, XLIX, no. 2 (2006) 577–92.

3. A. N. Porter, ed., *The Oxford History of the British Empire, Vol. 3. The Nineteenth Century* (Oxford: Oxford University Press, 1999), pp. 222–46.
4. From the Oxford History of the British Empire Companion Series: *Missions and Empire*, ed. Norman Etherington (Oxford: Oxford University Press, 2005); *Ireland and the British Empire*, ed. N. Kenny (Oxford: Oxford University Press, 2004), pp. 90–122; H. M. Carey, 'Religion and Society', in *Australia's Empire*, ed. D. Schreuder and S. Ward (Oxford: Oxford University Press, 2008), pp. 186–210.
5. The best short introduction to missions in the empire is A.N Porter, 'An Overview: 1700–1914', pp. 40–63, in *Missions and Empire*. See also J. Wolffe, *Religion in Victorian Britain, Vol. 5. Culture and Empire* (Manchester: Manchester University Press, 1997).
6. A. Porter, *Bibliography of Imperial, Colonial, and Commonwealth History since 1600* (Oxford: Oxford University Press, 2002), pp. 505–29 lists 646 items relating to missions which indicates the depth of the historiography.
7. A. N. Porter, *Religion Versus Empire?: British Protestant Missionaries and Overseas Expansion, 1700–1914* (Manchester: Manchester University Press, 2004); Brian Stanley, 'Commerce and Christianity: Providence Theory, the Missionary Movement, and the Imperialism of Free Trade, 1842–1860', *The Historical Journal*, XXVI (1983) 71–94; B. Stanley, *The Bible and the Flag: Protestant Missions and the Imperialism of Free Trade, 1842–1860* (Leicester: Inter-Varsity Press, 1990). See also the chapters by Breitenbach, MacKenzie and Brouwer below.
8. For an account of colonial India and a critique of the consensus argued by Andrew Porter and Brian Stanley that the connection between missions and imperialism was weak, see I. Coplan, 'Christianity as an Arm of Empire: The Ambiguous Case of India under the Company, *c.* 1813–1858', *The Historical Journal*, XLIX, no. 4 (2006) 1025–1054.
9. P. N. Miller, *Defining the Common Good: Empire, Religion, and Philosophy in Eighteenth-Century Britain* (Cambridge: Cambridge University Press, 1994), W. G. Mills, 'Victorian Imperialism as Religion', in *The Man on the Spot: Essays on British Empire History*, ed. R. D. Long (Westport, Conn.: Greenwood, 1995).
10. E. Said, *Orientalism* (London: Vintage, 1978). For more discussion of Said's view of religion, see W. D. Hart, *Edward Said and the Religious Effects of Empire* (Cambridge: Cambridge University Press, 2000).
11. J. Comaroff and J. L. Comaroff, *Of Revelation and Revolution: Christianity, Colonialism and Consciousness in South Africa*, 2 vols. (Chicago, IL: University of Chicago Press, 1991, 1997).
12. S. Thorne, 'Imperial Pieties', *History Workshop Journal*, LXIII, no. 1 (2007) 327.
13. For examples of this range of responses, see B. Stanley, ed. *Missions, Nationalism and the End of Empire* (Grand Rapids, MI: William B. Eerdmans, 2003).
14. L. Colley, *Britons: Forging the Nation, 1770–1837* (New Haven, CT: Yale University Press, 1992), p. 11.

15. S. J. Brown, *Providence and Empire: Religion, Politics and Society in the United Kingdom, 1815–1914* (Harlow: Pearson Longman, 2008).
16. H. McLeod, 'Protestantism and British National Identity 1815–1945', in *Nation and Religion*, ed. P. Van der Veer and H. Lehman (Princeton, NJ: Princeton University Press, 1999), pp. 44–70.
17. Sir John Seeley, in *The Expansion of Empire* (London: Macmillan, 1883), p. 208.
18. J. Samson, 'Are You What You Believe? Some Thoughts on Ornamentalism and Religion', *Journal of Colonialism and Colonial History*, III, no. 1 (2002), n.p.
19. S. Thorne, 'Imperial Pieties', *History Workshop Journal*, LXIII, no. 1 (2007) 319–27; S. Thorne, 'Religion and Empire', in *At Home with the Empire: Metropolitan Culture and the Imperial World*, ed. C. Hall and S. Rose (Cambridge: Cambridge University Press, 2006); Jeffrey Cox, *Imperial Fault Lines: Christianity and Colonial Power in India, 1818–1940* (Stanford: Stanford University Press, 2002). See review by S. Thorne, *Victorian Studies*, XLVII, no. 2 (2005) 295–7.
20. C. Hall, *Civilizing Subjects: Metropole and Colony in the English Imagination* (Oxford: Polity Press, 2002); *Cultures of Empire*: A Reader, ed. C. Hall (Manchester: Manchester University Press, 2000).
21. Hall and Rose, *At Home with the Empire*.
22. See Chapter 10 by McAleer in this volume.
23. S. Thorne, ' "The Conversion of Englishmen and the Conversion of the World Inseparable": Missionary Imperialism and the Language of Class in Early Industrial Britain', in *Tensions of Empire: Colonial Cultures in a Bourgeois World*, ed. F. Cooper and A. L. Stoler (Berkeley, CA: University of California Press, 1997); Idem, 'Religion and Empire'.
24. P. van der Veer, *Imperial Encounters: Religion and Modernity in India and Britain* (Princeton, NJ; Oxford: Princeton University Press, 2001).
25. A. Johnston, *Missionary Writing and Empire, 1800–1860* (Cambridge: Cambridge University Press, 2003).
26. Though for a recent synthesis, see E. Elbourne, 'Religion in the British Empire', in *The British Empire: Themes and Perspectives*, ed. Sarah Stockwell (Oxford: Blackwell, 2008). Religion is also a significant unifying theme for Jane Sampson, ed. *The British Empire* (Oxford: OUP, 2001).
27. F. Cooper and A. L. Stoler, *Tensions of Empire: Colonial Cultures in a Bourgeois World* (Berkeley, CA: University of California Press, 1997).
28. An adage credited to the jurist Joachim Stephani (1544–1623).
29. J. Wolffe, *God and Greater Britain: Religion and National Life in Britain and Ireland, 1843–1945* (London: Routledge, 1994); Idem, *Religion in Victorian Britain*.
30. S. J. Brown, *The National Churches of England, Ireland, and Scotland 1801–1846* (Oxford: Oxford University Press, 2001).
31. McLeod, 'Protestantism and British National Identity 1815–1945', pp. 44–70.
32. Porter, *Religion Versus Empire?*, pp. 158–62. See also E. D. Daw, *Church and State in the Empire: The Evolution of Imperial Policy, 1846–1856* (Canberra: Dept. of Government, Faculty of Military Studies, University of New South Wales, 1977).
33. For the SPG's Emigrants Spiritual Aid Fund which opened in 1849, see C.F. Pascoe, *Two Hundred Years of the* SPG (London: SPG, 1901), pp. 818–20

34. *Proceedings at a Meeting Held the Egyptian Hall, Mansion House,* pp. 26–7.
35. For the agonizing process leading to his decision see the coolly unsympathetic biography by E. S. Purcell, *Life of Cardinal Manning, Archbishop of Westminster,* 2 vols (London: Macmillan, 1896), especially pp. i, 593–628.
36. B. Stanley, 'Commerce and Christianity', 71–94.
37. For the significance of these, see R. Strong, 'A Vision of an Anglican Imperialism: The Annual Sermons of the Society for the Propagation of the Gospel in Foreign Parts 1701–1714', *Journal of Religious History,* XXX (2006) 175–98. See also H. Le Couteur, 'Anglican High Churchmen and the Expansion of Empire', *Journal of Religious History,* XXXII (2008), 193–215.
38. J. Gascoigne, *The Enlightenment and the Origins of European Australia* (Cambridge: Cambridge University Press, 2002), pp. 26–34, D. Pike, *Paradise of Dissent: South Australia, 1829–1857* (Melbourne: Melbourne University Press, 1967).
39. R. Strong, *Anglicanism and the British Empire c.1700–1850* (Oxford: Oxford University Press, 2007), ch. 4.
40. *Proceedings of a Meeting ... For the Purpose of Raising a Fund towards the Endowment of Additional Colonial Bishoprics* (London: Rivingtons, 1841). Fully discussed by Strong in Strong, *Anglicanism and Empire,* pp. 198–221.
41. H. Cnattingius, *Bishops and Societies: A Study of Anglican Colonial and Missionary Expansion, 1698–1850* (London: SPCK, 1952), p. 204.
42. T. E. Yates, *Venn and Victorian Bishops Abroad* (London: SPCK, 1978).
43. S. Thorne, *Congregational Missions and the Making of an Imperial Culture in Nineteenth-Century England* (Stanford, CA: Stanford University Press, 1999).
44. J. Wesley, *Journal of John Wesley* (Chicago, IL: Moody Press, 1951), ch. 3.
45. K. Condon, *The Missionary College of All Hallows 1842–1891* (Dublin: All Hallows College, 1986); For objections to the ultra-Irish priests from All Hallows, see J. J. McGovern and P. J. O. Farrell, 'Australia', in *A History of Irish Catholicism, Vol. 6,* ed. P. J. Corish (Dublin: Gill and Macmillan, 1971), pp. 59–60.
46. Wright, D. F. 'Chaplaincies, Colonial', in *Dictionary of Scottish Church History and Theology,* ed. Nigel M. de S. Cameron (Edinburgh: T. & T. Clark, 1993), p. 163.
47. J. Moir, 'The Settlement of Clergy Reserves, 1840–1855', *Canadian Historical Review,* XXXVII, no. 1 (1956) 46–62.
48. A. M. G. Stephenson, *The First Lambeth Conference, 1867* (London: SPCK for the Church Historical Society, 1967).
49. For the theology of empire, see D. Bebbington, 'Atonement, Sin, and Empire, 1880–1914', in *The Imperial Horizons of British Protestant Missions, 1880–1914,* ed. A. Porter (Grand Rapids, MI: William B. Eerdmans, 2003), pp. 14–31.
50. A. R. Pagden, *Lords of All the Worlds: Ideologies of Empire in Spain, Britain and France c.1500–c.1800* (New Haven, CT: Yale University Press, 1995).
51. A. Hastings, 'Christianity and Nationhood: Congruity or Antipathy? '*Journal of Religious History,* XXV, no. 3 (2001) 247–60; Idem, *The Construction of Nationhood: Ethnicity, Religion and Nationalism* (Cambridge: Cambridge University Press, 1997).

52. Barbara Fields, Christian Missionaries as Anti-colonial Militants', *Theory and Society* 11 (1982) 95–108.
53. For overview of these arguments, see introduction to B. Stanley, ed., *Christian Missions and the Enlightenment* (Richmond, Surrey: Curzon Press, 2001), pp. 1–21; B. Stanley and A. M. Low, eds., *Missions, Nationalism, and the End of Empire* (Grand Rapids, MI: William B. Eerdmans, 2003).
54. N. Erlank, '"Civilizing the African": The Scottish Mission to the Xhosa', in *Christian Missions and the Enlightenment*, pp. 146–7. See also Chapter 6 by John MacKenzie in this volume.
55. J. G. A. Pocock, *Barbarism and Religion, Vol. 2. Narratives of Civil Government* (Cambridge: Cambridge University Press, 1999). For Gibbon and imperialism, see also R. McKitterick and R. Quinault, eds. *Edward Gibbon Wakefield and Empire* (Cambridge: Cambridge University Press, 1997).
56. Despite the surprising elimination of Rev. Richard Johnston from the First Fleet by A. N. Porter, ed., in *The Oxford History of the British Empire, Vol. 3. The Nineteenth Century*, p. 225.
57. Edward Gibbon, *Decline and Fall of the Roman Empire*, ed. J. B. Bury, 7 vols. (London: Methuen, 1909–1914), vol. 4, ch. 38, p. 181.
58. Gascoigne, *The Enlightenment and the Origins of European Australia*, pp. 19–32.
59. L. B. Wright, *Religion and Empire: The Alliance between Piety and Commerce in English Expansion, 1558–1625*, repr. edn (New York: Octagon Books, 1965), p. 85. [First published by University of North Carolina Press, 1943.]
60. Stanley, 'Commerce and Christianity: Providence Theory, the Missionary Movement, and the Imperialism of Free Trade, 1842–1860', pp. 71–94; A. Porter, '"Commerce and Christianity": The Rise and Fall of a Nineteenth Century Missionary Slogan', *Historical Journal*, XXVIII (1985) 597–621.
61. Stanley, 'Commerce and Christianity', pp. 71–94.
62. S. Gilley and B. Stanley, *World Christianities, c.1815 – c.1914, The Cambridge History of Christianity, Vol. 8* (Cambridge: Cambridge University Press, 2006), H. McLeod, *World Christianities c.1914–c.2000, The Cambridge History of Christianity, Vol. 9* (Cambridge: Cambridge University Press, 2006).
63. C. Midgley, *Gender and Imperialism, Studies in Imperialism* (Manchester: Manchester University Press, 1998), A. Woollacott, *Gender and Empire* (Basingstoke: Palgrave Macmillan, 2006).
64. Citing Rev. W. J. Mayers, 'My Tour in "Brighter Britain"', *Night and Day*, XVI, no. 165–8 (1892).
65. See J. Belich, *Paradise Reforged: A History of the New Zealanders* (Albany, NY: Allen Lane, 2001) for the idea of New Zealand as 'Better Britain'.
66. E. Elbourne, 'The Sin of the Settler: The 1835–36 Select Committee on Aborigines and Debates over Virtue and Conquest in the Early Nineteenth-Century British White Settler Empire', *Journal of Colonialism and Colonial History*, IV, no. 3 (2003) n.p.
67. P. A. Buckner and R. D. Francis, ed. *Rediscovering the British World* (Calgary: University of Calgary Press, 2005), pp. 187–91.
68. See the Chapter 13 by Bateman in this volume.

Part I
Religious Metropoles

2
The Consolidation of Irish Catholicism within a Hostile Imperial Framework: A Comparative Study of Early Modern Ireland and Hungary

Tadhg Ó hAnnracháin

The religious persecution of the Irish in the Early Modern period for their fidelity to Rome was a core belief of nineteenth- and twentieth-century Catholic Irish nationalism. Nor was it merely confined to the often surprisingly rich and source-based, if highly slanted, historiography of figures such as M. J. Brenan, Thomas Walsh, D. P. Conyngham and M. V. Ronan.[1] On the contrary, detailed disquisitions on the 'horrors of the long, black night of the Penal Laws' could emerge in the most surprising of media, such as a newspaper like *An Camán*, which was primarily funded by a sports organisation, the Gaelic Athletic Association (GAA).[2] For several of the journalists who worked on this paper, the religious history of the early modern period was a crucial backdrop to their understanding of the struggle between British imperialism and Irish nationalism. In a special Christmas 1933 issue of *An Camán*, for instance, one of its occasional contributors, 'Dalcassian', declared: 'England's perversion from the faith created a further gulf between the two civilizations [of Ireland and England] and realizing that a national creed rooted on a glorious past would perpetuate national memories ... she determined to destroy them.'[3] Persecution thus represented a twin assault on both religion and national identity but, providentially, the defence of the first would allow for the revival of the second: 'Dalcassian' noted: 'One aspect alone of the life of the nation came unscathed throughout the long night of persecution – the ancient faith'.[4]

To other Irish nationalists who contemplated this picture of past persecution under British rule, a further providential pattern was discernible within the framework of Irish incorporation within the empire, and even in the loss of language, for it was this very process which had allowed the evolution of a shadow Irish spiritual empire, so ably considered by Fiona Bateman in the present volume (Chapter 13), of missionaries to be diffused throughout the world in the footsteps, indeed within the very bloodstream, of English temporal and linguistic expansion.[5] As Thomas Walsh and D. P. Conyngham put it:

> Our national church is the bond of our national existence; though the political arrangements of Ireland with the sister country have almost annihilated the political interest of the former, still that system has given to the church of Ireland an imperial character: for she is the mistress of religion in the British empire, gives her an imperial voice, by which the bigotry of England is branded with universal reprobation, and secures to Ireland an imperial importance.[6]

Such conceptualizations were undoubtedly attractive for they could mobilize very deeply embedded tropes of Christian triumph through abnegation and sacrifice providing a seductive reversal of a narrative of humiliation and defeat. Faith did not merely renew the nation that otherwise would have been submerged but miraculously allowed for the process of subjugation to serve as a springboard to a greater harvest in the vineyard of the Lord.

Ireland's Catholicism, therefore, at a stroke both allowed for the nation's resistance to incorporation into the British imperial framework and for its spiritual colonisation of British expansion throughout the world. Not only were such notions pleasurable but they also carried additional conviction because the central assumption on which the paradigm was based was probably correct. Christopher Hollis's 1927 assertion, for instance, that 'the Irish question only persisted because it was a religious question' remains a sustainable position eight decades later, even if his supplementary claim that 'the whole purpose of English policy was ... to destroy the Catholicism of the Irish' now appears hysterically overstated.[7] From a modern perspective it seems evident that the Catholicism of the Irish population was the single most important reason for the nationalism which ultimately abstracted the Irish Free State/Republic not only from the United Kingdom but eventually also from the British Empire/Commonwealth.

For the nationalist historiography which helped to provide the intellectual underpinning of the fledgling state the reason for this development was relatively unproblematic. As M. J. Brenan crisply noted, 'the superintending power of the almighty has been visibly displayed in the protection of the Church'.[8] The obscure workings of divine providence were undoubtedly convenient in this regard for they absolved historians of that era from having to engage too deeply with the factors which allowed for the survival and consolidation of Irish Catholicism. Instead comfortable notions of Catholicism as the elemental faith of the Irish people surviving miraculously in the face of English persecutions took root in a particular conception of Ireland's unique historical path.

The effort to replace such convenient recourse to a providential model has been at the heart of an ongoing debate within modern Irish historiography over the past three decades. If the protective hand of God is removed as a causal mechanism then how can the survival of Irish Catholicism within the framework of a profoundly anti-Catholic Imperial state be explained? As several of the chapters in the current volume confirm, the ramifications of this particular development were enormous. Without it, not merely would the twentieth-century Catholic spiritual empire discussed by Fiona Bateman, Chapter 13, not have existed but the nature of the Irish-influenced anti-Catholicism at the heart of imperial expansion analysed by John Wolffe would also necessarily have evolved in a fundamentally different way. Wolffe's analysis chiefly concerns itself with the renewal of anti-Catholicism following the hiatus of the late eighteenth century. The present discussion, however, focuses instead on the late sixteenth and seventeenth centuries when the most important developments in the religious partition of the archipelago took place.

By the early seventeenth-century hostility to Catholicism had been established as arguably the single most important element of national identity within the central English element of the British imperial matrix.[9] Popular disturbances of that era, even when motivated by the most evident economic conditions, almost inevitably acquired a patina of anti-Catholic rhetoric and paranoia about Popery was a crucial force linking Protestants of all social levels.[10] Nor was Scottish prejudice any less visceral, despite the decline in the numbers of Scottish Catholics to no more than 2 per cent of the seventeenth-century population.[11]

Yet parallel to what Patrick Collinson has described as the surprising process in which one of the most Catholic countries in Europe, namely England, became one of the least,[12] is the even more startling consolidation of one the most anomalous branches of medieval Catholicism, namely Ireland, as a bastion of the Church of Rome.[13] Until the acquisition of Quebec in the late eighteenth century, this fact rendered Ireland unique within the British imperial framework and continues to prompt the question of how such a paradoxical development could have come about in the era of *cuius regio eius religio*. Much of the literature in the sporadic debate which has addressed this topic has concentrated on reasons for the failure of the reformation rather than on the linked but by no means identical topic of Catholic success.[14] What this chapter proposes is a historiographical review of this problem which will bring to bear a comparative perspective on the process of Catholic consolidation, with a particular reference to the example of Habsburg central Europe.

A number of salient facts and conclusions have been produced by the existing scholarship on this problem of the past two decades. In terms of periodization, it can be stated reasonably confidently that at the outset of Elizabeth's reign in 1558 neither Protestant nor Catholic currents of reform had made any significant headway in Ireland among the general population of either Gaelic or English descent. There was little evidence of any lay enthusiasm for religious change but, on the other hand, relatively little overt opposition either, and such resistance as there was largely derived from political disaffection rather than religious conviction. The first Jesuit mission to Ireland in the 1540s was frankly pessimistic,[15] and the passage of Elizabeth's reformation legislation through the Irish parliament two decades later in 1560 seemed indeed to indicate that such pessimism had been justified.[16] Clerical conservatism certainly did represent a significant barrier to religious change in the existing colonial community, particularly because of the degree to which English cultural identity in Ireland had articulated itself in terms of medieval canonically orthodox Catholicism. As James Murray has demonstrated, for the clerical leadership of the older colonial community the historical justification of the English presence in Ireland derived from their mission to extend the canon law of the Western church to Gaelic Ireland and thereby civilize it.[17] Such sentiments represented a potential reservoir of resistance to attempts to alter the fabric of the medieval English Catholic legacy in Ireland. The strength of the Observant Franciscans in English and Gaelic Ireland represented another potential obstacle to the spread of the Reformation. Nevertheless, such clerical foci of resistance were not necessarily more formidable than those which the Tudor reform process ultimately surmounted in England during the course of the sixteenth century. But whereas in England the process of Reformation ultimately resulted in the cultivation of a deeply seated anti-Catholicism,[18] in Ireland the precise opposite occurred, a process rendered very visible by the events of the 1640s. Following a botched rising in Ulster in October 1641, sectarian conflict erupted all over the island. In every county, Protestants became the subject of attack by their Catholic neighbours who pillaged their property and stripped and expelled them from the localities. The following year, the formation of the confederate Catholic association, effectively a Catholic state occupying much of the island, offered even clearer testimony of the deep-seated and popular identification with Catholicism which had occurred in the vast majority of the population, both Gaelic and English, which traced its origin from the pre-Elizabethan period.[19] Crucially the social disorder and violence sparked by the original rebellion was contained by the creation of structures which emphasized the unity of the Catholic population and which derived their legitimacy in considerable measure from the formidable moral authority of the Catholic clergy. In a provincial synod of Armagh in March and in the first post-Tridentine synod of the entire Irish Catholic Church in May 1642, the Irish hierarchy threw its weight massively behind order, property and social hierarchy and actively solicited the support of the Catholic gentry of Ireland in controlling the pillaging, theft and violence which followed the rebellion. The clergy also actively participated in the governmental structures which ultimately developed and by administering an oath of association, which all Catholics were compelled to take, they played a critical role in the stabilization of new political structures among the general population.

The roots of this profoundly important development undoubtedly lay in the alienation of the Irish population by the expansion of the geographical reach of the English state in Ireland in the late sixteenth century. From 1541, all the inhabitants of Ireland had been accepted as subjects of the newly constituted kingdom of Ireland. In principle, therefore, Ireland was not a colony, although in practice it would come to enjoy, or suffer, an intermediate halfway status somewhere between colony and kingdom during the Early Modern period.[20] Yet arguably the most important rationale underpinning the later colonial treatment of the island was the failure of the state to inculcate religious loyalty to the established church, not only, and in contrast to Scotland, in the Gaelic population, but even in the existing colonial population, the Old English, those who traced their descent to settlers of Anglo-Norman origin. Thus the failure of the Reformation in Ireland was not merely the failure to convert, Anglicize and civilize the putatively barbarian and linguistically distinctive Gaelic population but also to build on the substantial English colonial presence established in the island for centuries prior to the Reformation.

The chief vehicle of religious alienation which has been identified with regard to this older pre-Elizabethan English colony was the economic damage which their community sustained in the course of the series of wars fought between Elizabeth's viceroys and various Gaelic and Gaelicized lordships. Such wars not only created massive grievances among the over-burdened English of Ireland but also hampered the state church by siphoning off the resources which might have been used to evangelize the population in the new religion.[21] Moreover, the scarcity of resources also hampered the state in its capacity to use instruments of coercion to enforce obedience in the ecclesiastical sphere. The military and financial strain occasioned by the various wars made state officials reluctant to enforce the requirements of the religious legislation for fear of creating a rebellious backlash. Nevertheless, from the late 1570s criticism was increasingly directed at the suspect religious practices of the Old English by frustrated administrators. This was at times a tactic consciously deployed as a means to delegitimize the accusations of governmental abuse which the older colonial community attempted to bring to the queen's attention. But it was tied also to wider discourse of loyalty in the latter part of the sixteenth century in Ireland. Increasingly in the Irish localities, the captains, seneschals and presidents who represented the expansion of the English state came in their casual violence and arbitrary justice to resemble the Gaelic and Gaelicized warlords whom they were intended to replace.[22] What distinguished them was their self-proclaimed loyalty to their sovereign, which was expressed as a principal aspect of their honour.[23] Such concepts of loyalty were extraordinarily enabling from an administrative perspective for none of the established elites in the island whether Gaelic or Old English could afford to subordinate their interests to the demands of state policy to the same degree as incoming New English, who might gain substantially by the overturning of the *status quo*, but who lacked an existing stake in it. Competitive loyalty vis-a-vis the Gaelic lordships with their assumed cultural inferiority and propensity to rebel was relatively easy to establish but it was religion which offered the principal grounds for the marginalization of the older colonial community. Criticisms of this nature in turn heightened the religious defensiveness of the Old English community who increasingly saw themselves as the marginalized victims of governmental innovation. It was in this context that the scions of established colonial families began to forsake the universities of England, where they would have been exposed to a Protestant education, opting instead for continental academic training in Catholic establishments. From the early 1590s this process was accelerated by the foundation of the first of a string of continental Irish colleges which ultimately came to out-perform the state's institution at Trinity College, in terms of providing educated clerical personnel for deployment in a pastoral context in the Irish localities.[24]

A series of brief comparisons by Karl Bottigheimer and Ute Lotz-Heumann have been enormously helpful in clarifying many of the processes just mentioned. In particular, their focus on Norway and Gaelic Scotland has been particularly enlightening in reassessing the obstacles posed by the Gaelic language to effective inculcation of Protestantism.[25] Yet the most under-theorized aspect of their work, as with other participants in the historiographical debate, concerns the mechanisms of Catholic success, which is presented as more or less predetermined on the basis of Protestant failure.[26] Consequently they offer relatively little analysis concerning the events of the seventeenth as opposed to the sixteenth century. It is in this context that central Europe reveals itself as a comparison of very considerable interest for it was precisely in the seventeenth century that the Habsburgs were able to bring about a considerable measure of religious change in their various imperial domains.

Bohemia forms the most celebrated example of the successful reimposition of Catholic rule. Taking advantage of Czech rebellion, after 1620 Emperor Ferdinand II instituted a determined attempt to re-Catholicize his newly re-conquered Bohemian kingdom.[27] In addition to the wide-scale confiscations in the wake of the victory at the Battle of the White Mountain, which were of especial value to the indigenous Catholic minority of Czech nobility,[28] large numbers of Protestant Czech nobles and burghers who refused to convert to Catholicism were forced into exile following a decree of 1627, often after having sold their property to Catholic relatives as the law allowed. This sustained programme of coercion was supplemented by an active process of evangelization, which included the foundation of 19 new Jesuit colleges in the lands of the Bohemian crown and it enjoyed, on the whole, a significant measure of success.[29]

In Ireland, where Catholicism was also successful but for quite different reasons, the closest equivalent to the Bohemian experience was the decade of the 1650s, which in folk historical memory came to represent the Irish equivalent of the Czech *Temno* or 'darkness'.[30] The *de facto*

Catholic state of the Confederates which had emerged from the wreckage of the 1641 rebellion was dissolved in 1649. In the same year an unprecedentedly successful conquest of Ireland was launched by the new republican regime in London which, by virtue of its unconditional nature, conferred an extraordinary freedom of action on the saints of the English Revolution in their dealings with the Catholic population, who were effectively deprived of either political or moral purchase on the process of government. The conquest and the ensuing settlement marked the high point in the level of persecution of the Catholic Church in Ireland. The initial tone was set at Drogheda when Catholic religious were, as Cromwell himself admitted, 'knocked on the head promiscuously';[31] in the early years of the war many priests, including three bishops, were executed summarily on capture. But from 1651 and increasingly after 1654 the preferred option became the shipping abroad of captured members of the priesthood. It has been estimated that at least a thousand priests were exiled during this period, while others were given permission to remain on the condition that they did not exercise their ministry. Catholic school teachers also faced a formidable level of harassment and were frequently arrested and transported to a life of indentured servitude.[32] Before the beginning of the conquest the Irish Catholic episcopacy was arguably the closest hierarchy in all Europe to the ideal Tridentine template, since they were universally seminary trained, overwhelmingly resident in their dioceses and appointed to their sees effectively at the behest of the Pope with only marginal interference from secular interests. By 1655, of the 27 bishops in place before Cromwell's invasion only one resident remained, Eugene MacSweeney of Kilmore, who was too old and infirm to represent any religious irritation to the regime. The effective destructive of the ecclesiastical apparatus of Catholicism in Ireland during these years, added to the punitive policy of the state towards Catholics in general, did lead to a significant number of conversions to Protestantism; the chief evidence for this derives from the details of their reconversion to Catholicism by newly active priests in the 1660s following the restoration.[33]

There are at least two reasons for the limited and short-term nature of Protestant advances in Ireland and the failure of Protestantism to hold its own in Bohemia against what would appear to be similar attempts to impose religious conformity by force. First, coercion was applied less systematically and for a far shorter period in Ireland than in Bohemia. Even in the later years of the 1650s the level of religious persecution faded considerably and the restoration of the monarchy in 1660 effectively reduced it to the level of the earlier seventeenth century, when the Catholic Church had made significant advances. Second, in marked contrast to Bohemia, active evangelization of the Irish population during the 1650s was remarkably limited.[34] During the Cromwellian period, the state actually deployed fewer Protestant ministers than during the 1630s, even if they do appear to have been somewhat more adequately provided for in material terms. At first glance Habsburg success in Bohemia would appear to offer a simple explanation of Protestant failure and Catholic success in Ireland: the English state simply failed to apply sustained coercion against the practice of a prohibited religion. A belief in the effectiveness of such reasoning may well have acted as the background assumption to the relative neglect of the seventeenth century in the debate about the processes of confessional change in Ireland. In effect the reasons why none of the Stuart monarchs were prepared to make use of stringent anti-Catholic legislation becomes the first port of call for an analysis of the island's peculiar religious evolution. Yet two problems can be identified with simply accepting such an explanation. First, while offering some explanation of the causes of Protestant failure, it still more or less neglects the causes of Catholic success. Second, it ignores the fact that the Bohemian, or perhaps more properly the Styrian, model (since it was in Styria that Ferdinand first refined the approach later to be used in the Czech lands)[35] was not the only strategy utilized by the Habsburgs to strengthen the position of Catholicism in their imperial domains.

In this regard Early Modern Hungary is of particular interest as an area of comparison. Following the disastrous battle of Mohács and the extinction of the Jagellon dynasty in 1526, a rough tripartheid division of the lands of the crown of St Stephen occurred: Royal Hungary, an arc of territory stretching from modern Slovakia and present-day Hungary west of Lake Balaton to Northern Croatia, fell under the control of Habsburgs who enjoyed the title of kings of Hungary and claimed the temporal and ecclesiastical prerogatives and *ius patronatus* which had pertained to that title; an independent principality under the suzerainty of the Ottoman *Porte* emerged in mountainous Transylvania to the East, while the central and southern portions of the medieval kingdom of Hungary, including the ancient capital, Buda, came under direct Turkish control. Prior to 1671 religious coercion was effectively not an option for the Habsburgs even in the Royal Hungarian portion of the kingdom which they actually controlled. The dynasty's hands were tied in this regard by the peace of Vienna in 1606 which guaranteed the orders of the kingdom of Hungary the freedom to practice the Protestant religion. Moreover, the frequent assembly of the Hungarian Diet helped enforce compliance with the provisions of the treaty on their monarch.[36] The Diet in turn was disposed to be vigilant in this regard because of the overwhelming Protestant majority in the country, a majority which at the most conservative estimate amounted to at least three-quarters of the Hungarian population at the close of the sixteenth century.[37] Underpinning the necessity for the regime to renounce programmes of religious coercion was the constant Turkish threat. Yet, despite this highly unfavourable situation, to a significant extent it was the re-Catholicization process of the seventeenth century which ultimately permitted the consolidation of a Catholic-dominated Hungarian kingdom, boosted by huge immigration, in the eighteenth century following the Turkish expulsion.

Of particular interest with regard to the religious failure of the state in Ireland was the Habsburg utilization of Hungarians as the basic element of its re-Catholicization strategy. This was most crucially evident in the magnate order of the Diet which was transformed into a Catholic bastion by 1649 through an interlocking process of conversion and creation of new magnates from native Catholic families.[38] In this process, the covert inducements towards Catholicism did not emanate merely from the foreign Habsburg court but were supplemented and brokered by native Hungarian figures, in particular during the long palatinate of Miklós Eszterházy (1625–1645). Despite an ultimately unsuccessful interlude of more absolutist government in the 1670s and 1680s, the Habsburg monarchy in Hungary depended enormously on a coterie of ten or eleven native Magyar and Croatian magnate houses.[39] The contrast with Ireland here is much greater than might appear at first glance. In Ireland also the Stuarts secured a solid majority of co-religionists within the peerage. Thus superficially by the late seventeenth century the two kingdoms would appear broadly comparable with the monarch's religion having triumphed among a privileged landed minority. However, in Ireland to a much greater extent than in Hungary, the religiously compliant aristocracy represented the replacement of the existing elite rather than its winning over. By the early eighteenth century three-quarters of the peerage active in the House of Lords of the Irish parliament were the products of families which had entered the kingdom since the beginning of the Irish Reformation in the 1530s. Only five of these active parliamentary families traced their ancestry to the indigenous Gaelic population, despite the fact that the majority of the island's population were still Irish speaking at this juncture.[40]

As with the magnate order, the Hungarian Catholic and Irish Protestant ecclesiastical establishments differed radically. For it was largely natives of the kingdom of Hungary who occupied the key positions on the seventeenth-century episcopal bench. There are some striking similarities in terms of intellect and literary output between the Protestant primate James Ussher in Ireland and his Catholic equivalent Péter Pázmány in Hungary. But whereas Ussher's membership of a pre-Elizabethan Old English family made him a minority figure in the Irish Protestant episcopacy, Pázmány, on the other hand, typified the native provenance of the Catholic clerical elite in Hungary during his era and he became one of the chief architects of a vernacular Catholic literature in Hungarian.[41] The contrast with Ireland in this regard, where the linkages between Protestantization and anglicization increased in strength over the course of the seventeenth century, and where conversion to the religion of the monarch involved a concommitant attachment to an anglicized church, is particularly evident.[42]

In these several respects, Pázmány and his Magyar and Slav colleagues in the Hungarian hierarchy are much more readily identifiable with the contemporaneous Irish Catholic episcopate who emerged as the leadership of a native and national, if extra-legal, church in the period after 1618. Drawn from and serving both the Old English and Gaelic populations of the island, the Catholic episcopal network covered the entire island and offered a mechanism to integrate both sections of the Catholic population within the framework of Catholic reform.[43] All the bishops of the new hierarchy had been trained in continental seminaries but the pattern of their appointment generally brought them back to their native dioceses or to one in the immediate vicinity.[44] Critically, their extraction from local elite groupings equipped them to act as formidable 'multipliers' of religious conviction.

The Catholic bishops of Royal Hungary were less successful than their Irish counterparts in terms of winning and confirming a majority of the population for their church. Protestantism survived in the Hungarian localities, although increasingly under pressure. Nevertheless, in conjunction with the reliably Catholic magnate order, the ecclesiastical establishment did manage to create a native Catholic tradition which posited itself as both organically Hungarian (and Croat) *and* loyal to the Church of Rome. Ultimately, this was of profound importance in the evolving religious complexion of the Hungarian kingdom, both before and after the re-conquest and re-peopling of the Turkish occupied areas. In sharp contrast with Ireland, the religion of the rulers was not shared exclusively by those communities which has been shaped by processes of plantation and immigration patronized and overseen by the central authority but could successfully stake a claim also to an indigenous character and heritage. Indeed, from the political perspective of Habsburg power, this process of Catholic indigenization was potentially a double-edged sword. The Catholic magnates and prelates of Hungary were the dynasty's most important and reliable collaborators in maintaining control of the throne of St Stephen but their loyalty was significantly more conditional than, for instance, the attachment of the Early Modern Protestant community of Ireland to the English connection. Gábor Bethlen's invasions of Royal Hungary in earlier decades of the seventeenth century attracted Catholic as well as Protestant support, although his often reluctant Catholic partisans were, it is true, significantly swifter to return to Habsburg allegiance. Later in the century, however, figures such as Miklós Zrinyi and Ferenc Wesselényi actively plotted rebellion against the Emperor Leopold, as indeed did György Lippay the Catholic primate of the kingdom. Later still the rebellion of Rákóczi Ferenc saw a Catholic leader with a significant coterie of co-religionist advisors patronizing a movement of resistance largely motivated by Protestant grievance in a fashion more reminiscent of the changed circumstances of late eighteenth- and nineteenth-century Ireland than the confessional confrontations of the Early Modern period.

Habsburg Royal Hungary thus offers an interesting religious counterpoint to seventeenth-century Ireland helping to reveal a number of factors which weakened the hand of the state church. Nevertheless, in seeking the additional reasons for the consolidation of Catholicism, as opposed to the mere weakness of the reformation, it is the *Hódoltság*, that part of the ancient kingdom of Hungary which fell under Turkish control after 1526 which offers the best evidence. Like Ireland the *Hódoltság* was perceived in Rome as lying *in partibus infidelium*, that is under the jurisdiction of a non-Catholic power. In some respects the official position of Catholicism in the *Hódoltság* was more favourable than in seventeenth-century Ireland. The Ottoman empire was erratic in its attitudes towards its Christian subject population, and generally more alert to the possibility of extracting cash than conversions but it upheld officially notions of Islamic tolerance of Christianity as a religion of the book. Christians were forbidden to attempt to convert Muslims, to extend or acquire new churches but retention of those they possessed was officially permitted. Officially, therefore, Turkish attitudes towards Catholicism were far more lenient than those of the Irish state. Moreover the Habsburgs of Royal Hungary continued to stake a claim to the area of the *Hódoltság*, professing concern with the plight of the Catholic population of the area and royal bishops drew some revenues from their titular *Hódoltsági* sees. Yet, despite these apparent advantages, Irish Catholicism proved incomparably more vibrant that its equivalent in the *Hódoltság*. Indeed the Hungarian Catholic establishment proved even less successful as an evangelical institution in this field of missionary endeavours on its doorstep than the Irish Protestant church.

Of critical importance in this regard was the difference in the number of Catholic clergy deployed on the ground. In utter contrast with the *Hódoltság* seventeenth-century Ireland was astonishingly well endowed with trained Catholic priests, despite the official prohibited status of the church. In 1622, Pázmány estimated the total number of secular priests serving in the *Hódoltság* at less than 20, who were supplemented by a handful of Jesuits and Franciscans.[45] As primate he was in a position to know. His estimate is confirmed by contemporary Jesuit reports. Everywhere Jesuit missionaries penetrated they were shocked at the extent of the shortage of priests, a shortage which seems to have been matched by a lack of Protestant preachers in many areas.

What priests people had come into contact with seem to have been survivors from an older era, probably consecrated by Franciscan bishops from Bosnia and frequently lacking any education whatsoever. So acute was the lack of trained clerical personnel that the Jesuits reported their great difficulty in leaving any locality, with even the Protestant population desperate for them to remain, offering to convert in return for continued spiritual supervision. The Croatian superior of one mission, Marino de Bonis, reported to Rome in 1617: 'To the preaching which was done in the country, flocked not only the Catholics, but also the Turks themselves and the heretics, many of whom were greatly edified by what they heard in the sermons.'[46] By early 1620, Marino de Bonis had encountered offers of this nature on so many occasions that he wrote rather sorrowfully to Rome: 'If here there were workers, great good would be accomplished. Wherever I go they say to me, if you stay with us, we will all be of your religion. And what can I alone do?'[47]

In Ireland, written within little more than a decade of Pázmány's report to Propaganda Fide, there is a report extant concerning the isolated, mountainous and largely Gaelic combined dioceses of Ardfert and Aghadoe. From this it can be deduced that there was a pool of perhaps 50 seminary educated priests in the dioceses who included among their numbers at least six doctors of theology, Richard O'Connell, Maurice O'Connell, Donncha Falvy, Florence O'Mahony, James Pierce and John O'Connor and three doctors of canon law, Edmund Pierce, Malachy O'Connell and another unnamed individual.[48] In other words, in both number and quality the Catholic clergy resident in one isolated Irish county far outstripped the entire Hodoltság. The diocese of Ardfert was the landing site of the Italian nuncio, GianBattista Rinuccini, and his party in 1645. Rinuccini's most important subordinate, Dionysio Massari, wandered away from the landing camp and recorded with amazement that despite the persecution of their heretical enemies 'amidst mountains and barren places, I found the knowledge of the holy Catholic faith flourishing' and that he did not encounter a single child who could not recite the 'Our Father' and the 'Hail Mary'.[49] A pattern of numerous Catholic clergy at work on the ground was not confined to merely isolated and mountainous regions far from central authority. The bishop of Ossory, David Rothe, believed that his centrally located and accessible diocese was overstaffed by the 1630s.[50] As noted previously, the historiographically neglected key to this phenomenon was the burgeoning network of continental Irish colleges which collectively outperformed the Protestant university at Trinity College in terms of supplying educated clergy to staff the shadow Catholic Church in Ireland. Practically all the trainees in the colleges were either Old English or Gaelic Irish natives who sought priestly training on the continent and who generally returned to their place of origin having completed their education. Thus the process of Catholic consolidation was largely accomplished by means of a native and not a foreign clergy. By comparison, however, Catholicism in the Hungarian *Hódoltság* was simply not strong enough to generate a sufficient number of would-be ordinands who were prepared to go abroad for priestly training and that factor outweighed the mission possibilities and even the institutional position of the Catholic Church within neighbouring Royal Hungary.

Taken together, therefore, these several areas of Hungarian comparisons suggest that the absence of effective persecution of Catholicism was not necessarily the only reason for the relative failure of the Protestant church in Ireland during the seventeenth century. In even more unpropitious circumstances than those facing the House of Stuart in Ireland, the Habsburgs managed to stimulate the development of a native Hungarian Catholic tradition but they did so by dint of a difficult process of persuasion and cooperation with the native Magyar and Slav elites within the kingdom rather than by their replacement. Turkish Hungary, on the other hand, offers a completely different level of comparison for it indicates that the strength of the position which Irish Catholicism gained was largely generated by a process of elite engagement with the Catholic Church which found a critically important expression in the creation of the network of Irish continental colleges. The interlocking of these two differing processes form the background to the historical connundrum that Ireland became the only country in Europe where the counter-reformation succeeded against the wishes of the state thus placing a Catholic problem at the very heart of the nascent British empire.

Notes

1. M. J. Brenan, *An Ecclesiastical History of Ireland* (Dublin: J. Coyne, 1840); T. Walsh and D. P. Conyngham, *Ecclesiastical History of Ireland* (New York: Kenedy, 1898); M. V. Ronan, *The Reformation in Dublin 1536–1558 (from original sources)* (London: Longmans, Green & Co., 1926).
2. *An Camán The Organ of Irish-Ireland Incorporating 'An Claidheamh Soluis' agus 'Fáinne an Lae'; A Review of National Affairs: Athletics, Language, Literature, Art, Industry*, 30 July 1932, p. 98. *An Camán* was the official newspaper of *Conradh na Gaeilge* (The Gaelic League) as well as of the Gaelic Athletic Association, but the vast majority of the expenses associated with the project were borne by the latter.
3. *An Camán*, 9 December 1933.
4. Ibid.
5. T. Browne, *Ireland: A Social and Cultural History, 1922–1979* (Glasgow: William Collins, 1981), pp. 35–9.
6. Walsh and Conyngham, *Ecclesiastical History of Ireland*, p. vii.
7. C. Hollis, 'Religious Persecution' in *European Civilisation: Its Origin and Development* ed. E. Eyre, 7 vols (Oxford, 1934–9), pp. iv, 723.
8. Brenan, *Ecclesiastical History of Ireland*, p. v.
9. J. Gregory, 'The Making of a Protestant Nation: "success" and "failure" in England's Long Reformation' in *England's Long Reformation 1500–1800*, ed. N. Tyacke (London: UCL Press, 1998), pp. 307–34.
10. D. Underdown, *Revel, Riot and Rebellion: Popular Politics and Culture in England 1603–1660* (Oxford and New York: Oxford University Press, 1987), p. 129.
11. T. Ó hAnnracháin, 'Catholicism in Early Modern Ireland and Britain', *History Compass*, III (2005) BI 143: 1–17.
12. P. Collinson, 'Comment on Eamon Duffy's Neale Lecture and the Colloquium' in *England's Long Reformation 1500–1800*, ed. N. Tyacke (London: UCL Press, 1998), pp. 71–86.
13. K. W. Nicholls, *Gaelic and Gaelicized Ireland in the Middle Ages*, 2nd edn (Dublin: Lilliput Press, 2003), pp. 105–30.
14. Among the more notable contributions to this debate have been N. Canny, 'Why the Reformation Failed in Ireland: une question mal posée', *Journal of Ecclesiastical History*, XXX (1979) 423–50; K. Bottigheimer, 'Why the Reformation Failed in Ireland: Une Question Bien Posée' in *Journal of Ecclesiastical History*, XXXVI (1985) 197–200; S. G. Ellis, 'Economic Problems of the Church: Why the Reformation Failed in Ireland', *Journal of Ecclesiastical History*, XLI (1990) 239–65; K. Bottigheimer and U. Lotz-Heumann, 'The

Irish Reformation in European Perspective', *Archive for Reformation History,* LXXXIX (1998) 313–53; J. Murray, 'The Diocese of Dublin in the Sixteenth Century: Clerical Opposition and the Failure of the Reformation' in *History of the Catholic Diocese of Dublin,* ed. J. J. Kelly and D. Keogh, (Dublin: Four Courts Press, 2000), pp. 92–111.

15. P. Corish, *The Irish Catholic Experience: A Historical Survey* (Dublin: Gill and MacMillan, 1985), p. 70.
16. H. R. Jefferies, 'The Irish Parliament of 1560: The Anglican reforms authorized', *Irish Historical Studies,* XXVI (1988) 128–40.
17. Murray, 'Diocese of Dublin', p. 99.
18. Underdown, *Revel, Riot and Rebellion,* p. 129; Gregory, 'Making of a Protestant Nation', p. 309.
19. For the best analysis of the sectarian violence which engulfed Ireland in 1641–1642 see N. Canny, *Making Ireland British 1580–1650* (Oxford: Oxford University Press, 2002), pp. 461–550; the religious basis of the confederate Catholic association is examined in T. Ó hAnnracháin, *Catholic Reformation in Ireland: the Mission of Rinuccini* (Oxford: Oxford University Press, 2002), especially pp. 16–81; see also M. Ó Siochrú, *Confederate Ireland: A Constitutional and Political Analysis* (Dublin: Four Courts Press, 1999).
20. In this regard see, for instance, *Political Thought in Seventeenth-Century Ireland: Kingdom or Colony* (Cambridge: Cambridge University Press, 2000), ed. J. Ohlmeyer, especially pp. 1–34 and 35–55.
21. C. Brady, *The Chief Governors: The Rise and Fall of Reform Government in Tudor Ireland, 1536–1588* (Cambridge: Cambridge University Press, 1994), especially pp. 209–44.
22. D. Edwards, 'Ideology and Experience: Spenser's *View* and Martial Law in Ireland' in *Political Ideology in Ireland,* ed. Hiram. Morgan (Dublin: Four Courts Press, 1999), pp. 127–57.
23. W. Palmer, 'That Insolent Liberty: Honour, Rites of Power and Persuasion in Sixteenth century Ireland', *Renaissance Quarterly,* XLVI (1993) 308–27.
24. This remains an under-researched phenomenon: for an introduction see T. J. Walsh, *The Irish Continental College Movement: The Colleges at Bordeaux, Toulouse, and Lille* (Dublin and Cork, 1973).
25. Bottigheimer and Lotz-Heumann, 'Irish Reformation in European Perspective', pp. 313–53.
26. Ibid., p. 336.
27. J. Pánek, 'The Religious Question and the Political System in Bohemia before and after the Battle of the White Mountain' in *Crown, Church and Estates: Central European Politics in the Sixteenth and Seventeenth Centuries,* ed. R. J. W. Evans and T. V. Thomas (London and New York: Palgrave MacMillan, 1991), pp. 129–48.
28. V. L. Tapié, *The Rise and Fall of the Habsburg Monarchy* (London: Pall Mall Press, 1971), pp. 93–9.
29. O. R. Chaline, *La Reconquête Catholique de L'Europe Centrale XVIe*-XVIIe siècle (Paris: Les Editions du Cerf, 1998), pp. 47–8, 76.
30. J. R. Palmitessa, 'The Reformation in Bohemia and Poland' in *A Companion to the Reformation World,* ed. R. Po-chia. Hsia (Oxford: Blackwell Publishing, 2004), p. 200.

31. Oliver Cromwell to Speaker Lenthall, 17 September 1649, in *The Writings and Speeches of Oliver Cromwell* ed. W. C. Abbott, 3 vols (Oxford: Clarendon Press, 1989), pp. ii, 128.
32. P. Corish, 'The Cromwellian Regime, 1650–60', in *A New History of Ireland III,* ed. T. W. Moody, F. X. Martin and F. J. Byrne (Oxford University Press, Oxford, 1976), pp. 375–86.
33. T. C. Barnard, *Cromwellian Ireland: English Government and Reform in Ireland 1649–1660* (Oxford: Oxford University Press, 1975), pp. 171–82; Corish, 'The Cromwellian Regime', pp. 375–86.
34. Barnard, *Cromwellian Ireland,* pp. 175–80.
35. Chaline, *Reconquête Catholique,* pp. 37–9.
36. In the immediate background to the rebellion of the 1640s were strong feelings of resentment at the monarch's failure to convene a Diet within the mandatory three-year period: see *Esterházy Miklós Nádor Iratai I. Kormányzattörténeti Iratok: Az 1642 Évi Meghiúsult Orszaggyulés Idõszaka (1640 December - 1643 Március)* ed. I. Hajnal (Budapest: Esterházy Pál Herceg Kiadása, 1930), pp. xiv–xv.
37. I. G. Tóth, 'Old and New Faith in Hungary, Turkish Hungary, and Transylvania' in *A Companion to the Reformation World,* p. 206.
38. P. Schimert, 'Péter Pázmány and the Reconstitution of the Catholic Aristocracy in Habsburg Hungary, 1600–1650', (Unpublished PhD thesis, University of North Carolina, Chapel Hill, 1989).
39. For a succinct English language analysis of this governmental system, see R. J. W. Evans, *The Making of the Habsburg Monarchy 1550–1700: An Interpretation* (Oxford: Oxford University Press, 1979), pp. 235–74.
40. T. Barnard, *A New Anatomy of Ireland: The Irish Protestants, 1649–1770* (New Haven, CT and London: Yale University Press, 2003), pp. 23–4.
41. M. Őry and F. Szabó, 'Pázmány Péter (1570–1637) in *Pázmány PéterVálogatás Műveiből,* ed. M. Őry, F. Szabó, P. Vass, 3 vols (Budapest: Szent István Tarsulat, 1983), pp. i, 11–84.
42. A. Ford, *The Protestant Reformation in Ireland, 1590–1641,* 2nd edn (Dublin: Four Courts Press, 1997), p. 194.
43. Ó hAnnracháin, *Catholic Reformation,* pp. 39–81.
44. D. Cregan 'The Social and Cultural Background of a Counter-Reformation Episcopate, 1618–1660', in *Studies in Irish History Presented to R. Dudley Edwards,* ed. A. Cosgrove and D. MacCartney (Dublin: University College Dublin, 1979), pp. 85–117.
45. I. G. Tóth, 'A Propaganda megalapítása és magyarország', *Történelmi Szemle,* XLII (2000) 1–2, 43.
46. 'Alle prediche che si facevano per le campagne, concorrevano non solo i cattolici, ma anche li turchi stessi et heretici, li quali molto s'edificavono dalle cose udite nelle prediche.' Brief relation of the state of the Christians subject to the Great Turk in Hungary and Slavonia by Marino de Bonis S. J., Superior of the mission, 1617, in *Erdélyi és Hódoltsági Jezsuita Missziók 1–2 1617–1625,* ed. B. Mihály, F. Ádám, L. László and M. István (Scriptum KFT, Szeged, 1990), p. 299.
47. 'Se qua ci fussero operarii, gran bene si faria. Dovunque vado mi dicono, se voi starete con noi, tutti saremo della vostra fede, E che cosa posso far io solo?' Marino de Bonis to Marino Gondalán (1 January1620); ibid., p. 378.

48. B. Jennings, 'Miscellaneous documents II 1625–40' in *Archivium Hibernicum,* xiv no. 9 (1949); A.P.F. (Archivio Storico della Sacra Congregazione per l'Evangelizzazione dei Popoli o de 'Propaganda Fide') 'S.O.C.G.' (Scritture Originali referite nelle Congregazioni Generali) 140, ff. 69r–77r.
49. S. Kavanagh, ed., *Commentarius Rinuccianus, de sedis apostolicae legatione ad foederatos Hiberniae Catholicos per annos 1645–9,* 6 vols (Dublin: Irish Manuscripts Commission, 1932–1949), 13; APF., 'Miscellaneae. Varie', 9, p. 56.
50. Ó hAnnracháin, *Catholic Reformation,* p. 58.

Further reading

Bottigheimer, Karl and Ute Lotz-Heumann, 'The Irish Reformation in European Perspective', *Archive for Reformation History,* LXXXIX (1998) 313–53.

Canny, N. *Making Ireland British 1580–1650* (Oxford: Oxford University Press, 2002).

Corish, P. *The Irish Catholic Experience: A Historical Survey* (Dublin: Gill and MacMillan, 1985).

Ford, A. *The Protestant Reformation in Ireland, 1590–1641.* 2nd edn (Dublin: Four Courts Press, 1997).

Gregory, J. 'The Making of a Protestant Nation: 'Success' and 'Failure' in England's Long Reformation'. In *England's Long Reformation 1500–1800,* ed. N. Tyacke (London: UCL Press, 1998), pp. 307–34.

Ó hAnnracháin, T. *Catholic Reformation in Ireland: The Mission of Rinuccini* (Oxford: Oxford University Press, 2002).

Palmitessa, J. R. 'The Reformation in Bohemia and Poland'. In *A Companion to the Reformation World,* ed. R. Po-chia Hsia (Oxford: Blackwell Publishing, 2004).

3
Anti-Catholicism and the British Empire, 1815–1914

John Wolffe

In 1897, Queen Victoria's Diamond Jubilee year, Mr A. C. Howe, a prominent local merchant, was moved to deliver a powerful speech at a public meeting in Victoria, British Columbia. He proclaimed that

> Providence has thrown in our way this opportunity to declare for the principles which are the foundation of the British Empire; principles which have made the Briton's name respected from the rising of the sun to the going down of the same; and if we do not take up the challenge of the Romish Church, we are not worthy to take our stand beneath the banner of St George.

The circumstances that stirred such high stakes rhetoric might seem relatively trivial. Victor M. Ruthven, a convert from Roman Catholicism, had been delivering a series of Protestant lectures, and selling a book entitled *Crimes of Romish Priests*. Local Catholics had instigated proceedings against Ruthven on the grounds that this work was indecent. The magistrates had dismissed the charge, but Howe maintained that the prosecution was an attack on free speech and an attempt to reverse the work of 'Latimer and Ridley, who spoke for liberty' and of 'Drake, Hawkins and Frobisher who fought for liberty'. Here 'in our own fair city of Victoria', Rome was showing 'the same old cloven hoof which trod upon the neck of Europe for ten sad centuries' and 'the same old spirit which built the fires at Smithfield'.[1] Howe's speech concisely illustrates a central theme of this chapter, the existence of a strong link between imperial and anti-Catholic rhetoric, and the insistence that both were founded in a love of liberty. Such language also served to assert the maintenance of strong historic links between metropole and periphery. Granted that the ties between anti-Catholicism and empire were less uniform and more complex than Mr Howe perceived them to be, his speech suggests that the subject is one well worthy of investigation.

The association between Protestantism and the British Empire merits scholarly exploration at two different levels. At the first more generalized level, Brian Stanley and Andrew Porter have ably investigated the sometimes ambivalent relationship between institutional Protestant missions and British overseas expansion.[2] Such missionary activity was seldom explicitly anti-Catholic, only becoming so when there was specific rivalry or confrontation with Roman Catholic missions, as for example in Tahiti in 1844.[3] This chapter, however, is concerned with the second level of organized explicit anti-Catholicism, a phenomenon primarily apparent in white settler colonies and which has yet to receive the scholarly attention it deserves. There have indeed been significant but not exhaustive studies of anti-Catholicism in a national and regional context, notably on Britain and Ireland in the mid-nineteenth century, and on Canada, particularly in the late nineteenth century.[4] There are surveys of the role of Protestantism in British national identity in the nineteenth and early twentieth centuries, and of the long-term development of the organized British anti-Catholic movement.[5] However, coverage of Australia, New Zealand, South Africa and other parts of the Empire is more limited and, as Stuart Piggin observes in relation to Australia, 'has been much better told from the Catholic side than the Protestant'.[6] Above all, there has not yet been any systematic exploration of the topic in an overall imperial context, as a divisively cohesive 'empire of religion'. This chapter constitutes a preliminary foray into this wider field. It will begin with discussion of some aspects of anti-Catholicism and empire over the nineteenth century as a whole. Attention will then turn to the activities of the Imperial Protestant Federation (IPF) in the 1890s and 1900, the most significant attempt to make explicit links between militant Protestantism and the British Empire. The IPF's history effectively illustrates both the considerable short-term potentialities of an anti-Catholic vision of empire, and its longer-term fragility and divisiveness.

Defending the Protestant Empire 1815–1890

The early nineteenth-century expansion of settlement in British North America and Australia – and eventually New Zealand and southern Africa – coincided with a revival in anti-Catholic activity in Britain. The late eighteenth century represented something of a hiatus in the

British tradition of anti-Catholicism, partly because of reaction against the savagery of the Gordon Riots of 1780; partly because events in France from 1789 onwards weakened the previously axiomatic equation of Protestantism and Francophobia. Nevertheless, the preconditions for its revival were established by the Union of the British and Irish Parliaments in 1800. Protestants were liable to perceive the presence of a substantial Catholic minority within a single unitary state as a dangerous anomaly. The passing of Catholic Emancipation in 1829, by removing constitutional barriers to Catholic participation in political life, made it seem doubly important that national Protestant identity should be clearly reasserted. These constitutional and political developments were paralleled by the emergence of increasingly explicit anti-Catholic tendencies in the expanding evangelical movement. Evangelicals were concerned to convert Roman Catholics as well as to limit their political and social influence. Their efforts were initially centred on Ireland, in the so-called 'Second Reformation' movement of the 1820s, which revived in the Irish Church Missions to Roman Catholics of the 1850s.[7] As the mid-century Irish diaspora, of both Protestants and Catholics, gathered momentum such tensions and aspirations were carried to other parts of the empire.

In the early to mid-nineteenth century reference to the concept of 'empire' was quite common in anti-Catholic literature. In 1828, an anonymous pamphleteer related it to current interest in the application of the prophecies of the Apocalypse. He perceived the hand of providence in England's triumph over the 'Atheistic Empire' of France and in the 'increased dominion of the Church of England over the remote dependencies of the Empire'.[8] In 1829, Richard Warner, rector of Great Chalfield, Wiltshire, argued that Catholic Emancipation was incompatible with the 'liberty, laws and Protestant succession of the British Empire'.[9] Around 1840, the Protestant Association published its 'National Standard for the Maintenance of Civil and Religious Liberty in the British Empire', arguing that a new parliamentary oath to renounce the allegedly intolerant and persecuting doctrines of Rome was necessary to prevent 'papal demagogues and…bishops' from dismembering the empire.[10] During the repeal campaign of 1843, the Irish Protestant agitator Robert M'Ghee perceived a trial of strength between Queen and Pope for rule 'in the British empire'.[11] In the late 1840s and early 1850s the National Club, which sought to promote the anti-Catholic cause in parliament, made its successive published statements under the overall title of *Addresses to the Protestants of the Empire*.[12]

The concept of empire expressed in such texts reflected the enduring influence of the language of the Act of Appeals of 1533, which had asserted that 'this realm of England is an empire', in other words an independent state that should be free from papal interference.[13] The unity of that 'empire' seemed to depend on the successful promotion of a common Protestant religious profession. Although initially the focus was primarily limited to Britain and Ireland, by mid-century a wider perspective was developing, for example in a pamphlet published in 1851 urging a robust legislative response to the recently established Roman Catholic episcopal hierarchy. The author, a naval officer, argued that it was essential that the mother country should set an example to the colonies 'of sound loyalty and the Protestant faith; applicable, whenever they may require it, to their respective communities as integral portions of the Empire.'[14]

In the 1820s tensions in Ireland were already stirring echoes in other parts of the globe. In 1823 William Parker, a recent settler in the Cape of Good Hope, published a pamphlet in two parts, with one of those splendidly cumbersome early nineteenth-century titles that serves to sum up the content of the whole work:

> *The Jesuits Unmasked; Being an Illustration of the Existing Evils of Popery in a Protestant Government, Duly Exemplified in Letters from the Cape of Good Hope, where the English Settlers have been Exposed to Great Distress and Oppression from the Practices and Influence of Popish Emissaries; and where a Deep-Laid Plan, Originating at the Jesuits Institution at Stonyhurst, in Lancashire, of Extirpating Protestantism had Commenced.*[15]

It is both amusing and suggestive to set Parker's title against the judgement of the most recent authoritative history of Christianity in South Africa that 'A history of the Christian Church in South Africa before 1860 might have ignored the Roman Catholic Church altogether, so limited was its contact with other denominations and so small the number of its adherents in South Africa.'[16]

The pamphlet is significant though, not only in providing a striking illustration of the disjunctions between perception and reality that run through the subject-matter of this chapter, but also in illustrating how anti-Catholicism both drew upon and reinforced geographically widespread imperial networks. Parker was a Protestant from Cork, who in 1819 led a party of settlers from Ireland to the Cape.[17] On arrival in the colony, he met Lieutenant-Colonel Christopher Bird, the chief secretary, and during their conversation, claiming that he was unaware Bird was a Catholic, he denounced 'Popery' as 'the bane of Ireland'.[18] With this unprepossessing beginning it seems that the two men took a strong dislike to each other. During the next few years Parker developed a catalogue of grievances against the colonial government, which he believed had allocated his party land that was manifestly of too poor a quality to support them all. He also perceived the government, which in 1821 made an annual grant of £75 to Father Scully, who served the tiny Catholic community in Cape Town, to be giving undue favour to the Roman Catholic Church.[19] In Parker's mind all these circumstances were attributable to the alleged Jesuitical machinations of Colonel Bird, who he discovered had been educated at Stonyhurst and was a brother of the alleged 'chief-priest of the [Jesuit] Order, at Preston'.[20] Parker, whose perceptions of Roman Catholicism had been formed in his native Ireland and who was aware of the anti-Protestant Pastorini prophecies currently circulating at home, saw the infant South African Church as the outpost of a massive global conspiracy.[21] He transmitted his allegations to London, in petitions to parliament, letters to the influential, from the King downwards, and eventually through his published pamphlet. His attack on Bird led to the unfortunate chief secretary being required formally to take the anti-Catholic oath and, when he conscientiously refused, to him being superseded in his post.[22] Parker's publications also gave an added stimulus to anti-Catholic feeling in Britain at a time of ongoing debate over Emancipation.

In Australia too the early development of the local Catholic community rapidly stimulated a perception of papal conspiracy. In 1840, the laying of a foundation stone for a Catholic chapel in Harington Street, Sydney and a public meeting to form a branch of the Catholic Institute of Great Britain provoked the Protestant response:

> Let Papists be honest, and let them not endeavour to throw us off our guard by disavowing or concealing the principles of intolerance and persecution on which their Church, whatever individuals among them might do, would act towards us the moment it had the power.[23]

During the 1840s John Dunmore Lang, the leading Presbyterian minister in New South Wales, feared that Australia would be transformed into 'a province of the Popedom'.[24] He advocated Scottish Protestant emigration to counteract Catholic influence. Lang travelled to Britain to promote his campaign, and in 1847 in Edinburgh, at a time of considerable anti-Catholic excitement in England and Scotland, appealed directly to British and Irish Protestants in a pamphlet entitled *Popery in Australia and the southern hemisphere, and how to check it effectually.* In Victoria in the early 1860s conspiracy theories were rife in the Protestant press and the Pope's forces were thought to be aiming at the dismemberment of the empire. In this atmosphere an initially consensual proposal to reduce the Governor's salary came to be perceived as a Jesuitical plot.[25] In 1864, one Thomas Slater published a pamphlet in Melbourne recounting his earlier travels around the globe in which he perceived Jesuits to be at work everywhere, notably in North America and the Pacific islands. Slater had settled in Australia in the mid-1850s and claimed to have 'observed the secret organisations of the Jesuits permeate through all [Australian] society'. They were, he believed, responsible for numerous violent or unexplained deaths.[26]

Elsewhere too small-scale local Catholic activity was readily perceived as a manifestation of much more powerful global forces. When in 1847 an Irish crowd in Halifax Nova Scotia burnt an effigy of the British prime minister a local Presbyterian newspaper believed that the incident 'reminded our citizens of the burnings and robberies and murders of Ireland, and the treatment they might expect to receive should the Catholics unfortunately gain an ascendancy in this place'.[27] In Grahamstown, South Africa, an anti-Catholic pamphlet by the Rev. William Sargeant, published in 1868, was provoked by the activities of the local Catholic Church, led by the energetic Bishop Patrick Moran, but developed into a much broader attack on the alleged political influence of the papacy.[28] It is hardly surprising that in such an environment Catholics could themselves develop a converse belief in Protestant global conspiracy against them, evident in Moran's own subsequent ministry from 1869 as Bishop of Dunedin in New Zealand.[29] Consequent defensiveness was liable further to stimulate Protestant suspicions.

Following the formation of organized Protestant societies and the commencement of anti-Catholic periodicals in Britain and the United States from the late 1820s onwards, there were similar developments in Australia and Canada during the 1840s and 1850s. The *Sydney Protestant Magazine* began publication in April 1840.[30] A Canadian Protestant Association was formed in 1854, 'to support and defend their liberties from the designs and intrigues of Popish mercenaries, and as for the maintenance of the public peace and tranquility.' Although the qualifications for membership emphasized spiritual credentials, the purpose thus appeared more political than religious.[31] The objects of the Protestant Defence Alliance of Canada, formed in Montreal in 1875, reflected a consciousness of vulnerability in the face of the Catholic majority in Quebec, and aimed to resist 'all efforts on the part of the

Roman Catholic hierarchy to violate the principles of civil and religious rights and liberties.'[32] The Orange Order was also a major force in Canada and Australia, as in Scotland and Ireland but in contrast to England, where its presence was more limited and localized. Its influence extended beyond its own formal lodge structures, being evident in bodies such as the Canadian Protestant Association, which referred to its members as 'brothers', initiated 'candidates' into membership, and provided for sanctions against those who divulged 'the private affairs of the Association'.[33]

From the middle of the century onwards, a number of anti-Catholic orators travelled widely around British possessions and the United States, thus strengthening a sense of common cause and identity. In the 1850s, John Orr, a black man calling himself the Angel Gabriel, provoked riots in the west of Scotland, in several North American cities and then in his native British Guyana.[34] In the same period the Italian Alessandro Gavazzi toured Britain, Ireland, the United States and Canada, where in 1853 his meetings led to riots in Quebec and Montreal.[35] The most energetic traveller of them all was the Canadian Charles Chiniquy, a former Catholic priest, who following his excommunication in 1858 joined the Presbyterian Church. In 1860, Chiniquy made his first visit to Britain and Ireland where he spoke in 85 different localities. He returned there in 1874, visited Australia in 1878, New Zealand in 1880 and Britain again in 1883. His final visit to Britain was made in 1896–1897, when, now in his late eighties, he impressed his hosts with his continuing vigour.[36]

The Imperial Protestant Federation

Given the evident strength of anti-Catholicism around the empire throughout the nineteenth century, it might at first sight seem surprising that no attempt to organize it at an imperial level was made until the 1890s. The formation of the IPF needs, however, to be seen in the context both of contemporary imperialist enthusiasm and of *fin de siècle* Protestant insecurity. From around 1880 onwards some of those who in an earlier generation would have been staunchly anti-Catholic were becoming more ambivalent towards ultra-Protestantism.[37] As with the initial formation of domestic anti-Catholic societies around the time of the passing of Catholic Emancipation 60 years before, the impulse to organize was stirred by a consciousness that Protestant identity was becoming more vulnerable and contested. Immediate stimuli came from the continuing political repercussions of the Home Rule struggle in Ireland, and the long-running evangelical campaign against Anglican ritualism, which was becoming a significant issue in colonial churches as well as in England itself.

The main architect of the IPF was Walter Walsh (1847–1912), one of the select group of Victorian Anglican evangelicals who made successful careers as anti-Catholic writers and organizers. Walsh gained experience in working for the Irish Church Missions and the Protestant Reformation Society, service as political agent for the anti-convent MP Charles Newdegate, and journalism for the *Press and St James Chronicle* and the *English Churchman*.[38] In 1889, he founded the *Protestant Observer*, a monthly defined by its anti-Catholicism, which he edited for the rest of his life. From the outset he gave extensive coverage to colonial developments, and claimed that the *Protestant Observer* was widely circulated across the empire. A feature of Walsh's propaganda was a readiness to draw strong parallels between diverse geographical contexts: thus the Salisbury administration was accused of 'reprehensible' inconsistency when it handed Protestants in Madagascar over to the rule of a Catholic power, France, while it continued to uphold Ulster Protestant resistance to Home Rule.[39] The impact of Catholic marriage laws in Quebec was seen as a foretaste of what would happen in Ireland if Catholics ever gained political control there.[40] Walsh's journalistic skills attained their fullest and most influential flowering in 1897, with the publication of *The Secret History of the Oxford Movement*, a masterpiece of anti-Anglo-Catholic propaganda which was widely read and circulated on account of its fluent literary style and quasi-scholarly credibility.

In the meantime, Walsh sought to progress his vision of imperial anti-Catholic vigilance through promoting closer associations between the numerous fragmented Protestant societies. In September 1896, he articulated his strategy in *The Protestant Observer*:

> The Colonies will give us their united support in maintaining the Protestant Constitution, the Protestant Coronation Oath, and the Protestant succession to the Throne, and in furthering such laws as shall render their subversion an impossibility, while such a Federation will, in its turn, greatly strengthen the Colonies by looking after their interests in the British Parliament[41]

Such hopes were reciprocated in Australia, where the *Victorian Standard*, an Orange newspaper, described the proposal as 'heroic, and…to be commended on account of the ceaseless activity of the Jesuits, priests, and Romanists, generally, to regain the lost power of the Papal Church.'

It thought such an organization would be especially valuable in mobilizing 'numbers of good Protestants who elect to remain outside the Orange Society for reasons which, to them, seem sufficient'. It might be termed 'universal Orangeism' and would render the 'Protestant British Empire' safe against the assaults of Rome.[42]

However, this first attempt to form a federation of Protestant societies foundered because of a lack of distinct purpose and objectives.[43] After protracted negotiations, the IPF was eventually constituted early in 1898, under the chairmanship of Colonel Thomas Myles Sandys (1837–1911), who following his military career, was elected Conservative MP for Bootle in 1885 and subsequently became Grand Master of the Loyal Orange Lodge of England.[44] The secretary was Edward H. Garbett (son of the prominent evangelical clergyman Canon Edward Garbett), assisted by Walsh's own son James. Initial growth was rapid and by October of 1899 20 organizations in Britain had affiliated.[45] Almost immediately, however, the IPF found itself encountering opposition from within the anti-Catholic movement itself. The underlying issue was the reluctance of other societies to allow the IPF to take political action on its own initiative rather than merely serving as a consultative body. As a consequence, the IPF found itself competing with a rival umbrella organization, the London Council of United Protestant Societies.[46] The dispute combined with a preoccupation with mobilizing Protestant forces for the general election of 1900 seems initially to have prevented the IPF from living up to its name by substantive activity outside Britain.

Then, in January 1901, Queen Victoria died. Inevitable and foreseeable though this event was, it heightened concern for the Protestant identity of the empire as symbolized by the monarchy. Four years before, celebration of the Diamond Jubilee had been overshadowed by recognition of the Queen's mortality, and fears that the Prince of Wales was subject to Catholic influence.[47] In its obituary leader, *The Protestant Observer* asserted that 'To the Protestants of the British Empire the relationship of our late Queen to the Protestant Religion has ever been one of vast interest and importance.' The article pointed out that she had sworn to maintain the Protestant religion, and claimed that her personal sentiments were apparent in her distaste for ritualism, her affection for the Church of Scotland, her marriage to the Lutheran Prince Albert and in her strong opposition to the 'Papal Aggression' of 1850.[48] Walsh was to develop this portrayal in a full-length book, *The Religious Life and Influence of Queen Victoria,* published in 1902.[49] Although in reality Victoria's Protestantism was much more liberal than that of the predominantly evangelical readers of the *Protestantism Observer*,[50] the image still seemed a credible one. Walsh hoped that her spirit would descend on her successor.[51]

Sectarian sensitivities impinged significantly on the funeral arrangements. The Bishop of Winchester, Randall Davidson, was alarmed when he heard that the royal family wanted the service to include the Russian Kontakion, which with its implicit prayers for the dead would seem to sanction Catholic tendencies. He recalled:

> I felt it my duty to point out to the King that the use of this Anthem on such an occasion would certainly hurt the feelings of very many and might do real harm. I had some difficulty in getting an interview with him about it, but the moment I explained the matter he saw it and felt that it must be altered[52]

Meanwhile in Canada there was a sharp private dispute between the Catholic prime minister, Sir Wilfred Laurier, and the Protestant governor-general, the earl of Minto, over arrangements for the memorial service in Ottawa. Laurier objected to Minto's plans to make the service in the Anglican cathedral a state ceremony, on the grounds that there was no established church in Canada.[53] Minto accepted Laurier's advice, but with a rather bad grace. In a letter to the Colonial Secretary, Joseph Chamberlain, he attributed Laurier's position to 'bigotted Roman Catholic' influence in the Canadian cabinet, and thought 'that I cannot think Sir Wilfred really in touch with the strong Imperial feeling here'.[54]

The most significant implication of the Queen's death was that the accession of Edward VII led to pressure for the revision of the explicitly anti-Catholic language of the Accession Declaration made when the new monarch first opened parliament, and the Coronation Oath to maintain the Protestant religion. The IPF sought to pre-empt moves for the revision of the Declaration by immediately circulating 'hundreds of thousands' of pamphlets and leaflets 'all over the British dominions'. It was though relieved when on 14 February the new King 'in a clear ringing voice' made the Declaration in its existing form. Their rejoicing, however, was short-lived as the following month parliamentary moves began to have the Declaration revised ready for the next accession. The IPF launched a vigorous agitation, circulating literature throughout the empire, giving permission to colonial societies to reprint it locally, placing a full page advertisement in the *Times* and telegraphing a statement to 'all the principal daily newspapers in Australia, Canada, and other parts of the Empire'. Efforts were particularly focused on Canada in order to counteract extensive Roman Catholic petitioning against the Declaration. The consequent wave of public meetings and petitioning appears to have influenced the government to decide quietly to drop its attempts to change the Declaration.[55]

The agitation over the Accession Declaration promoted the IPF as a genuinely imperial rather than primarily British organization. On 21 March 1902 its Imperial Council unanimously approved a 'Solemn Protestant League and Covenant for the British Empire', an interesting antecedent of the Ulster Covenant of a decade later. They gave thanks for the blessings of the Reformation, but believed the machinations of Jesuits and Ritualists were seeking to undo them. They pledged to resist the political influence of Rome 'by all lawful means', to resist 'if necessary, with our lives' any attempt to place a Roman Catholic on the throne, and to enter into 'a firm Bond for mutual defence and assistance' should there be an attempt 'made in any part of the British Empire to deprive our Protestant brethren of their civil and religious liberties'. On a more spiritual level they affirmed the authority of the Bible and resistance to 'the elevation of the traditions of men to a level with the Word of God'.[56]

The 1902 Annual Report contained extensive accounts of colonial activity. Its affiliated societies now included several in Australia and one in Canada. It claimed to be active in many other colonies, the West Indies, India, Burma, South Africa, Ceylon, Gibraltar, Hong Kong, Malta, Newfoundland, Singapore and the Straits Settlements.[57] It gave particular priority to distributing its literature throughout the Empire: for example, the Tasmanian Christian Colportage Association, locked in a bitter conflict with ritualists on the island, was supplied with publications, including 300 free copies of Walsh's *Secret History of the Oxford Movement*.[58] Meanwhile the International Colportage Association of Canada was reprinting the IPF's publications and requesting that a visiting lecturer should be sent out to them as soon as possible.[59]

In Canada, the IPF did something to fill a vacuum in Protestant organization left by the collapse of the Protestant Protective Association, which had had an important if short-lived political impact in the 1890s.[60] This organization had initially been an offshoot of the American Protective Association, which despite determined efforts to adapt to Canadian political conditions, was still open to the charge of being an alien republican import into Canada at odds with imperial ties.[61] The IPF, however, offered a strong linking of Protestantism and imperialism. It appears though to have worked through existing bodies, particularly the Orange Order, rather than to have promoted new ones.[62]

Stimulated by the IPF, anti-Catholic activism also gathered momentum in Australia and was initiated in New Zealand. The Australian Protestant Defence Association was constituted in June 1902 and held a large public meeting in Sydney Town Hall that October. Its primary objectives were the return of men who would uphold its principles to parliament and local government, and the upholding of Protestant interests in education and the public service. Its detailed printed provisions for the establishment of local branches indicated that it was a well-organized and ambitious body.[63] A Protestant Union of Victoria was also in operation by the end of 1902, with the objectives of 'the enlightening of public opinion and the uniting of Protestants'.[64] The political strength of Protestantism in New South Wales became apparent in subsequent elections and was celebrated in a 'Great Protestant Demonstration in Sydney' in November 1904. It was attended by a crowd estimated at 16,000, including many women, hailed by one speaker as 'of much or even greater important to the politicians than men, because ... with the mothers of Australia rested the Protestantism of the nation'.[65] Meanwhile in 1902 an 'influential gentleman' in New Zealand had asked for advice on forming an 'undenominational Protestant organization which may be able to effectually counteract the alarming growth of the Papal power in all departments of the state in the Legislature', and in 1903, the Protestant Defence Association of New Zealand was formed in Dunedin.[66] Its first annual meeting in May 1904 reported steadily increasing membership and a regular programme of lectures. There was a branch in Auckland by 1907.[67]

In South Africa, the relationship of Protestantism and empire was more ambivalent, because the irreproachably Protestant Boers had recently fought a bitter war in the vain attempt to resist integration into that very empire. For whatever reasons, no anti-Catholic organizations of the kind developing in Canada, Australia and New Zealand were formed at this period, and in March 1903, the *Protestant Observer* noted 'a great need for Protestant work of a distinctively controversial character in South Africa'.[68] In the meantime propaganda from London was clearly helping to mould local perceptions of Roman Catholicism, as revealed in the rather disarming enquiry of a lady from Woodstock, Cape of Good Hope:

> May I ask if there is any other ground for believing that convents in this colony are the same habitations of cruelty as those in

> England and elsewhere, than that they are all conducted by the same system.[69]

The IPF was concerned not only with the growing Roman Catholic presence in the region but also with High Church ritualistic tendencies in the (Anglican) Church of the Province of South Africa, which it believed were compromising the Protestant character of the empire. In May 1903, the *Protestant Observer* approvingly reprinted an article alleging that Anglicanism of this kind was 'working ceaselessly and silently to prevent the union of Boer and Briton because Anglicanism is the drudge of Rome'.[70] It therefore supported the small rival Church of England in South Africa which, despite its origins in the deposition in 1865 of the liberal John William Colenso from the bishopric of Natal for alleged heresy, was now staunchly evangelical.[71]

Decline and realignment

From 1905 onwards, the IPF developed a more specific preoccupation with supporting the cause of Irish Protestants and loyalists, expressed particularly in the commencement of a monthly publication, *Grievances from Ireland*, which was distributed to MPs, peers and newspaper editors. It compiled information regarding the alleged intimidation and oppression of Protestants, while professing to stand 'above the turmoil of ordinary Party strife'. It was though staunchly opposed to Home Rule, 'not, however, because of its merely political aspect, but in consequence of the religious interests involved'.[72] A strong emphasis on Irish issues and denunciation of Home Rule as anti-Protestant also characterized the series of flysheets the IPF published in the 1906 General Election campaign.[73] Increasingly the IPF's conviction that 'The Imperial greatness of the British Nation is built upon Protestantism'[74] seemed to be primarily a reflection on the Irish situation rather than on more far-flung connections. While the strong anti-Home Rule sentiments of the Irish Protestant diaspora nevertheless guaranteed continuing colonial support for the IPF, its original vision had significantly contracted.

In the later years of the Edwardian decade, the IPF appeared to be in decline. The *Protestant Observer* continued to report anti-Catholic activity around the empire, but there was little evidence that much effort was being made to coordinate and stimulate it.[75] Increasingly the IPF came to be overshadowed by the London Council of Protestant Societies, which, as its name suggests, had a weaker sense of the imperial dimension of anti-Catholic efforts.[76] The general weakening of anti-Catholic influence was strikingly demonstrated following Edward VII's sudden death in May 1910. This again brought to the boil the simmering question of the Accession Declaration, with the Liberal government and the new king, George V, both strongly antipathetic to the historic wording.[77] The IPF rallied its forces and despatched 'vast quantities of literature to all parts of the Empire'.[78] Numerous protests were received from the colonies and forwarded to the government. A letter from Ontario affirmed that 'Canada is Protestant and British to the core' and that changing the Declaration would 'do more to loosen the ties that bind Canadians in loyalty to the throne than anything else'. A meeting in Johannesburg was convinced that any alteration in the Declaration would endanger the Protestant succession.[79] Nevertheless, the Asquith government was able, even in the midst of wider constitutional crisis, to get both Houses to agree to change the new King's Declaration to a simple statement that he was a 'faithful Protestant'. In the eyes of the Protestant Reformation Society 'the Vatican and its sympathizers within the Church of England have snatched a triumph, aided thereto by the dependence of the government upon the Nationalist vote, and alas! by the lazy ignorance of to-day in regard to Rome and its doings.'[80] Under the banner headline 'SOLD FOR A MESS OF PAPAL POTTAGE', the *Protestant Observer* urged its readers solemnly to resolve 'never again to support the Liberal or Conservative party in any way whatever', a counsel of deep frustration rather than of effective political strategy.[81]

For the IPF the wound was indeed grievous. At the end of 1910 it moved its London offices from Southampton Street, off the Strand, to 325 Clapham Road, a step justified on grounds of economy and convenience but still a significant retreat from the centre of the imperial metropolis.[82] *Grievances from Ireland* also ceased publication in 1910.[83] Whether from despair or by coincidence, the IPF's two leading figures barely outlived the unrevised Accession Declaration, and with Sandys's death in November 1911 and Walsh's in February 1912, it lost first its valuable figurehead and then its essential ideologue.[84] Although the *Protestant Observer* enjoyed a growth in circulation in the context of the Irish Home Rule crisis, and there was particular demand for its September 1914 issue which contained a leader on 'Rome's hand in the war', appeals for funds for the IPF became increasingly desperate in tone.[85] Moreover, in the medium term the war's impact was much less positive, and the journal ceased publication in 1917. A legacy to the IPF was reported in *The Times* as late as 1926,[86] but the organization made little further impact on the historical record.

Nevertheless, despite lack of leadership from London, organized anti-Catholicism continued to thrive in some colonial settings. During the early twentieth century the southern hemisphere dominions developed further Protestant organizations of the kind that had emerged in Britain and Canada during the nineteenth century. The First World War years saw significant sectarian strife in New Zealand and the formation in 1917 of the Protestant Political Association, which was an influential force until the mid-1920s.[87] In Australia, in 1919, the Victorian Protestant Federation headed its constitution 'For God, King and Empire'.[88] The Protestant Association of South Africa was founded in 1923 and United Protestant Associations were active in New South Wales and Queensland during the inter-war period.[89] In a booklet published in the 1930s, the Protestant Truth Society surveyed the current state of the empire, and discerned the hand of Rome in Malta, Ireland, Australia, New Zealand, South Africa and Canada. Italian fascism was perceived as 'the new arm of the Papacy' while the Vatican was seen as instrumental in bringing the Nazis to power in Germany. 'Rome', it alleged, 'never sleeps, and while our statesmen are almost blind to her ambitions the Vatican plots to thwart us in every part of the world.'[90]

Above all there was Ireland. It was, however, a revealing paradox that the IPF declined at the very period when intransigent Ulster Protestant opposition to Home Rule was gathering momentum, as symbolized by the signing of the Covenant in September 1912. The present analysis though confirms from a different perspective James Loughlin's judgement that the 1910 Accession Declaration controversy had been 'a litmus test of the political relevance of popular Protestantism – one that had decisively demonstrated its diminishing significance.'[91] Despite the religious rhetoric of the Covenant, Sir Edward Carson and the other Ulster Protestant leaders do not appear to have had any direct links with the IPF and the British Protestant movement. Their imperialism, moreover, was a consequence rather than a cause of their Unionism.[92] For its part the *Protestant Observer* interpreted the Covenant in narrowly religious terms as an expression of opposition to Roman Catholicism.[93] In June 1913, a large Protestant gathering in the Royal Albert Hall sought to present the Home Rule crisis in its religious aspect and pledged 'every lawful' support for their Irish co-religionists. The event, at which the speakers were churchmen rather than parliamentarians, appeared, however, somewhat outside the mainstream of Unionist political action.[94] The subsequent enduring polarities of twentieth-century Northern Ireland ensured continued prominence for anti-Catholicism, but the primary stimulus was the specific Irish context rather than the imperial Protestant vision of the founders of the IPF.

The appeal to Protestantism had a significant impact on the religious, social and cultural fabric of the British Empire. That impact was at its most powerful when religious concerns, such as missions to the Irish poor or the struggle against ritualism, could combine effectively with more political and constitutional ones, whether resistance to Home Rule, the defence of the Accession Declaration or above all the abstract but emotive appeal to 'liberty'. At the turn of the twentieth century, however, the rhetoric of the IPF could not obscure the reality that militant Protestantism was too narrow a basis for a cohesive overall imperial Protestant ideology. Its legacy lay rather in its contribution to the development of the narrower sectarian construction of empire that developed in Ulster, and among the Orange Lodges and other loyalist sympathizers in Britain and the colonies, notably in Rhodesia and Natal.[95] Such a vision was always a politically divisive one, and its potential to fragment rather than unite the empire became painfully apparent as events in Ireland unfolded between 1916 and 1923. Back in 1897, Protestants more realistic than Mr Howe of British Columbia had already recognized, however reluctantly, that the expansion of the empire in the Victorian era had been grounded in religious pluralism rather than exclusive anti-Catholicism.[96]

Notes

1. *Protestant Observer,* October 1897, 146; November 1897, p. 174.
2. B. Stanley, *The Bible and the Flag: Protestant Missions and British Imperialism in the Nineteenth and Twentieth Centuries* (Leicester: Apollos, 1990); A. N. Porter, *Religion versus empire? British Protestant missionaries and overseas expansion, 1700–1914* (Manchester: Manchester University Press, 2004).
3. G. B. A. M. Finlayson, *The Seventh Earl of Shaftesbury* (London: Eyre Methuen, 1981), pp. 208–9.
4. E. R. Norman, *Anti-Catholicism in Victorian England* (London: George Allen and Unwin, 1968); D. Bowen, *The Protestant Crusade in Ireland 1800–70* (Dublin: Gill and Macmillan, 1978); J. Wolffe, *The Protestant Crusade in Great Britain 1829–1860* (Oxford: Clarendon Press, 1991); D. G. Paz, *Popular Anti-Catholicism in Mid-Victorian England* (Stanford: Stanford University Press, 1992); J. R. Miller, 'Anti-Catholic Thought in Victorian Canada', *Canadian Historical Review,* LXVI (1985) 474–94.
5. H. McLeod, 'Protestantism and British National Identity 1815–1945', in *Nation and Religion,* ed. P. van der Veer and H. Lehmann (Princeton, NJ: Princeton University Press, 1999), pp. 44–70; J. Wolffe, 'Change and Continuity in British Anti-Catholicism, 1829–1982', in *Catholicism in Britain and France since 1789,* ed. F. Tallett and N. Atkin (London: Hambledon Press, 1996), pp. 67–83.

6. S. Piggin, *Evangelical Christianity in Australia: Spirit, Word and World* (Melbourne: Oxford University Press, 1996), p. 34.
7. For detailed accounts of these developments see Wolffe, *Protestant Crusade*, pp. 1–64, 162; Bowen, *Protestant Crusade*, pp. 208–56.
8. R. H. M., *The Englishman's Polar Star!! Or A Deeply Interesting and Highly Important View of Certain Historical Facts, As Connected with the Honour and Safety of the British Empire; being a Preface to a New Interpretation of Apocalypse of St John by the Rev. G. Croly* (Preston, 1828), pp. 43–5.
9. R. Warner, *Catholic Emancipation, Incompatible with the Safety of the Established Religion, Liberty, Laws and Protestant Succession of the British Empire: An Address to the Protestants of the United Kingdoms* (London, 1829).
10. *A National Standard for the Maintenance of Civil and Religious Liberty in the British Empire*, Publications of the Protestant Association 33 (*c.*1840).
11. R. J. M'Ghee, *The Pope and Popery Exposed in the Present Power & Plots Against the Religion, Laws & Liberties of the Empire: In a Speech Delivered at Exeter Hall, May 10th 1843*, Publications of the Protestant Association 40 (1843).
12. Wolffe, *Protestant Crusade*, p. 217.
13. G. R. Elton, ed., *The Tudor Constitution: Documents and Commentary* (Cambridge: Cambridge University Press, 1965), p. 344.
14. A. Davies, *Proposals for Uniting the British Colonies with Their Mother Country* (London, 1851), pp. 54, 72.
15. (London, 1823).
16. J. Brain, 'Moving from the Margins to the Mainstream: The Roman Catholic Church', in *Christianity in South Africa: A Political, Social & Cultural History*, ed. R. Elphick and R. Davenport (Cape Town: David Philip, 1997), p. 195.
17. Parker, *Jesuits Unmasked*, Part I, xi.
18. Ibid., Part II, pp. 40–1.
19. Ibid., Part I, pp. xviii–xx; W. E. Brown, *The Catholic Church in South Africa from its Origins to the Present Day* (London: Burns and Oates, 1960), pp. 9–10.
20. Parker, *Jesuits Unmasked*, Part I, pp. xiv–xv. See also W. Parker, *Proofs of the Delusion of His Majesty's Representative at the Cape of Good Hope, and of the Iniquity of the Public Officers Acting under His Excellency's Orders, during the Absence of the Right Hon. Lord C. H. Somerset* (Cork, 1826).
21. Ibid., pp. i–iii.
22. Brown, *Catholic Church*, p. 26.
23. *Sydney Protestant Magazine*, September 1840, p. 188.
24. D. W. A. Baker, *Days of Wrath: A Life of John Dunmore Lang* (Melbourne: Melbourne University Press, 1985), pp. 193–4, 222, 247–8.
25. M. M. Pawsey, *The Popish Plot: Culture Clashes in Victoria 1860–1863* (Sydney: Catholic Theological Faculty, 1983), pp. 103–23.
26. T. Slater, *Popery and Despotism Unmasked, by a Light Being Thrown on Some of the Insidious Crimes of Secret Societies at Present Existing in the British Empire and the United States of America* (Melbourne, 1864), p. 40 and *passim*.
27. *Guardian*, 26 March 1847, quoted by A. J. B. Johnston, 'Nativism in Nova Scotia: Anti-Irish Prejudice in a Mid-Nineteenth-Century British Colony' in *The Irish in Atlantic Canada 1780–1900*, ed. T. P. Power (Fredericton: New Ireland Press, 1991), p. 27.

28. W. Sargeant, *A Check to Popery in South Africa or a Sermon Delivered by Dr Ricardo, RCC in St Patrick's Cathedral, Grahamstown, Cape of Good Hope January 1st 1868, Examined and Refuted, by an Appeal to Scripture, Antiquity and Common Sense* (Grahamstown, 1868), pp. v–vii, 117–20.
29. H. Laracy, 'Paranoid Popery: Bishop Moran and Catholic Education in New Zealand', *The New Zealand Journal of History*, X (1976) 51–62.
30. Available on-line at http://www.nla.gov.au/ferg (accessed November 2007).
31. *Constitution and By-Laws of the Canadian Protestant Association* (Kingston, 1854).
32. *Protestant Defence Alliance of Canada. Constitution & c.* (Montreal, 1876).
33. Dr David Fitzpatrick has extensively studied the relationship between Orangeism and Empire in as yet unpublished research. For a survey of Orangeism in Canada see C. J. Houston and W. J. Smyth, *The Sash Canada Wore: A Historical Geography of the Orange Order in Canada* (Toronto: University of Toronto Press, 1980) and for a significant case study S. W. See, *Riots in New Brunswick: Orange Nativism and Social Violence in 1840* (Toronto: University of Toronto Press, 1993). See has observed that 'Loyalty to the British Empire constituted a core element of Canadian nativism.' (S. S. W. See, '"An Unprecedented Influx": Nativism and Irish Famine Immigration to Canada', *American Review of Canadian Studies*, XXX (Winter 2000) 431.
34. *Liverpool Mercury*, 23 May 1854; *Caledonian Mercury*, 21, 27 March 1856. I am indebted to Prof. David Killingray for information on Orr's background and activities.
35. R. Sylvain, *Clerc, Garibaldien, Prédicant des Deux Mondes: Alessandro Gavazzi (1809–1889)* 2 vols (Quebec: Tours, 1962).
36. Richard Lougheed, *La Conversion Controversée de Charles Chiniquy* (Quebec: Editions La Clairière, 1999), pp. 8, 104, 138–9, 147–8.
37. MacLeod, 'Protestantism and National Identity', pp. 56–7.
38. I. T. Foster, 'Walsh, Walter' in *Oxford Dictionary of National Biography*.
39. *Protestant Observer*, September 1892, p. 141.
40. Ibid., June 1893, p. 86.
41. Ibid., September 1896, pp. 136–7.
42. Ibid., February 1897, pp. 18–19.
43. *The Imperial Protestant Federation, Report for 1899–1900 and a History of the Formation and Progress of the Federation* (London, 1900).
44. *The Times*, 19 October 1911.
45. *The Protestant Observer*, October 1899, p. 154.
46. *Report for 1899–1900*, pp. 86–138.
47. *Protestant Observer*, June 1896, 96; February 1897, pp. 18–19.
48. Ibid., February 1901, p. 25.
49. Walsh, *Queen Victoria*, pp. 11, 13–15, 261.
50. For an assessment see W. L. Arnstein, 'Queen Victoria and Religion', in *Religion in the Lives of English Women 1760–1930*, ed. G. Malmgreen (London: Croom Helm), pp. 88–128.
51. *Protestant Observer*, February 1901, p. 26.
52. Lambeth Palace Library, Davidson Papers, Vol. XIX, no. 101, pp. 25–6.
53. National Library of Scotland, Minto Papers (4th Earl), MS 12563, fos. 58–9, Laurier to Minto, 28 January 1901.

54. Ibid., MS 12557, Minto to Laurier, 29 January 1901; Minto to Chamberlain, 15 February 1901.
55. *Protestant Guide* (London: Marshall Bros., 1902), pp. 90–122; *The Times,* 27 June 1901.
56. *Protestant Guide,* pp. 149–52.
57. Ibid., p. 286.
58. Ibid., pp. 21–4.
59. Ibid., p. 25.
60. J. T. Watt, 'Anti-Catholic Nativism in Canada: The Protestant Protective Association', *The Canadian Historical Review,* XLVIII (1967) 45–58.
61. J. R. Miller, 'Bigotry in the North Atlantic Triangle: Irish, British and American influences on Canadian anti-Catholicism, 1850–1900', *Studies in Religion,* XVI (1987) 299–300; *The Protestant Protective Association in Ontario, Being Extracts from Speeches and Two Letters of Mr J. D. Edgar MP* (1893?).
62. *Protestant Observer,* June 1904, p. 90.
63. *Australian Protestant Defence Association: Constitution and By Laws for the Use of Branches* (Sydney, 1902); *Protestant Guide,* pp. 16–17.
64. *Protestant Observer,* November 1902, p. 175.
65. Ibid., April 1904, pp. 63; October 1904, pp. 155; February 1905, pp. 22, 26.
66. *Protestant Guide,* pp. 17–18; *Protestant Observer,* July 1903, p. 106.
67. Ibid., September 1904, pp. 142; August 1907, pp. 122–3.
68. Ibid., March 1903, p. 50.
69. *Protestant Guide,* p. 20.
70. Ibid., May 1903, p. 78.
71. Ibid., August 1908, p. 125. On CESA see Anthony Ive, *A Candle Burns in Africa: The Story of the Church of England in South Africa* (CESA: Gillitts, Natal, 1992).
72. *Grievances from Ireland,* 11 (1905) 9–11.
73. *General Election Protestant Literature* (1906); *The General Election and National Prosperity* (1906).
74. Ibid.
75. *Protestant Observer,* 1908, pp. 24, 125, 138, 162, 170, 181; 1909, pp. 106, 124, 147.
76. On the LCUPS, which became the United Protestant Council, see *The Protestant Dictionary,* ed. C. S. Carter and G. E. A. Weeks, new edn (London: The Harrison Trust, 1933), p. 552.
77. The King had informed Asquith that he would refuse to open Parliament, when he would have to make the Declaration, unless the wording was changed (H. Nicolson, *King George the Fifth: His Life and Reign* (London: Constable, 1952), p. 162). His personal position was, of course, not public knowledge so the IPF and its allies posed as defenders of the Protestant integrity of the monarchy against compromising ministers.
78. *Protestant Observer,* August 1910, pp. 121–2.
79. Ibid., July 1910, p. 108; August 1910, pp. 118, 121–2.
80. *Work and Witness,* new series, IV, no. 3 (October 1910), p. 1. Cf *The Times,* 29 July 1910.

81. *Protestant Observer*, September 1910, p. 135.
82. Ibid., April 1912, p. 60.
83. Ibid., May 1912, p. 80.
84. Ibid., November 1911, p. 169; February 1912, p. 57.
85. Ibid., September 1914, 134; October 1914, 142; February 1915, 13; June 1915, 72.
86. *The Times*, 29 May 1926.
87. P. S. O'Connor, 'Sectarian Conflict in New Zealand', *Political Science* (Wellington, NZ), XIX (1967) 3–16.
88. National Library of Australia, on-line catalogue.
89. Ibid.; www.protestantbooks.co.za/association.asp (accessed 22 December 2006).
90. W. H. Peace, *The Vatican and the British Empire* (London, n.d.), p. 5 and *passim*.
91. J. Loughlin, *Ulster Unionism and British National Identity Since 1885* (London: Pinter, 1995), p. 59.
92. A. Jackson, 'Irish Unionists and the Empire, 1880–1920: Classes and Masses' in K. Jeffery, ed. *'An Irish Empire'? Aspects of Ireland and the British Empire* (Manchester: Mancherster University Press, 1996), pp. 123–48.
93. *Protestant Observer*, October 1912, pp. 153–4.
94. Ibid., July 1913.
95. Cf. D. W. Lowry, '"The Ulster of South Africa": Ireland, The Irish and the Rhodesian Connection', in *Southern African-Irish Studies*, ed. D. P. McCracken and D. Lowry (Durban, 1991), pp. 122–45.
96. C. Stirling, *The Decline of England; or the Other Side of the 'Diamond Jubilee' Shield: An Appeal to the Protestants of the Empire* (London, 1897).

Further reading

Bowen, D. *The Protestant Crusade in Ireland 1800–70* (Dublin: Gill and Macmillan, 1978).

Loughlin, J. *Ulster Unionism and British National Identity Since 1885* (London: Pinter, 1995).

McLeod, H. 'Protestantism and British National Identity 1815–1945'. In *Nation and Religion*, ed. P. van der Veer and H. Lehmann (Princeton, NJ: Princeton University Press, 1999), pp. 44–70.

Miller, J. R. 'Anti-Catholic Thought in Victorian Canada', *Canadian Historical Review*, LXVI (1985) 474–94.

Miller, J. R. 'Bigotry in the North Atlantic Triangle: Irish, British and American Influences on Canadian anti-Catholicism, 1850–1900', *Studies in Religion*, XVI (1987) 289–301.

Norman, E. R. *Anti-Catholicism in Victorian England* (London: George Allen and Unwin, 1968).

O'Connor, P. S. 'Sectarian Conflict in New Zealand', *Political Science* (Wellington, NZ), XIX (1967) 3–16.

Pawsey, M. M. *The Popish Plot: Culture Clashes in Victoria 1860–1863* (Sydney: Catholic Theological Faculty, 1983).

Paz, D. G. *Popular Anti-Catholicism in Mid-Victorian England* (Stanford: Stanford University Press, 1992).

Watt, J. T. 'Anti-Catholic Nativism in Canada: The Protestant Protective Association', *The Canadian Historical Review,* XLVIII (1967) 45–58.

Wolffe, J. *The Protestant Crusade in Great Britain 1829–1860* (Oxford: Clarendon Press, 1991).

Wolffe, J. 'Change and Continuity in British Anti-Catholicism, 1829–1982'. In *Catholicism in Britain and France since 1789,* ed. Frank Tallett and Nicholas Atkin (London: Hambledon Press, 1996), pp. 67–83.

4
An Empire of God or of Man? The Macaulays, Father and Son

Catherine Hall

Zachary Macaulay and his son Thomas Babington Macaulay were both very well known figures in their time. Both were buried in Westminster Abbey, Zachary memorialized for his contribution to the campaigns against the slave trade and slavery, Thomas for his history writing. In this chapter, I explore the ways in which these two men thought about the relation between religion and empire, the contribution each made to specific imperial projects in their own lifetimes and the visions they had of the peoples and places of empire. Thinking across the period from the late eighteenth century, when Zachary entered public life, to the mid-nineteenth century, for Thomas died in 1859, draws attention to shifts in racial and imperial assumptions across a half century. Both men spent their adult lives, in very different ways, preoccupied with questions of nation and empire. Here I focus only on specific moments in those careers.[1] But the similarities and differences between father and son remind us of the historical specificity of thinking about race and empire, the particular conjunctural moments out of which people come, the distinction between evangelical and secular discourses, the relative salience of religion at different moments in the history of empire and the importance of factors other than the discursive.

The stories of Macaulay Senior and Junior cross nations and continents: their family was indeed an imperial family. Zachary Macaulay was born in Scotland in 1768, spent time as a young man in Jamaica and Sierra Leone and then settled in England in 1799 until his death in 1838. He remained a Scot, marked, for example, by his capacity to speak Gaelic. Three of his brothers spent time in India, including Colin who served at Seringapatam, was imprisoned by Haidar Ali for four years, and later became Wellington's *aide-de-camp* in Indian campaigns. Another brother was a naval captain, while yet another was employed by the Sierra Leone Company. These sons of a Scottish Presbyterian minister of very modest means, like so many Scots, found empire to be very successful way of making a living or even a fortune, one which in the case of Zachary Macaulay was made and lost.[2] Furthermore, cousins and nephews worked and settled in India and Sierra Leone. Thomas Babington Macaulay was born in England and unlike his father was immensely proud to be an Englishman. Apart from his years as an undergraduate in Cambridge and the four years in India between 1834 and 1838, he lived all his life in London. After 1838 he was able to live as an independent literary man because of the very large salary he received in India, most of which he saved, and the inheritance that came from his uncle Colin.[3] He had numerous relatives across the empire, especially in India and Sierra Leone. His beloved sister Hannah married Charles Trevelyan who spent years as an Indian civil servant. Yet Macaulay's reputation as a historian was made by his *History of England* which banished the empire to the very margins of his 'island story'.[4]

Zachary was one of twelve children, and came from a line of Presbyterian ministers. He grew up in the west of Scotland in the 1760s and 1770s, a time when that country was far from the full economic and political transformation and the 'assimilation' into England later to be so powerfully evoked by his son in his *History*.[5] As the oldest boy at home, for his elder brother Colin had gone into the army, he was involved in the education of his younger siblings and this gave him, he recorded in his autobiographical fragment, the habits of authority, self-confidence and impatience, not to speak of a dogmatic and magisterial style in writing and speaking. At nine he suffered a serious accident to his right arm which occasioned many operations, and this, together with his blindness in one eye, meant that he was more bookish and spent more time with adults than other children. He loved the classics, a love that he was to pass on to his son. At 14 he was sent to Glasgow to pursue a mercantile career and while there 'was continually laying the plan of wonderful adventure'.[6] Two years later, having got into some unspecified trouble, he was intending to go to the East Indies, but a relative offered him patronage in Jamaica, a place where many Scots were seeking their fortunes. Aged 17 he became a bookkeeper and under-manager on a plantation. He was initially shocked by slavery but decided that he would have to harden himself to it if he were to survive. He was in Jamaica for at least five years and then in 1789, aged 21, he sailed for England.[7]

Zachary Macaulay's imperial career has to be understood in the context of the aftermath of the American War of Independence, the loss of the 13 colonies, the trial of Warren Hastings and the attempts to make empire respectable and reclaim British associations with particular definitions of liberty and freedom. The year 1789, when he returned to England, was the year of the French Revolution. Both the formal establishment of the colony of Sierra Leone and the start of the trail of Warren Hastings had taken place in the previous year: key moments in the attempt to reconstruct empire in the wake of the loss of the 13 colonies. The American War of Independence had seen the appeal by British commanders to enslaved Africans to fight for Britain in return for their freedom. Black loyalists were now constituted as subjects of empire. Slavery, as Christopher Leslie Brown argues, could no longer be defined as the problem of American and Caribbean colonists: it was becoming a British issue.[8] At the end of the war some of those African-Americans ended up in London and contributed to the categorization of poor black people on the streets of London as 'a problem'. Granville Sharp conceived of the idea of a settlement for freed Africans, a Province of Freedom, and the first colonists arrived in Sierra Leone in 1787. Since the end of the Seven Years War the whole issue of how to integrate new populations into the empire had become a concern: Were these Africans, Native Americans and Indians British subjects? While Quakers and evangelicals focused on the position of Africans, taking up the slave trade and slavery, Burke prepared his impassioned critique of Warren Hastings: India must be freed from corruption. Burke facilitated a regeneration of the imperial ideal alongside his conservative polemic against the revolution in France.[9] He raised issues of accountability and this chimed with other conceptions of imperial trusteeship being formulated in the late eighteenth century. As the evangelical poet Cowper put it,

> That where Britain's power
> Is felt mankind may feel her mercy too.[10]

This was the moment at which Zachary Macaulay returned to England, and this was the political conjuncture in which he was formed. On arriving in England in 1789 he went to stay with his sister Jean who had married Thomas Babington, a leading Evangelical. Deeply impressed by the domestic peace of their country home in Leicestershire, Rothley Temple, he was drawn to their faith and experienced conversion. Babington introduced him to other leading Evangelicals and he was taken up as a promising young man. By the late 1780s the group around Henry Thornton, the banker, were deeply enmeshed in questions about the slave trade. After the failure of the first settlement in Sierra Leone, they established the Sierra Leone Company in 1791, intending to expand commerce in West Africa and demonstrate that there was an alternative to slavery. Thomas Clarkson had recently published his *Essay on the Impolicy of the Slave Trade* which emphasized the huge costs of the trade in British as well as African lives and argued that Africa's rich resources represented a plethora of commercial opportunities.[11] The mission of the Company was defined as 'the honourable office of introducing to a vast country long detained in barbarism the blessings of industry and civilisation'.[12] Managed by 13 Directors elected annually by shareholders, Henry Thornton was the chair and Thomas Clarkson one of the directors. Sharp's initial dream of a self-governing 'Province of Freedom' had been reconfigured: it was now a colony ruled from London.

Zachary Macaulay, fresh from Jamaica and seen as having the requisite knowledge of Africans and the plantation system, was sent by the Company to Sierra Leone to report on the state of the colony in the wake of its first failure in 1789. Black loyalists in Nova Scotia who had been deeply disappointed not to receive promised land grants had petitioned the British government for justice. Might they settle in West Africa, rejuvenate the colony and provide a base for the hoped for transformation of African commerce – from bodies to other commodities? In 1792, John Clarkson, naval lieutenant and brother to Thomas captained the journey of 1,200 plus from Nova Scotia to Sierra Leone and supervised the establishment of a new settlement. Appointed by the Company as the first Superintendent, he became completely convinced of the legitimacy of the black settlers' claims for land, an attitude which disturbed the Directors. On his return to England because of ill health he was effectively forced to resign. In 1793, Macaulay, seen by the Directors as a much safer bet, was sent out as a member of Council and was soon to become Governor. He remained in Sierra Leone until 1794 and returned to England in 1795 when he became engaged to Selina Mills, a protégé of Hannah More and her sisters. In 1796, he returned to Sierra Leone where he stayed until 1799.[13] On his return he married Selina. In 1800 their first son Tom was born and in 1802 they settled in Clapham joining their evangelical friends on the Common.

On his return to England, Macaulay, now 31, was appointed as Secretary to the Sierra Leone Company. He became a key adjutant of Thornton and Wilberforce in their multiple schemes for both nation and empire – their efforts to create a more virtuous and religious world in a conservative idiom.

Macaulay's public reputation is that of the man behind the scenes who provided the charming and charismatic Wilberforce with the material he needed to make the case against the slave trade and slavery. With his 'insatiable appetite' for business, he was a rather severe figure, constantly correcting himself and others.[14] A deeply committed evangelical, his central concern was to lead a Christian life and do all he could to make others live likewise. His conversion took place in the wake of his years in Jamaica when he had had to accustom himself to the brutal and dissolute nature of white society and indeed make himself anew. Jamaica, dominated by the plantation and the whip was a living hell, in total contrast to the domestic felicity of Rothley Temple under the watchful patriarchal eye of his brother-in-law Babington. His autobiographical memoir, written in Sierra Leone, had as its epigraph the words of John Newton, the one time captain of a slave ship who experienced conversion:

Thou didst once a wretch behold
In rebellion blindly bold
Scorn Thy grace, Thy power defy
That poor rebel, Lord, was I.[15]

His model was Thornton, a resolutely conservative figure with clear views of social and political hierarchy, a man with a 'uniform and abiding impression of his accountableness to God for every moment of his time, and every word he utters'.[16] Macaulay was constantly anxious about his own self-discipline – and that of his wife and children. Blind in one eye and with a right arm affected by his childhood accident he did not cut an elegant figure. Many respected him, few seem to have felt great warmth for him. Thornton saw him as 'solid, well informed, very resolute, clearheaded and sensible... well read and well instructed... in all manner of colonial subjects... extremely zealous in the cause'. He went to Sierra Leone 'well understanding that the point to be laboured is to make the colony a religious colony'.[17] Babington was convinced that Zachary had been 'selected by the Lord, in a manner rather remarkable, as His instrument in a great work'.[18]

To make the colony a religious colony: this was the key to Macaulay's vision and that of the Clapham Sect. And more than that, it was to make a religious world. It is his years in Sierra Leone and it is this period, and the insights it gives us into his attitudes to race, religion and empire in the 1790s, that I focus on here.

The preferred evangelical representation of 'the negro' was of the passive and docile victim of white barbarism, most memorably portrayed in the Wedgewood cameo of the kneeling man raising his eyes to the heavens and asking 'Am I not a man and a brother?' Macaulay may well have found this hard to reconcile with his time on the plantation: but there is little evidence about this. In 1795, he voyaged with a slave ship from West Africa to Barbados so that he would know the middle passage for himself. It was a horrific journey, one that could be narrated in familiar anti-slavery rhetoric: 'their cup is full of pure, unmingled sorrow, the bitterness of which is unalloyed by almost a single ray of hope' he wrote.[19] Yet such a picture of abjection and misery did not sit easily with his experiences in Sierra Leone. The black settlers who were the subjects of his small kingdom had struggled to escape slavery in the Americas and had left Nova Scotia because of their disillusionment with what they saw as the betrayal of British promises. They believed in their own rights and entitlements and many were unwilling to be told how to conduct themselves – in their working lives, their social and domestic responsibilities and indeed their religious practices. A substantial proportion belonged to dissenting groups, particularly the Methodists and Baptists and Macaulay as a pious Anglican with a strong attachment to the established church found this a constant challenge. He was especially horrified by the Methodists, who accounted 'dreams, visions and the most ridiculous bodily sensations as incontestable proof of their acceptance with God and their being filled with the Holy Ghost', and worse still were practising 'a pure democracy'.[20] Indeed to his mind they were fast sliding into 'the wretched state of barbarism of their African forefathers'.[21]

From his arrival in Sierra Leone Macaulay was convinced that the right tone was one of authority. While at home he was deferential to Thornton, Babington, and Wilberforce, always aware of his inferior social position. In Freetown, the heart of the settlement, he was certain that 'strong language and a decided pre-emptory tone are absolutely necessary'.[22] Here he may have drawn on his West Indian experience. While he was severely critical of the behaviour of planters in the Caribbean and white traders and merchants in West Africa, he did not doubt that he was the commander and guide of 'his' family, as he construed it, regularly lecturing them in preacherly mode on their duty to the Company that had provided for them so generously. Revolution in France, radicalism in England, rejection of social hierarchy and constituted authority: these were evils to be dealt with. While a 'new man' himself, he had no time for claims from uppity black settlers. When he judged their expectations improper he had no hesitation in using his authority to quash them. John Clarkson thought him 'illiberal', approving as he did of arbitrary power.[23] Anna Maria Falconbridge, a white resident of Sierra Leone, regarded him as a 'canting parasite'.[24] While the Nova Scotians saw themselves as partners in an enterprise with the Company, the Directors and their employees were prone to treat them in a markedly unequal manner. The Company coinage designed by Thornton and executed at Boulton and Watt's Soho foundry celebrated a black and white hand clasped together, yet the reality of the power relation was rather different. When two of the settlers, Isaac Anderson and Cato Perkins, took a petition to London in 1793 detailing their grievances, Dawes, the then Governor, and Macaulay wrote letters explaining away their issues. The petitioners were furious at the way they were treated: 'We did not come upon a childish errand, but to represent the grievances and sufferings of a thousand souls', yet they were met, 'as if *Slaves*, come to tell our masters, of the cruelties and severe behaviour of an *Overseer*'.[25]

The evangelicals, including Macaulay, believed that all human beings were descended from Adam and Eve and belonged to the same human family. The differences that were encountered were to do with the particular circumstances of life – not any 'original faults in moral character' or 'natural inferiority in understanding'.[26] But they also believed in natural order and hierarchy. All could be civilized – just as the Scottish Highlanders had been rescued from barbarism. In order to define the task which they saw themselves as needing to do, and for which they needed support of varied kinds from the mother country, they detailed the 'barbarisms' of Africa and other colonial sites on which they worked. In the process they succeeded in fixing representations of African difference and inferiority and disseminating them more widely than ever before.[27] They claimed universalism but practiced forms of racial hierarchization alongside those of class and gender: this was indeed a rule of difference.[28]

In the Company phase of settlement the power was held by their appointed Governor and Council (all white), but the already existing system of local self-government was maintained to some degree as were black juries: an extraordinary phenomenon in the British Atlantic world. In 1796 militias were established in the colony and white men had the hitherto unheard of experience of serving under black officers. Macaulay's determination on some occasions to treat white and black the same under the law roused the ire of other Europeans. The idea of white men being punished for offences against black settlers was beyond the bounds of the imagination as far as many ships' captains were concerned. Yet despite these important marks of a belief in African potential and a resolute opposition to the slave trade Macaulay had conflict with the settlers from the beginning of his tenure of office. He was determined to maintain the authority of the Company against the claims of the settlers, particularly in relation to land. Violence erupted in 1794 when he dismissed two Company porters for threatening the captain of a slave ship. After the devastating French attack on the colony in the same year there was fury at the way in which he attempted to claim back property saved from the assault and insist on new oaths to the Company without which no employment, medical care, or schooling for children would be permitted, not to speak of the right to vote. 'We wance did call it Freetown', wrote Moses Wilkinson and his fellow Methodists to John Clarkson, now 'we have a Reason to call it a town of Slavery'.[29] Macaulay complained of the 'malcontents' and when challenged as to why no black men had been appointed to offices and were paid less than whites he declared that these were privileges to be won.

> Write as well, figure as well, Act as well, think as well as they do and you shall have a preference. I have anxiously sought among you for men to fill offices, nor is there at this moment an office in the Colony filled by a White which a Black could fill.[30]

Tempers exploded around the elections for local government offices in 1796. Macaulay complained bitterly to his superiors of the 'wayward humours, the perverse disputings, the absurd reasonings, the unaccountable prejudices, the everlasting jealousies, the presumptuous self-conceit, the gross ignorance and insatiable Demands of our settlers'. He lectured them in offensive terms, attacking them for listening to 'every selfish or base deceiver who...would abuse or revile your Governors'.[31] The proposal by some of the settlers that white men should not be allowed to stand for office provoked him to laughter: 'there was something so unique in making a white face a civil disqualification'.[32] His descriptions of those elected ranged from 'a noisy factious fellow, exceeding griping and selfish', 'devoid of principle', 'plausible and specious among ignorant people' to 'a pestilent fellow – factious, noisy, busy, bold and blind'.[33] Women householders, of whom there were a significant number, had been granted the vote in this election but this was now withdrawn as was the constitution. The only place of 'absolute peace' in the colony was his household where 25 African boys and girls, the sons and daughters of local chiefs and headmen, were accommodated and taught the ways of civilization. There 'my will is generally the law to all within our pales'.[34] It was the 'poison of the age of Reason' Macaulay believed, which resulted in such disruptive spirit, fermented by seditious Methodists.[35] By his last weeks in the colony in 1799 he had to keep a loaded gun by his bed and a light burning through the night. Soon after his departure open rebellion broke out in 1800, only to be put down by a company of Maroons from Jamaica who had been brought in to change the demographic balance of the colony. The attack on the settlers was led by his brother Alexander. The published report of the Company in 1801 regretted the tragic failure of hopes for Africa while Wilberforce wrote privately to Dundas that the Nova Scotians 'have made the worst possible subjects, as thorough Jacobins as if they had been trained and educated in Paris'.[36]

Macaulay's influence in Sierra Leone persisted, however. As Secretary to the Company he was in a key position to intervene on appointments and provide the information from which the annual reports were penned by Thornton. After the abolition of the slave trade Sierra Leone became a centre for recaptives, as they were named, those who had been freed from the slave traders of other nationalities by the British squadrons that patrolled the Atlantic waters. It was Macaulay who conceived the idea of apprenticeship, a way of teaching those who had been enslaved to learn to be free to labour, and this was superintended by his second cousin Kenneth M. Macaulay who was to become a celebrated figure in the colony for rather different reasons from his pious relative.[37]

Macaulay had encountered what he saw as the barbarisms of empire, both black and white, in Jamaica and Sierra Leone. Enslaved Africans, white planters, African princes, white traders, black settlers all came in for his criticism. These were worlds remote from the domestic, from virtuous white women presiding over 'home', and an England where there was at least a band of evangelicals to struggle together over vice and irreligion. He was horrified by the indiscriminate violence of the French attack on Sierra Leone in 1794 and had the cruelties of the revolution in San Domingue graphically described to him by a travelling Frenchman. The world was indeed a dangerous place and only 'real religion' and a constant attention to God's presence, His Word, and His Law, held the unquenchable passions of man in check. While firmly believing in the potential of Africans to be civilized, the problem was that it was so difficult to create the conditions in which this could happen. And there were so many false prophets, dissenters and radicals in particular, leading the gullible astray. The 25 African children who he had taken into his household accompanied him back to England and were settled in Clapham evoking admiration as a successful experiment in civilization.[38] There was no question of equality between African and English for the foreseeable future. The Company might have to negotiate with African princes for land and trading rights, rely on Maroon troops to defeat a settler rebellion, depend on African men and women to labour, but none of this meant that Africans and Europeans were the same. Difference and inequality were firmly entrenched in the evangelical imagination, alongside a profound belief in the right of all persons to have access to Christian teaching and salvation.

The dream of a religious colony as defined by conservative Evangelicals had not been a success, though according to all observers Sierra Leone was a profoundly religious place. Efforts were turned to missionary work, with the foundation of the Church Missionary Society, to the possibilities for evangelical work in India, and to the struggle to counter vice and irreligion at home. The *Christian Observer* was founded, which Macaulay edited, to provide a counterweight to the expanding secular press. Until 1807 efforts were concentrated on the campaign against the slave trade. The abject figure of the enslaved on the middle passage or on the plantation was an easier one to manage than the feisty settlers of Sierra Leone. Once the British slave trade was abolished Sierra Leone was brought formally into the empire and the Company disbanded. From 1808 he acted as Secretary for the new African Institution, committed to monitoring the trade and expanding legitimate commercial opportunities in Africa.[39] He went into business trading with Sierra Leone and the East Indies. Within a few years it was clear that the hopes that the ending of the trade would lead to improvements on the plantations were hollow and Macaulay, together with James Stephen, led the attempts to expose brutalities and regulate slavery. Macaulay's extraordinary memory and range of knowledge, his analytical skills, his briefings of political leaders, his collection of evidence, his pamphlets and journalism, all made him an invaluable figure. In 1823 he was in the forefront of the new society pledged to work towards the gradual abolition of slavery. He established and edited the *Antislavery Reporter* a key source of information on the conditions of slavery across the empire. Pro-slavery forces kept up a series of vitriolic attacks on him, accusing him of malpractice both in Sierra Leone and Jamaica. In 1823 he gave his business over to his nephew and partner to devote himself to the struggle against slavery, with disastrous consequences. By 1833, when slavery was abolished, younger men were leading the struggle and having lost his money and, in 1831, his wife, he was a much less resilient figure. He himself died in 1838. Apprenticeship had been abolished that year and a monument was erected to him in Westminster Abbey: the bust incorporates the figure of a kneeling African with the motto 'Am I not a man and a brother?'[40]

If this was the father, what then of the son? Tom, the first born of Selina and Zachary, was two when the family moved to Clapham. His was a profoundly evangelical childhood, surrounded by the families of the Thorntons, Grants, and Wilberforces, his 'second mother' Hannah More. Educated privately because his father feared the unholy culture of a public school, his encounter with Cambridge was his first introduction to a more secular world, though his education in the classics was probably a key alternative influence from his earliest years. He never experienced conversion and while he maintained a virtual silence throughout his life on his own religious beliefs, largely it would seem from respect for his father, there is no evidence of serious spiritual reflection in his writings. What Tom learned from his Clapham childhood, however, was a love of family and domesticity, a love so deep that he was never able to leave his family of origin. He also inherited a fear of European-style revolution and an expectation of a place in the public world of politics, along with the confidence to claim acceptance among the political elite. While his father was renowned for his quiet, behind-the-scenes work, Tom liked to be in the limelight.

The children of 'the Saints' were taught to feel pity for 'the African'. But there were always discordant notes. Zachary's stories of Sierra Leone, according to Tom's nephew and biographer George Otto Trevelyan, left his uncle unable 'to entertain any very enthusiastic anticipations with regard to the future of the African race'. He disliked his father's passion for abolitionism, and in private letters, even as early as 1833, was referring to Africans as 'niggers' – a term that carried deeply derogatory meanings. He was impatient of what he saw as the excessive 'negrophilia' of the older generation. As he recorded many years later in his journal, 'I hate slavery from the bottom of my soul; and yet I am made sick by the cant and the silly mock reasons of the Abolitionists. The nigger driver and the negrophile are two odious thing to me.'[41] While his first public speech in 1824 was on an anti-slavery platform and his first published essay in the *Edinburgh Review* was on West Indian slavery neither went beyond a conventional abolitionist rhetoric and these were not subjects that he subsequently pursued. The campaigns to end the slave trade and slavery were driven by religious and moral imperatives. Tom's vision of imperial politics was rather different and here I focus on the ways in which that was elaborated in the 1830s.

By the end of the 1820s the Tory-Anglican hegemony of the post-war years was collapsing – conservatism at home and autocracy in the empire were being challenged.[42] Issues about reform across both nation and empire were pressing.[43] Ireland was on the brink of eruption over Catholic emancipation, Dissenters had been granted civil rights, radicals and reformers were demanding an extension of the franchise, Jamaica was on the verge of the most serious rebellion of the enslaved, a rebellion that was to make emancipation inevitable. This was the conjuncture in which Macaulay Junior came to political maturity, a time when once again British society was seriously at risk and might succumb to revolution. While the Evangelicals had turned to 'real religion' as the key to national and imperial regeneration, Macaulay Junior's preoccupation was in cohering the nation, making it whole and stable and ensuring that internal divisions were resolved. He was a supporter of Catholic emancipation, believing that religious affiliation could be superseded by national belonging and that religious minorities should be assimilated. His maiden speech in the House of Commons in 1830 was in support of the removal of civil disabilities practised against the Jews and was rooted in his conviction that Jews could be brought into the nation, could be Englishmen. In 1833 he again argued for an end to civil discrimination on religious grounds: all men should have the right to practise their religious beliefs. Since Jewish men owned substantial property, it was important that they should be patriots. They had wealth, and therefore power: they should also have the responsibilities and rights associated with citizenship. 'There is nothing in their national character', he argued, 'which unfits them for the highest duties of citizens.'[44] A Jew was not a 'Musselman' or a 'Parsee', or a 'Hindoo who worships a lump of stone with seven heads'.[45] The responsibility of rulers was to make men patriotic. 'If the Jews have not felt towards England like children', he argued, 'it is because she has treated them like a step-mother.'[46] Like his father he imagined society in a family idiom, but while for Zachary the test of belonging was religious, for his son it was nationalist. England must assimilate her potentially wayward children. While Evangelicals laboured to convert Jews and many were opposed to Jewish emancipation, Macaulay believed the nation came first rather than God.[47] He was emphatically an Englishman.

In 1830, having made his reputation as a Whiggish polemicist under the patronage of Brougham, he entered the House of Commons just in time to engage in the great theatrical debates over reform that occupied both Lords and Commons in 1831 and 1832. He became famous as a result of his oratorical triumphs, his capacity to thrill and enthrall the House by sheer force of argument and words. In his speeches on reform, he argued that societies could only be ruled by public opinion or the sword, and that public opinion was clearly in favour of reform. Enlarging the franchise had become a historical necessity: the government must be brought into harmony with 'the people', by which he meant middle-class men. His imagined nation, his England, was marked as the most civilized in the world, separated from the 'tattooed savages of the Pacific', from enslaved 'negroes', from 'Mohawks and Hottentots'.[48] It was also distinct from those others who were deemed closer in their level of development, yet far from the rational world of Englishmen – the backward Scots, the dangerous revolutionaries of France, and the desperate insurgents of Ireland. English superiority was marked by a belief in property, stability and a capacity to renew the constitution, to reform in time. Middle-class men must be brought into the political nation, just as Irish Catholics had been. The English people, he came to believe, were marked by their ability to assimilate others, to make others in their own image.

Jews and Catholics could be English provided they adopted English customs and habits and obeyed English law. If they did not they must take the consequences. While Catholics should have civil equality, the disturbances across rural Ireland were totally unacceptable. Granted a place in the Whig cabinet as a consequence of his contribution to the winning of reform, Macaulay made a key speech on the Coercion Bill for Ireland in 1833, a bill which repealed Habeas Corpus and gave the police the right to search for arms. Ireland was in danger of polluting the body politic and must be cleansed. In these same months negotiations were proceeding over the ending of slavery. The West India lobby drove a hard bargain with the government and Macaulay found himself caught between the expectations of the Whig government, of which he was now a junior member, and the concerns of his father and the abolitionists. The deal that was brokered, with 20 million compensation for the 'owners' of the enslaved and a system of apprenticeship, made it abundantly clear that freed Africans were imagined not as equals but as children who were to be taught the ways of freedom.

Macaulay's key role in these months, however, was to draft and steer through the new Charter Act for India. A novice on Indian affairs, his 1833 speech laid out the lines that he was to follow both in his subsequent term of office there, and in his essays on Clive and Hastings. The future for Indians was to become like the English. They could retain their own religious customs, just as Jews, Catholics and Dissenters could, but in language and thought, in education, commerce and law, the society must be anglicized. While the East India Company was to continue to rule for the British, a stronger government presence was established in India's new Supreme Council. 'We are trying', Macaulay argued, speaking volumes with his language, 'to bring a clean thing out of an unclean, to give a good government to a people to whom we cannot give a free government.' While a limited form of representative government was possible in Europe given a substantial middle class, such a system was utterly out of the question in India: an enlightened and paternal despotism was the only possibility. India was a dependency: the solution was 'to engraft on despotism those blessings which are the natural fruits of liberty'.[49] Indians were, as he was to put it to his Leeds constituents, 'a conquered race, to whom the blessings of our constitution cannot as yet be safely extended'.[50] The East India Company had established order. There was now a government 'anxiously bent on the public good'; 'bloody and degrading superstitions' were losing their hold, and there were signs that 'the morality, the philosophy, the taste of Europe' were beginning to have a salutary effect. The 'higher classes of natives' had become interested in the language and literature, the civilization and culture, which secured superiority to the English. It was conceivable that eventually Indians themselves would be able to enter high office, but this could only be by slow degrees. It was even possible that India might eventually be a nation and, 'to trade with civilised men', he argued, is infinitely more profitable than to govern savages.[51] Echoing Burke, and indeed Charles Grant, a leading evangelical of his father's generation, he argued that England had a responsibility to India.[52]

> As a people blessed with far more than an ordinary measure of political liberty and of intellectual light, we owe to a race debased by three thousand years of despotism and priestcraft. We are free, we are civilised, to little purpose if we grudge to any portion of the human race an equal measure of freedom and civilisation.[53]

For Macaulay, then, the task was to educate those subjects, to offer them enlightenment. Like his father he identified the superiority of Western culture with its intellectual power (the power which had enabled both generations, father and son, to rise themselves), but for him there was no religious element. Indian 'backwardness', in his view, was a result not of their 'heathenism' but of the absence of European civilization.

While the Charter Act legislated that no one should be barred from high office by virtue of their religion, their birth or their colour, Macaulay was quick to add the recommendation that natives must not be brought into high office hastily. His apparent universalism and belief in civil equality was always undermined, as it had been with his father in Sierra Leone, by the simultaneous production of racial, ethnic and other hierarchies. In theory Indians might be equal in the future once they had been educated and had transformed themselves into something else: in practice this moment was always deferred. Indians were stranded as Dipesh Chakrabarty puts it, in the 'waiting-room of history'.[54]

In the wake of his work on the revised charter for the East India Company, Macaulay was offered the newly created position of lay member of the Supreme Council that had been created to govern India. He left England in February 1834 to spend three and a half years in what he described as painful exile. The 'rulers of India' could never be more than 'pilgrims and sojourners', a tiny minority trying to command a country to which they did not belong, always looking to another 'home'.[55] They were strangers in a strange land and his tastes, as he told his sisters, were 'not oriental'.[56] He was shocked on arrival at Madras by the fact of blackness: the 'innumerable swarms of natives', 'the dark faces, with white turbans, the flowing robes: the trees not our trees: the very smell of the atmosphere that of a hothouse, and the architecture as strange as the vegetation'.[57] He was all too conscious that English power, as yet, could only be maintained by the military. But the hope must be to win consent through the construction of new colonial subjects, civilized subjects, educated into the ways of Englishmen.

While in India he worked with the Governor-General, Bentinck, to anglicize the system of government, hoping that he would 'be able to effect much practical good for this country'.[58] He was appalled by 'Hindu apathy' and the level of bribery and corruption, but also had little sympathy for the majority of the Anglo-Indian community, their philistinism and small mindedness. Like his father he was critical both of the colonizers and of the 'natives'. But his contact with 'natives', unlike Zachary, was minimal apart from the innumerable servants who surrounded him and his sister Hannah in their grand Calcutta household. He rewrote the penal code (though it was not to be enacted until the 1860s) seeing it as his especial responsibility to protect 'the natives' from abuses of power and arguing for uniformity in judicial administration. Arriving at a time when policy makers were deeply split between those in favour of providing government support to the teaching of

Arabic and Sanskrit to the elite and those who wanted to focus resources on the teaching of English, he threw his full weight behind the latter and composed his infamous Minute on Education. As Gauri Viswanathan has argued, the subsequent introduction of English literature as a central aspect of the curriculum resolved the problem of religious education – a secular and literary education would transform character: religion could be left as a matter of private belief.[59] Indeed Macaulay was convinced that 'false religions' would lose their hold, for as he wrote to Zachary in October 1836, no doubt partly in deference to his father's views, 'no Hindoo who has received an English education ever continues to be sincerely attached to his religion'. The reason was clear for the religion was 'so extravagantly absurd that it is impossible to teach a boy astronomy, geography, natural history, without completely destroying the hold which that religion has on his mind.' Moslems, he opined, were more resilient for they had much in common with Christianity. Nevertheless, they too would abandon their faith. 'The natural operation of knowledge and reflection', he concluded, would ensure the disappearance of these false beliefs amongst the respectable classes of Bengal within 30 years.[60] The eventual triumph of Western thought was assured.

Macaulay returned to England in 1838 with a secure income. While in India, he had been contemplating writing a history of England, giving up politics for literature. After a period in which he tried to combine the two, he decided to devote himself to his writing. His immensely popular essays on India told the dramatic stories of Clive and Hastings, their famous conquests, their flawed characters and their legacy: a country that could be benevolently ruled with secular principles. His *History of England*, in five volumes, was a spectacular success. True to its name, it was indeed a history of England. It told of the English as an imperial race, the successful assimilation of the Scots (in part because of the acceptance of the Scottish Presbyterian Church) and the continued problem of the Irish (intimately linked with the insistence on an established church). Its focus was on the making of the nation and a homogeneous Englishness which did not depend on religious belonging. In this narrative the empire was peripheral to England.[61]

Yet to both generations of this family the empire was crucial: their source of income and independence, that most valued of masculine attributes. They lived in different temporalities, dominated by different politics, and the sites of empire on which they operated were very different. Zachary was a demanding father and Tom had to struggle to separate himself and articulate his own opinions.[62] Their attitudes to nation and empire were both deeply connected and markedly different. Both believed that difference was a matter of culture, and was not decreed by nature or the environment. Neither liked their encounter with the otherness of the colonized or the colonizers. Both enjoyed the power of being colonizers themselves. Both assumed their beliefs were appropriate for others. But the father carried the image of God everywhere; for the son it was the nation that defined the parameters of belonging.

Notes

1. This essay is part of a much larger project on Macaulay and the writing of national and imperial histories. I am grateful to the Economic and Social Research Council for their support.
2. On Scots and empire see the Chapters 6 and 5 by John MacKenzie and Esther Breitenbach in this volume.
3. For biographical information see the *Oxford Dictionary of National Biography*; *Life and Letters of Zachary Macaulay*, ed. Viscountess Knutsford (London: Edward Arnold, 1900); G. O. Trevelyan, *The Life and Letters of Lord Macaulay* (London: Longman, Green and Co, 1881); J. Clive, *Thomas Babington Macaulay. The Shaping of the Historian* (London: Secker and Warburg, 1973).
4. C. Hall, 'At Home with History. Macaulay and the *History of England*', in *At Home with the Empire*, pp. 32–52.
5. T. B. Macaulay, *The History of England* [1st pub. in 5 vols (1848–59)] 3 vols (London: Dent, 1906). See particularly vol. I, chs. 13 and 18.
6. Knutsford, *Life and Letters*, p. 5.
7. For an account by a missionary wife of Jamaica at the time of slavery and emancipation see John McAleer's Chapter 10 in this volume.
8. C. L. Brown, *Moral Capital. Foundations of British Abolitionism* (Chapel Hill, NC: University of North Carolina Press, 2006).
9. N. Dirks, *The Scandal of Empire: India and the Creation of Imperial Britain* (Cambridge MA: Harvard University Press, 2006).
10. Cited in Brown, *Moral Capital*, p. 205.
11. E. G. Wilson, *Thomas Clarkson: A Biography 1760–1846* (Basingstoke: Macmillan, 1989).
12. C. Fyfe, *A History of Sierra Leone* (Oxford: Oxford University Press, 1962), p. 36.
13. There are a number of accounts of Macaulay's time in Sierra Leone all of which make use of his journals and letters, now in the Huntington Library. See particularly Knutsford, *Life and Letters;* Fyfe, *A History of Sierra Leone*; E. G. Wilson, *The Loyal Blacks* (New York, Capricorn, 1976); J. St. G. Walker, *The Black Loyalists: The Search for a Promised Land in Nova Scotia and Sierra Leone* (London: Longman, 1976); C. Pybus, *Epic Journies of Freedom. Runaway Slaves of the American Revolution and Their Global Quest for Liberty* (Boston, MA: Beacon, 2006) & '"A Less Favourable Specimen": The Abolitionist Response to Self-Emancipated Slaves in Sierra Leone' *Parliamentary History Supplement* (2007) 98–113; D. Coleman, *Romantic Colonization and British Anti-Slavery* (Cambridge: Cambridge University Press, 2005); S. Schama,

Rough Crossings. Britain, the Slaves and the American Revolution (London: BBC Books, 2005). See also S. Schwarz, ed. and Introduction, *Journal of Zachary Macaulay*, Parts 1 and 2, University of Leipzig Papers on Africa, History and Culture, Series no. 4 (Leipzig: University of Leipzig, 2000–2001).
14. Cited in Knutsford, *Life and Letters*, p. 3.
15. Ibid., p. 3.
16. Ibid., p. 202.
17. Cited in Wilson, *The Loyal Blacks*, p. 294.
18. Cited in Knutsford, *Life and Letters*, p. 24.
19. Ibid., p. 89.
20. Schwarz, ed., *Journal*, Part 1, pp. 60–1.
21. Cited in Pybus, 'A Less Favourable Specimen', p. 107.
22. Cited in Knutsford, *Life and Letters*, p. 171.
23. Cited in Wilson, *The Loyal Blacks*, p. 288.
24. *Maiden Voyages and Infant Colonies. Two Women's Travel Narratives of the 1790s*, ed. D. Coleman (Leicester: Leicester University Press, 1999)), p. 116.
25. Cited in Wilson, *The Loyal Blacks*, p. 297.
26. Cited in Schwarz, ed., *Journal*, Part 2, Introduction, p. xxiv.
27. Metcalf makes this argument in relation to India. T. R. Metcalf, *Ideologies of the Raj. New Cambridge History of India, Vol. III.4* (Cambridge: Cambridge University Press, 1994), p. 39.
28. On the concept of a 'rule of difference', an adaptation of Partha Chatterjee's 'rule of colonial difference', see C. Hall, 'The Rule of Difference: Gender, Class and Empire in the Making of the 1832 Reform Act', in *Gendered Nations: Nationalism and Gender Order in the Long Nineteenth century*, ed. I. Blom, K. Hagemann and C. Hall (Oxford: Berg, 2000), pp. 107–37.
29. Cited in Wilson, *The Loyal Blacks*, p. 321.
30. Ibid., p. 323.
31. Ibid., p. 328.
32. Cited in Knutsford, *Life and Letters*, p. 157.
33. Cited in Wilson, *The Loyal Blacks*, pp. 328–9.
34. Ibid., p. 344.
35. Cited in Wilson, *The Loyal Blacks*, p. 379.
36. Ibid., p. 388.
37. Fyfe, *History of Sierra Leone*, p. 146.
38. For an account of the school see B. L Mouser, 'African Academy – Clapham 1799–1806', *History of* Education, XXXIII, no. 1 (January 2004) 87–103.
39. W. Ackerson, *The African Institution (1807–1827) and the Antislavery Movement in Great Britain* (Lampeter: Edwin Mellen Press, 2005).
40. For a discussion of the monuments to abolitionists see J. Oldfield, *'Chords of Freedom': Commemoration, Ritual and British Transatlantic Slavery* (Manchester: Manchester University Press, 2007), ch. 3; M. Dresser, 'Set in Stone? Statues and Slavery in London', *History Workshop Journal*, LXIV (2007) 162–99.
41. Trevelyan, *Life and Letters*, p. 17.
42. C. A. Bayly, *Imperial Meridian. The British Empire and the World 1780–1830* (London: Longman, 1989).
43. Hall, 'The Rule of Difference'; see also Catherine Hall, 'Marxism and Its Others', in *Marxist History-Writing for the Twentyfirst Century*, ed. C. Wickham (Oxford: Oxford University Press, 2007).

44. T. B. Macaulay, *Speeches on Politics and Literature* (London: Dent, 1909), p. 93.
45. Ibid., p. 87.
46. Lord Macaulay, 'Civil Disabilities of the Jews', in *Literary and Historical Essays Contributed to the Edinburgh Review* (Oxford: Oxford University Press, 1913) Part 2, p. 90.
47. For a most interesting discussion of Macaulay see G. Viswanathan, *Outside the Fold. Conversion, Modernity and Belief* (Princeton, NJ: Princeton University Press, 1998), ch. 1.
48. Macaulay, *Speeches*, p. 16.
49. Ibid., p. 104.
50. *Leeds Mercury* 8 February 1834.
51. Macaulay, *Speeches*, pp. 124–6.
52. Grant's 1792 pamphlet on the degenerate state of India had been very influential, particularly in the charter debates of 1813. Charles Grant, *Observations on the State of Society among the Asiatic Subjects of Great Britain, Particularly with Regard to Morals; and the Means of Improving It.* Printed by the House of Commons, 15 June 1813.
53. Macaulay, *Speeches*, pp. 124–6.
54. D. Chakrabarty, *Provincializing Europe: Postcolonial Thought and Historical Difference* (Princeton NJ: Princeton University Press, 2000), p. 8.
55. Macaulay to Margaret Cropper, 15 June 1834, in *The Letters of Thomas Babington Macaulay, Vol. 3 January 1834–August 1841*, ed. T. Pinney (Cambridge, 1976), p. 39.
56. Macaulay to Selina and Fanny Macaulay, 19 October 1834, in *Letters, Vol. 3*, ed. Pinney, p. 97.
57. Macaulay to Margaret Cropper, 15 June 1834, in ibid., pp. 37–9.
58. Macaulay to Napier, 1 January 1836, in ibid., p. 163.
59. G. Viswanathan, *Masks of Conquest. Literary Study and British Rule in India* (London: Faber and Faber, 1990).
60. Macaulay to Zachary Macaulay, 12 December 1836 in *Letters, Vol. 3*, ed. Pinney, p. 193.
61. Hall, 'At Home with History', pp. 32–52.
62. Clive puts great emphasis on the oedipal relation in his account of Tom's development. Clive, *Macaulay*.

Further reading

Ackerson, W. *The African Institution (1807–1827) and the Antislavery Movement in Great Britain* (Lampeter: Edwin Mellen Press, 2005).

Clive, J. *Thomas Babington Macaulay. The Shaping of the Historian* (London: Secker and Warburg, 1973).

Coleman, D. *Romantic Colonization and British Anti-Slavery* (Cambridge: Cambridge University Press, 2005).

Hall, C. 'At Home with History: Macaulay and the *History of England*'. In *At Home with the Empire: Metropolitan Culture and the Imperial World*, ed. C. Hall and S. O. Rose (Cambridge: Cambridge University Press, 2006), pp. 32–52.

Schama, S. *Rough Crossings. Britain, the Slaves and the American Revolution* (London: BBC Books, 2005).

Trevelyan, G. O. *The Life and Letters of Lord Macaulay* (London: Longman, Green and Co, 1881).
Viswanathan, G. *Outside the Fold. Conversion, Modernity and Belief* (Princeton, NJ: Princeton University Press, 1998).
Walker, J. St. G. *The Black Loyalists: The Search for a Promised Land in Nova Scotia and Sierra Leone* (London: Longman, 1976).

5

Religious Literature and Discourses of Empire: The Scottish Presbyterian Foreign Mission Movement*

Esther Breitenbach

Introduction

That foreign missions played a not insignificant role within the British Empire has been increasingly recognized as the body of scholarship on missions grows. Many scholars in this field have aimed to evaluate the impact of foreign missions on the peoples with whom they worked, and to elucidate the nature of the encounter between missionaries and indigenous peoples in colonial territories. More recently the impact of foreign missions on people in Britain has been a focus of research, and there is growing evidence that the foreign mission movement was instrumental in shaping understandings of empire and of the other peoples governed by the imperial state, as well as in shaping the construction of identities in Britain.[1] This chapter discusses the ways in which missionary societies and the Presbyterian churches mediated an understanding of empire for people in Scotland.[2]

In Scotland, as in England, organized support for foreign missions first arose at the end of the eighteenth century. Inspired by the example of the Moravian missions, and the establishment in England of the Baptist Missionary Society in 1792 and the London Missionary Society (LMS) in 1795 (in the foundation of which several Scots were involved), the Glasgow Missionary Society (GMS) and Edinburgh-based Scottish Missionary Society (SMS) came into being in early 1796. These societies were non-denominational and led by evangelicals from both the Established and dissenting churches. The attempt in the same year by

Church of Scotland evangelicals to obtain the church's support for foreign missions was unsuccessful.[3] It took until 1824 till the evangelicals were able to persuade the Church of Scotland to change its mind, with missionary work being inaugurated by the appointment of Alexander Duff to Calcutta in 1829. By this time the GMS and SMS were supporting small numbers of missionaries in Jamaica, South Africa and India. Some of these missionaries were transferred to the Church of Scotland in the 1830s, while the remaining GMS missionaries were transferred to the Free Church of Scotland or the United Presbyterian Church in the aftermath of the Disruption of 1843.[4]

At the Disruption all but one of the Church of Scotland missionaries went over to the new Free Church, and the Church of Scotland was obliged to start afresh in building up a missionary presence abroad. The Free Church of Scotland, energized by the evangelical zeal which produced the Disruption, proved able to field more missionaries than its established rival, as did the similarly evangelical United Presbyterian Church of Scotland, formed in 1847 by the union of the United Secession and Relief churches. From the immediate post-Disruption period onwards the three main Presbyterian denominations were the major channel for missionary enthusiasm in Scotland. In 1900, the Free Church and United Presbyterian Church combined to form the United Free Church of Scotland, and in 1929 this body was reunited with the Church of Scotland. Thus throughout the second half of the nineteenth century, there were three major Scottish Presbyterian churches working in mission fields, the Established church being one of these. The nineteenth century in Scotland can be characterized as a time of both religious disputatiousness and religious zeal, and arguably foreign missions benefited from the churches' rivalry in demonstrating their zeal.[5]

A significant feature of support for foreign missions in Scotland, as elsewhere, was the creation of Ladies' Associations, first established in the 1830s.[6] These also split on denominational lines, maintaining informal links with their respective churches between the 1840s and 1880s.[7] In the 1880s, the position of women's missionary organizations became more formalized within the churches, reflecting both the growth of interest in the work of women missionaries and of numbers, both absolutely and as a proportion of the missionary workforce. Crucially, women were active supporters of the foreign mission movement from the outset, with their role as organizers at home and as mission workers abroad growing in importance as the nineteenth century advanced.

In nineteenth-century Scotland, the Presbyterian churches were dominant in the field of missionary endeavour, though dissenting and non-denominational groups also actively supported missionary work, most prominently the Scottish auxiliaries of the LMS, and the Edinburgh Medical Missionary Society (EMMS). Throughout the nineteenth and into the twentieth century Scots continued to make a significant contribution to the LMS, supplying a sizeable proportion of its missionaries, including some of its best known such as Robert Moffat, James Chalmers, and most famous of all, David Livingstone.[8] The EMMS, founded in 1841, funded the training of medical missionaries, who undertook their medical apprenticeship in the slums of Edinburgh's Cowgate. The society supported directly a small number of medical missionaries, but primarily placed them with the churches or other missionary societies. Among other denominations, the Reformed Presbyterian Church supported a small number of missionaries in the New Hebrides, taken into the Free Church in the union of the two churches in 1876. The Episcopal Church also supported missionary activity, with a Scottish Episcopalian missionary society being formed in 1846, but did not support missions directly until the 1870s.[9] The Catholic Church in nineteenth-century Scotland, as a poor church, was itself regarded as being the subject of a mission, and it was not until 1933 that the Catholic missionary organization, the White Fathers, established a foundation in Scotland.[10] In nineteenth- and early twentieth-century Scotland formal support for foreign missionary work was thus an exclusively Protestant phenomenon. It was also an overwhelmingly Presbyterian phenomenon.

It should be noted that foreign mission work was conducted in parallel with home mission work, and that effectively any group of people who were not Protestant were considered appropriate objects of attempts at conversion. There were thus missions to Jews, and to Catholics in France and Ireland, for example. Anti-Catholicism was an intrinsic part of Scottish Presbyterianism, and anti-Catholic sentiment often found vehement expression. Within the missionary discourses to be discussed in this chapter, such sentiments were usually implicit, since the key focus of missionary concern was the character of religions such as Hinduism, Islam or African belief systems. On occasion, however, missionaries turned their attention to denunciations of Catholicism.[11]

Missionary literature and discourses

Support for foreign missions grew throughout the nineteenth century in Scotland, and continued to grow in the early twentieth century. A corollary of this growing interest was the production of a voluminous body of literature by and about missionaries, both for missionary supporters and church members and for a wider public. By the late nineteenth century the readership of such literature was extensive reflecting the religious nature of Scottish society in this period. The argument of this chapter is that the dominant discourses within missionary literature were influential in shaping Scots' understanding of the experience of empire, their constructions of imperial 'others' and the construction of a Scottish imperial identity.

This literature was inherently religious, though not necessarily theological, in that it sought to inform Scottish Christians about the progress of Christianization in other parts of the world. Within this literature were enunciated explicit discourses about the nature of Christianity in relation to other religions, about behaviours which Christianity sought to inculcate and to regulate, and about the nature of the missionary encounter with others in colonial territories. Also present in this literature, refracted through a religious prism, were discourses on the meaning of political events, imperial expansion and imperial administration.

Such discourses were articulated through a variety of types of publications: pamphlets containing sermons, speeches and addresses; annual reports; periodical literature; collections of occasional papers; histories of missions; biographies, and the occasional memoir or autobiography. From the period of formation of the first missionary societies in the late eighteenth century, letters from missionaries were copied and circulated, and pamphlets of sermons and addresses were published, the latter remaining perennial throughout the nineteenth century. In the 1820s the first major missionary periodical was published, and from the 1840s, following the Disruption, such literature was published on denominational lines, including separate publications for women readers on women's missionary work.[12] Between the 1830s and 1870s there were a handful of memoirs and biographies of missionaries, with the latter taking off as a genre from the late 1870s onwards. Histories, too, began to be published more frequently from this period. These forms of literature, with the exception of some biographies or memoirs, might be defined as being 'official' literature, in the sense of being generated, published or sanctioned by missionary societies and churches.

The most prolific production of missionary biographies occurred between the 1890s and 1920s, though some were published both before and after this period. The earliest publication on the life of a Scottish missionary was John Wilson's memoir of his wife Margaret, published in 1838, and largely based on letters written by Margaret Wilson to her family and friends.[13] Between 1838 and the late 1870s, works by or about the following missionaries were published: Robert Moffat, John

Philip, Robert Nesbit and Hope Waddell.[14] After the late 1870s, with the publication of works on Robert Moffat and Alexander Duff, there was a steady flow of biographies.[15] Some of these were published in popular editions, or in special editions for children, for example *The White Queen of Okoyong,* on Mary Slessor.[16] Biographies might go through several editions, and some missionaries were the subject of multiple biographies, with Robert Moffat, Mary Slessor, and Alexander Duff coming into this category.[17] Livingstone is in a league of his own, with over 100 books written about him between the 1870s and the 1950s.[18] Some missionary biographies continued to be reprinted until at least the 1960s, and they remained popular as Sunday School prizes with Livingstone and Slessor being the most likely to feature here.[19]

By the late nineteenth century, the three main Presbyterian denominations between them were circulating around 250,000 copies of missionary periodicals, and by the early twentieth century, around a third of members of the Free Church of Scotland and the United Presbyterian Church were subscribers. There is also evidence of the active encouragement of church members by ministers to read and subscribe to periodicals, including free distribution of these to all members of the congregation. Indeed, Dow has argued that by the 1860s it was generally accepted that ministers should make pulpit readings of extracts from missionary periodicals, and that local Presbyteries frequently encouraged this.[20] Missionary societies and, above all, the Presbyterian churches in the post-Disruption period had the capacity to reach a wide audience as evidenced both by the levels of religious adherence in this period,[21] and by the popularity of literature by and about missionaries. Thus it can be argued that the continuous production and circulation of such types of publication in the nineteenth and early twentieth century meant that they played a significant role in shaping Scots' understanding of empire.

Prominent among the discourses within missionary literature were those of class, gender and nation, as well as race. In describing and communicating their experience of working with colonial peoples, missionaries and their supporters were also making statements about themselves and their own values. Representations of missionaries as educated and professionally skilled individuals, as upholders of the sexual morality and gender roles which were the dominant convention within Scottish society of the time, as typifying the Scottish character and making a contribution to the 'fair name of Scotland',[22] can be found in abundance in this literature. The focus in this chapter is, however, necessarily selective, and it concentrates on discourses which were central to the representation of the missionary enterprise in its encounter with other peoples in colonial territories.[23] Dominant discourses in missionary literature continuously emphasized the superiority of Christianity as a religion, as a system of knowledge and as a system of morality, with sexual morality and the position of women being central to the latter. A related dominant discourse was that of the 'civilizing mission' in which foreign missions regarded themselves as playing a key role, in particular as promulgators of Christian morality. This discourse, however, could conflict with the belief that all peoples were essentially equal in their humanity, a premise that underpinned the vision of Christianization of other peoples. Ambiguities were thus apparent in missionary positions with shifts occurring over time. Ultimately, however, such discourses were intrinsically linked with the development of a racialized world view. As well as shaping the attitudes of Scots to the 'others' of empire, missionary literature helped shape their attitudes to the empire itself. Their relationship to the empire and imperial administration was an explicit concern of missionaries and their supporters, as was the conceptualization of the role of the foreign mission movement within empire. These discourses also reflected ambiguities and shifts over time, but ultimately, they fostered the growth of imperialist sentiment in Scotland.

The superiority of Christianity

Given that the missionary enterprise was in essence a religious endeavour, explicit discourses of religious belief naturally permeated missionary literature. The superiority of Christianity was repeatedly reaffirmed as a counterpoint to other religions, consistently denounced as 'superstition', 'idolatry', false beliefs', and 'heathenism'. While some missionaries engaged in lengthy verbal and written debates over religious and philosophical systems, for example, with Indians of various faiths, and while some became scholars of languages, the periodical literature invariably summarily dismissed other religions and cultures. Attacks such as those on the systems of 'Parsis, Mohammedans and Brahmans' as 'horrid delusions' and 'multifarious idolatries' were typical, and indeed persistent throughout the nineteenth century.[24] Furthermore, missionary scholarship might be regarded with suspicion, as was evident in the criticism of William Miller for being too tolerant of Hinduism.[25]

Affirmation of the superiority of Christianity was often articulated through moral horror and revulsion at the beliefs and practices of other peoples, sometimes depicted in lurid terms. Hinduism, for example, was condemned for its depravity, the proof of which was afforded by 'crowds of females' prostrating themselves before their idol, and rolling about 'in the most indecent manner'.[26] While in India it was such irrational 'superstition' and 'idolatry' and the 'degradation' of women that were constantly denounced, in Africa it was the 'savagery' of 'heathenism' that attracted comment. This included a variety of practices perceived as cruel by missionaries, such as punishments for practising witchcraft, the sacrifice of widows or concubines, exposure of twins and female circumcision. Even Livingstone, praised for his fellow feeling for Africans and his 'sympathy with even the most barbarous and unenlightened' was, according to his biographer, Blaikie, disgusted by the 'painful, loathsome, and horrible spectacle' of certain forms of 'heathenism'.[27]

In challenging other religions and systems of belief, it was not only that Christianity was seen as providing religious truth in contrast to the 'false beliefs' of others, but also that European systems of thought, and above all science, had attained a rigour and potency that would necessarily defeat other ways of thinking.[28] Thus, the incorporation of Christian religious instruction in education in India could challenge the power of such beliefs over people, since 'their own faith, being largely founded on fable, cannot stand before the light of science'.[29] Indeed, the 'terrible power' of Indian religions could not 'stand the fierce light of European thought', which, in higher schools and colleges would destroy such beliefs.[30] The linking of Christianity to 'European thought', and in particular to science, was a notable feature of representations of Scottish missionary work and achievements. The view that science and religion complemented each other, actively advanced by Livingstone, for example, was widely shared by missionaries.[31] It has been commented that 'in Scotland the conflict between science and religion did not rage as bitterly as it did elsewhere', a situation attributed partly to the divisions in the church and in provision of education, which allowed a diversity of opinion and courses.[32] As MacKenzie has argued, the application of science presented a self-image of missionaries as 'people who controlled their natural and human environments with the help of technology, science and Western medicine'.[33] And it emphasized the difference between European and Indian and African societies.

While missionary work might often be thought of as involving first and foremost preaching and evangelism, for Scots missionaries education and medicine came to play an increasingly important role as the nineteenth century progressed. Both were seen as an effective route to

Christianization, and presented in the literature as such, whether the institutions of higher education established by Alexander Duff and others, or the teaching of literacy to orphan girls, such as those in Madras, praised for their 'proficiency in reading and their knowledge of Scripture, inculcated so kindly and zealously by their teacher'.[34] The education of girls in orphanages not only protected them 'from the countless corrupting and degrading influences of heathenism', but it also served to modify their disposition and to mould their character in a Christian form.[35] Medical missionary work at its inception was promoted as much, if not more, on the basis of its potential religious benefits. As EMMS founding member, Benjamin Bell, argued, 'the practice of the healing art may become a powerful auxiliary to the preaching of the Gospel', since 'Christian medical men... by gaining the confidence and goodwill of their patients, may open and smooth a pathway to their hearts for the saving truths of religion.'[36]

The position of women

Christianity was seen as superior to other religions which missionaries encountered, not just because of its greater rationality and promotion of scientific thought, but also because it was regarded as superior in its treatment of women. The regulation of gender relations was central to Christianity, as it was to other religions, and much criticism of the latter focused on those cultural beliefs and practices which governed such relations. This was a prominent part of missionary discourse from the earliest days of missionary activity, for example, in the campaign against *sati* in India,[37] and in campaigns against slavery, which was frequently condemned for its deleterious effect on sexual morality and for presenting obstacles to stable married life. Furthermore, the fact that some women in India lived in seclusion, prohibited from contact with white men, led to calls for more women missionaries, a call to which women responded in increasing numbers in the late nineteenth century as they gained access to higher education and professional careers.

Wives of missionaries and single women missionaries had worked in educating girls in India from the 1830s. The visiting of 'zenanas' (women's quarters where seclusion of women was practised) came later, and was regarded as a significant advance. Apparently first mooted in the 1840s by Church of Scotland missionary, Dr Thomas Smith, zenana visiting did not become a major part of missionary work until the 1860s, when it was reported back to the Church of Scotland as being 'like the discovery of a new continent'.[38] Male missionaries played an active role in supporting the development of this work, arousing interest at home with persistent denunciations of the position of women in India, which was said, more than any other country, to involve 'the greater degradation of women and her subjection to the man'.[39] As missionary work with women in India expanded later in the nineteenth century, the tone of moral shock and denunciation lessened, though Indian women's lives continued to be painted bleakly. *Our Indian Sisters*, recommended in 1898 to those in charge of mission work-parties or meetings, showed 'the state of degradation in which the women of India are held without the light of the Gospel'.[40]

Many denunciations of African 'savagery' and 'heathenism' focused on cultural practices affecting women, a theme present in accounts of missionary work in Nigeria from the 1840s onwards, for example. At the turn of the century, when missionary activity was initiated in Kenya, practices such as female circumcision appalled the missionaries.[41] This was not, however, explicitly described. Rather, allusion was made to the 'indescribable vileness of certain of the customs', which served as the justification for the mission to provide a 'place of refuge' for girls, who desired 'to leave evil Kikuyu customs, and to learn the "things of God"'.[42] Though a range of African practices were regularly condemned, generally speaking the position of women in Africa was less emphasized than in India as a reason for women to take up missionary work, since it was not only women who could work with women, though there was perceived to be an advantage in this. That gender roles were different in Africa, however, excited comment: 'According to native ideas the woman supports the family and the husband.'[43] But, 'European notions are different, and the missionaries have had to work in training native ideas into more civilised ways.'[44] Thus, much in the manner that middle-class philanthropists at home imposed their view of domestic ideology on working-class girls, women missionaries trained their African pupils for domestic labour and service.

As a counterpoint to the 'degradation' of women in imperial territories, missionaries stressed the freedom and equality of women in Scotland. Alexander Duff assured his audience at the Scottish Ladies' Association AGM in 1839 that Christianity had ensured 'as one of its inseparable fruits, the reinstatement of woman in all her privileges of original equality with man'.[45] Marriage, for women in Scotland, placed them in the position of 'helpmate', sharer of joys, comforter of sorrows and companion in pilgrimage, compared to the position of slavery that was the fate of married women in India.[46] Women writing for their own periodicals were less inclined to stress their equality in marriage, but commented on the lives of Indian women, for example, in ways that indicated belief in their own superior position. They made no rhetorical appeals to abstract ideas of equality and freedom as male missionaries were wont to do, more simply stating that they were glad, for example, to be able to take exercise and to enjoy the freedom of not being confined to the zenana, in contrast to the 'narrow lives' of the 'thousands shut up'.[47]

Given that regulation of gender relations is a central preoccupation of religion, it follows that discourses surrounding the position of women would be prominent in missionary literature. This emphasis functioned to engage the interest of women themselves as active supporters and as potential missionary agents, as well as reinforcing a belief in the superiority of Christianity and social systems governed by Christian beliefs (implicitly understood as Protestant) in their treatment of women. These discourses also functioned to galvanize support for the enactment of social reforms by the imperial administration, and in this were intrinsically linked to the idea of the 'civilizing mission'.

The civilizing mission versus the equality of humanity

When the resolution to support foreign missions was first put to the Church of Scotland in 1796, among the arguments which defeated this was the view that peoples had to be civilized before they could be Christianized.[48] Having resolved to support foreign missions in 1824, the Church of Scotland subsequently endorsed the view that 'the formative influence of Western knowledge' was required for the development of the form of rationality necessary to grasping the 'evidence' of Christianity.[49] In adopting this strategy, advanced by Alexander Duff, the Church of Scotland both enshrined the importance of education and the need for prior civilization of others if they were to be receptive to Christianity, and this set the framework for much missionary practice for the rest of the century. This strategy moved into the Free Church along with Duff in 1843. Thus missionaries were construed as bringers of 'civilization' as well as of Christianity. Most writers of missionary literature supposed that other peoples could approach this state of civilization. Yet there was an inherent tension between the view that other peoples needed to be civilized and a belief in human equality, a tension which was apparent in competing discourses of equality and inferiority.[50]

Some missionaries were depicted as passionate advocates of the rights of others, especially of African peoples. For example, concern about the status of Africans in South Africa was voiced by John Philip, active advocate of the rights of 'Hottentots', and his role as such was discussed in the pages of the *Missionary Register,* following the publication of his *Researches in South Africa* in 1828.[51] A universalistic belief in human equality was, however, most clearly expressed in writings on slavery. In 1839, on the first anniversary of emancipation in the West Indies, for example, white missionaries celebrated together with black people, rejoicing at the downfall of slavery.[52] Despite such celebrations, the consequences of emancipation continued to be a matter for debate, and missionaries felt it necessary to mount a defence of this. Mr Robb, of Jamaica, accused Thomas Carlyle of misrepresenting black people in his *Occasional Discourse on Negro Slavery.* To picture 'the black man' as a 'lazy animal', content with the small amount of labour to keep him in pumpkin was a 'coarse caricature'.[53] Such charges were false, and a whole people should not be condemned for the 'delinquencies of a portion'. What rights 'to torture and task the African', Robb asked, had the 'self-styled wiser white man'.[54]

Subsequent to emancipation, Scottish missionaries in Jamaica went together with freed slaves to West Africa, to set up the United Presbyterian mission at Old Calabar.[55] The impact of slavery in West Africa was kept in the public view both by the work of such missionaries and by accounts of explorers. By the middle of the century attempts were being made to create an organizational base at home for the exploration and evangelization of Central Africa, with the objects of furthering commerce, advancing the interests of geographical and other sciences and 'especially to abolish the horrid traffic in slaves'.[56] Livingstone's expeditions into central Africa brought this issue into even greater prominence, and it continued to be at the forefront in arguments for and accounts of the development of missions, particularly in Nyasaland [Malawi].

The images of 'savagery' so common in representations of Africa were seldom present in an Indian context, the Indian 'mutiny' of 1857 proving an exception. In this context Indians were characterized as 'murderous ruffians', and 'bloodthirsty mutineers', displaying 'diabolical fury' and 'Asiatic treachery'.[57] The more typical terminology used for India, however, was that of vice, degradation and squalor. India was recognized to have a 'civilization' albeit one that had declined. The social structures created by that civilization, such as the caste system, were seen to present serious obstacles to Christianization. Thus, Africans might be easier to convert, since Africa as 'a simple, untutored savage, who needed plain, practical teaching, ... was likely to turn to God far sooner than India would do'.[58]

Though competing discourses of inferiority and equality of colonial 'others' were present in representations of other peoples, the dominant themes were their 'superstitious' systems of belief, sometimes cruel customs, sexual laxity or immorality (contrasted with Christian monogamy and modesty), the lack of civilization particularly exemplified by the lack of literacy or a literate culture and the lack of a work ethic. Even where missionaries were presented as advocates of rights or champions of abolition, and as treating other peoples as 'brothers', their language was not always free from condescension. Furthermore, the generalizations offered as characterizations of other societies lacked complexity and, often, accuracy, and inevitably played their part in the formulation of ideologies of racial hierarchy, even if missionaries themselves seldom articulated these explicitly. MacKenzie has argued that 'The full panoply of social Darwinian notions, involving fundamental genetic difference and the inevitability of competition and extinction induced if necessary by war', does not appear in missionary writings, since this would have run counter to the idea of redemption.[59] This appears to be generally true of Scottish Presbyterian missionary literature, yet there are echoes of Social Darwinism in biographies of missionaries. Moreover, Social Darwinists were to be found among missionary supporters, the most notable case being Henry Drummond, a popularizer of Social Darwinist views, for which his account of his travels in central Africa served as a vehicle.[60]

Humanitarian and egalitarian discourses were most vocally articulated as an integral part of anti-slavery sentiment, and were intermittently expressed in relation to the rights of African peoples, especially in South and Central Africa. As elsewhere in Britain, confidence in an egalitarian stance declined in response to the difficulties following emancipation in the West Indies, to the Indian uprising of 1857 and 1858, to acts of resistance such as the Morant Bay rebellion in 1865 and to the growing perception of the difficulties of making conversions, especially in India.[61] As Peter Clayworth has pointed out, the belief that others could become civilized Christians was also eroded by settler lobbies in various parts of the empire and by the work of ethnological theorists.[62] On the other hand humanitarian views were somewhat revived by the challenge to the Arab slave trade represented by the Scottish missions in Nyasaland. Competing discourses of the inferiority of other 'races' and of human equality are thus observable in missionary literature throughout the nineteenth and in the early twentieth century. Over time, however, the former came to dominate over the latter. This shift in attitudes coincided with imperial expansion and the rise of imperialist enthusiasm in Britain. Missionary discourses also played a role in this process, and it is to how their relationship to empire was articulated that I now turn.

Attitudes to empire

There is an argument that the ambitions of British foreign missions were not coterminous with empire.[63] That neither evangelicals' vision of global Christianization nor their foreign mission operations were confined to the territories of empire is indeed true. Nonetheless, the existence of empire was crucial to the foreign mission enterprise. It was the experience of empire in the American colonies in the eighteenth century that provided the training ground for missionary work,[64] and the territories in which most missionaries were active were part of the British Empire or within the sphere of imperial influence. Missionaries were seldom to be found providing a critique of the concept of empire or the right of Britain to establish imperial rule as such, and indeed tended to see the existence of the empire as an act of Providence enabling Christianization. What they were intermittently critical of were aspects of imperial administration and policy, and on occasion they actively lobbied the imperial government to protect and further their interests, coming to see these as increasingly coinciding, as John MacKenzie has argued of South Africa.[65] However other British Protestant missionaries may have positioned themselves, within the Scottish context this widely perceived coincidence of interests manifested itself in the claim that the Presbyterian foreign mission movement was a Scottish contribution to empire of which the nation could be proud.

While the relationship of missionaries and the churches to the sphere of imperial politics was an ambiguous and sometimes uncomfortable one, there was nonetheless an explicit engagement with the political sphere on a number of occasions. Such engagement in politics sometimes excited disapproval, but this was not always the case. In the controversies over slavery in the Caribbean in the early nineteenth century, the Directors of the SMS actively intervened to constrain their agent, Mr Blyth, from expressing his views. Mr Blyth was criticized for his friendship with Mr McQueen, a determined defender of slavery, and appears to have also been criticized for carrying out missionary work with slaves, thereby giving credence to slavery. The Directors of the Society issued a statement endorsing the view that missionaries should not express political opinions, that is, views on slavery.[66] Following the emancipation of slaves in the West Indies in 1838, missionaries were able to speak out freely in celebration of emancipation and in condemning the continuing slavery in Cuba and North America.[67] This evidence suggests that Scottish missionaries in Jamaica were opposed to slavery, but given their dependence on the support of planters to carry out their work did not necessarily give vocal expression to these views.[68]

By contrast John Philip's advocacy of the rights of Africans in South Africa was openly claimed as political by his supporters, and he was defended against accusations of rashness and recklessness in his stance on the rights of 'Hottentots' and 'Caffres', since 'The African Colonists have no more right to hold the Lands of the Caffres than the West Indian Planters had to enslave the Negroes.'[69] While Philip's radical views excited opprobrium in some quarters, Alexander Duff's political lobbying in favour of education in English in India seems to have met with universal approval. Duff was described as having heroic stature as a promoter of the missionary cause, as a preacher and as an educationalist and lobbyist who had shown 'genius' in putting into practice his educational scheme and in influencing the government.[70]

The above examples illustrate missionaries' criticism of aspects of imperial administration, and their efforts to persuade the government to change laws. The Indian uprising of 1857 provides a clear case of criticism of imperial administration being voiced by missionaries and their supporters, but not of the concept of empire as such. This episode stands out particularly as the cause of much soul-searching as to its religious as well as political meaning. The 'calamity' prompted Christians to think about 'the causes of the Lord's controversy with us, and the duty which the churches and the nation owe to that country'.[71] Though Britain had brought benefits to India, God did not make 'this little island' the mistress of India simply 'that India should become a mine of wealth to the sons of Britain, or the means of her national aggrandisement'.[72] Rather it was for the higher end of Christianizing India. Thus a critique of both the East India Company and the government was offered, as well as of the Scottish people for their lukewarm support of missionary work. In 1857 the public debate in Scotland, as elsewhere in Britain, about the role of missionaries in helping to provoke the Mutiny, was resolved in favour of support for further missionary expansion. Thus, subsequent to the Indian Mutiny there was greater confidence in a strategy of missionary expansion, a confidence undoubtedly increased by the change in imperial governance which followed the Mutiny, and brought to an end the power of the East India Company, which had been antipathetic to missionary interventions.

In the same year Livingstone made his dramatic appeal for missionaries to aid Africa along the path of Christianity and commerce, meeting an enthusiastic reception in Scotland, if not immediate action to set up new missions. While the evidence from missionary literature indicates that interest in missionary work continued to increase in Scotland in the decade subsequent to this, it was not until after Livingstone's death in 1873 that there was a further surge of public interest in the missionary cause. The launch in 1875 of the Livingstonia expedition, jointly supported by the three main Presbyterian churches, and its subsequent progress maintained a high level of public interest. Though the choice of Nyasaland as a site for Scottish missions was in fact fortuitous, as other sites, including Somalia had been under consideration by the Free Church at this time, it clearly chimed with the desire to pay homage to Livingstone as a great Scot, and to emulate his example.[73] It was the development of the missions in Nyasaland that was to lead to a much closer alignment of missionary interests with those of empire, though this pattern was also apparent elsewhere.

As well as being covered in detail in missionary periodicals, the expedition to establish the Livingstonia mission received a great deal of attention in the press, with letters and reports appearing regularly in Scottish newspapers, indicating the progress being made on the journey to Lake Nyasa and incidents and encounters on the way.[74] The growing concern for the position of the Scots missionaries and traders in Nyasaland in the late 1880s led to the formation of an organized lobby, urging the government to intervene to protect its sphere of influence in the light of attacks by Arab slave-traders and encroachments by the Portuguese, who as participants in 'the scramble for Africa' were laying claims to territories in the area surrounding Lake Nyasa. There was a vigorous campaign by missionaries in 1887 and 1888, and by 'their powerful supporters in Britain' for action against the Arabs and Portuguese in the north of Nyasaland. This included a series of public meetings in Aberdeen, Glasgow, Edinburgh and Dundee, and culminated in 'a monster petition signed by over 11,000 ministers and elders of the Scottish churches'.[75] Furthermore, this campaign involved joint action between members of the 'Free and Auld Kirks' in a way that had not happened since the Disruption. A number of the meetings were chaired by Balfour of Burleigh, a prominent member of the Church of Scotland, then in Lord Salisbury's cabinet and he also led the delegation which presented the petition.[76] The subsequent declaration of a Protectorate over part of Nyasaland in 1889 was followed by the establishment of British rule over the whole of Nyasaland in 1891.[77]

As these examples illustrate, the disavowal of engagement with politics frequently reiterated in missionary literature was belied by active lobbying when the occasion demanded. It seems clear that the 'official' line of the churches was that religious life should be kept separate from the political sphere, and that it was not appropriate for churches to intervene in the politics of empire. The relationship between religion and politics was, however, sometimes an uncomfortably close one, and it was not always possible to maintain this separation.[78] In the early days of foreign missions, expressions of radical and political views tended to come from amongst the evangelicals, in non-denominational societies, in dissenting churches and within the Church of Scotland itself, with this latter group seceding to form the Free Church of Scotland at the Disruption of 1843. Subsequent to the Disruption, however, the main Presbyterian churches made political interventions from time to time. In general, such interventions were in the interests of furthering missionary expansion or of protecting missions and missionaries from the encroachments of on the one hand, Arab slave-traders, and on the other, rival European imperialists. Thus there was a common cause for the churches to espouse, and one that was intrinsically allied to imperial expansion, which the churches too came to more explicitly embrace.

A Scots contribution to empire

That there was an increasing coincidence between religious and imperial interests is perhaps best illustrated by the way in which the Presbyterian foreign mission movement came to be represented as a Scots contribution to empire. Such claims were typically couched in terms of the civilizing impact of missions, and these claims were made predominantly in relation to Africa, though Duff's role in advancing education in the English language in India was also praised. In constructing the claim of a Scots contribution to empire, it was not only prominent missionaries from the Presbyterian churches who personified this achievement but also Scots serving with other societies, most notably LMS missionaries, such as Robert Moffat and David Livingstone.

Missionaries such as Robert and Mary Moffat were described as having succeeded in making the country safe for Europeans, through their Christian influence on the 'heathen Bechwanas'.[79] Similarly, it was claimed that Dr Laws in his 'missionary enterprise' achieved the 'civilizing of the Ngoni', a people described as imposing, warlike and brave.[80]

The missionaries at Blantyre, in the 16 years they had been there, had 'changed the very soul of the place – the habits, the character, the life of the people'.[81] Thus, 'philanthropy and Christian Missions' were seen as having the capacity to restore 'the trend of the negro' from its backward movement to a forward evolutionary path.[82]

While the moral, religious and peaceful aspects of this contribution to empire were emphasized, it could also be tied to territorial expansion. This was stated most explicitly in the secular press, where references to Nyasaland as a 'Scotch colony' were found, together with tributes to the missionaries for achieving this.[83] But missionary literature itself also generated such a view. This was less the case in the missionary periodicals, which, although presenting an idealized view of the missionary, focused on descriptions of missionary life in carrying out professional duties, whether as ministers, doctors, nurses or educators, and on indications of success such as the numbers being educated or converted. Biographies, however, energetically fashioned the image of the heroic missionary, with Livingstone, of course, providing the best example of the iconization of missionary as hero.[84] One factor in this iconization, as MacKenzie has argued, was that Livingstone, as a Congregationalist, could be appropriated by all the main Presbyterian churches in Scotland, as well as churches elsewhere.

This creation of Livingstone as a heroic figure is well illustrated by William Garden Blaikie's biography and his claims for the impact of Livingstone's work.[85] Many of the subsequent 'extraordinary numbers of popular biographies' were reworkings of Blaikie's biography,[86] the purpose of which was 'to make the world better acquainted with the character of Livingstone', since he himself was modest and concentrated on his discoveries and researches.[87] 'As a man, a Christian, a missionary, a philanthropist, and a scientist, Livingstone ranks with the greatest of our race', and his life was telling both as 'a plea for Christian missions and civilisation' and as an illustration 'of the true connection between religion and science'.[88] Blaikie claimed for Livingstone a 'posthumous influence' that had entirely 'changed the prospects of Africa', through the steps taken to suppress the slave trade, commercial undertakings, exploration and the 'marvellous expansion of missionary enterprise'.[89]

Though Livingstone featured most prominently as representing the Scottish missionary contribution to empire, later missionaries were also acclaimed as 'empire builders'. The development of Scottish missions in Central Africa fostered the most explicit claims for national achievement, sometimes situated in the context of Scotland's history of colonial enterprise. It was thus seen to redeem the failure of Darien, with the

Livingstonia Mission being 'the greatest national enterprise ... since Scotland sent forth the very different Darien expedition'.[90] Its national character was attested to by support from 'all Christian Scotland', with subscriptions coming in 'from every class and quarter, from city merchant prince and Highland crofter'.[91] Though missionaries themselves had no thought of empire building, they could be claimed as 'moral' empire builders, adding to the 'grandeur of the British Empire'.[92]

In her turn, Mary Slessor was presented as a hand-maiden of empire, facilitating the expansion of British jurisdiction and assisting in its administration.[93] Her recognition by government in this role and the official approval afforded by government honours may have been crucial to her selection as material for idealization, despite her somewhat maverick character as a missionary.[94] Her 'heroic pioneer work' was undertaken with 'dauntless courage' as she single-handedly set about 'putting down the cruel and barbarous superstitions and customs that were everywhere rampant'.[95] With her 'infinite knowledge' of the language, 'and by her shrewdness and adaptability in understanding the native character' she was able to dispense justice over a wide area, gaining recognition to do so from the British authorities. As a representative of the Native Court, 'her work as a missionary was linked up with the systematic pacification of the country which the Government had entered after she began her labours'.[96]

Such assertions of the missionary contribution to empire became commonplace, continuing to be made into the twentieth century, and it was the iconization of Livingstone as the typical Scot that above all symbolized this representation of the missionary enterprise as a Scottish contribution to empire. Over time the emphasis on Livingstone's Scottishness became more pronounced. The centenary of his birth in 1913 provided ample opportunity for public celebration of the Scottishness of his 'ancestry' and 'blood', the 'staunchness of his convictions' and the 'probity of his character',[97] while the 1920s witnessed the establishment of the Livingstone memorial at Blantyre.[98] Thus the imagery and construction of the missionary as 'empire builder' served to simultaneously reinforce nationalist sentiment and imperialist fervour. The churches and missionary supporters at home had both facilitated and actively participated in this process.

Conclusion

This chapter has outlined the growth of support for foreign missions in nineteenth- and early twentieth-century Scotland, which came to be channelled mainly through the three major Presbyterian denominations after the Disruption of 1843. This movement functioned through Scottish networks of locally based committees and associations, which raised funds and disseminated information about foreign missions. Over time a voluminous missionary literature was produced, with an extensive readership. Thus, though numbers of missionaries as such were not necessarily large, institutional mechanisms existed to amplify and broadcast the meaning of the missionary experience. Through mechanisms such as these, as Hall has argued of Birmingham, cities and towns across Scotland became 'imbricated with the culture of empire'.[99] The missionary enterprise of Presbyterian Scotland, as Thorne has argued of the English Congregationalists, 'provided the empire with one of its more enduring ideological legitimations', through focusing the attention of church-goers on the empire 'on a regular and passionate basis'.[100]

My argument is that the dominant discourses articulated within missionary literature inevitably fed the creation of a racialized world view and fostered enthusiasm for imperialism. This does not necessarily mean that it was what missionaries intended. There was a degree of disjuncture, perhaps significant, between what missionaries may have privately thought and what they publicly expressed,[101] and a disjuncture between how missionaries perceived their work and their relations with the peoples with whom they worked and how readers at home understood this. Yet missionaries themselves contributed to the recycling of stereotypes and persistent tropes of missionary discourse, such as the superiority of Christianity, the savagery and superstition of others, the 'equal' treatment of Christian women compared to their degradation within other religious and cultural systems and the role they were playing in the 'civilizing mission'. Against the weight of this weary repetition, the more intermittent strains of humanitarianism struggled to be heard.

The typical stereotypes and tropes of missionary literature, in their representation of the 'others' at whom Christianization was aimed, were not unique to Scotland. Indeed, as chapters in this volume show, they functioned as a 'stock of metaphors and generalizations' shared by English-speaking communities on both sides of the Atlantic,[102] and were circulated not just by Protestant missionaries in an imperial context, but could be used in relation to the 'heathen' at home, to Catholics in Ireland and by Irish Catholic missionaries in relation to their missions abroad in the twentieth century.[103] However, their formulation by Scots for a Scottish audience meant that they furnished simultaneously representations of 'others' of empire and of Scots, in the same way that missionary societies and churches elsewhere articulated their local or national identities in counterpoint to the peoples of empire.[104] Of particular significance within the Scottish context was the role played by the institutional autonomy of the church, following the Union of 1707, in carrying constructions of Scottish national identity,[105] and in providing the basis for a distinctly Scottish foreign mission movement. The Disruption of 1843 appears, paradoxically, to have strengthened both the movement and its Scottish identity, since the split energized the missionary movement as well as evangelicalism and church extension at home, while at the same time motivated the different Presbyterian denominations to stress their character as Scottish churches. Despite the rivalry between the main Presbyterian denominations, evident in disputes at home, the foreign mission movement created the opportunity for the churches to find common cause in fulfilling Livingstone's legacy in Africa. It was the experience in Africa in particular that was to give rise to the claim of missions as a Scots contribution to empire, though the work of Indian missions was also incorporated in this claim. This projection of the foreign mission movement as a Scottish contribution to empire in itself helped to foster the growth of imperialist sentiment. Missionary literature, in its various forms, as a vehicle for representations of empire and the role of Scots within it, though inherently religious in its intentions and embedded in institutional religious life, was thus instrumental in building popular support for imperial power in its secular form.

Notes

* I am grateful to the anonymous reader from Palgrave for the helpful comments which have assisted with revisions of this chapter from the original conference version. I am also grateful for the award of an ESRC postdoctoral fellowship (Award no. PTA-026–27-144), which has enabled this work to be carried out.

1. See, for example, S. Thorne, *Congregational Missions and the Making of an Imperial Culture in Nineteenth Century England* (Stanford: Stanford University Press, 1999); C. Hall, *Civilising Subjects: Metropole and Colony in the English Imagination 1830–1867* (Cambridge: Polity Press, 2002); A. Johnston, *Missionary Writing and Empire, 1800–1860* (Cambridge: Cambridge University Press, 2003).
2. This chapter resonates with John MacKenzie's Chapter 6 in this volume.
3. See *Account of the Proceedings and Debate in the General Assembly of the Church of Scotland, 27th May, 1796* (Edinburgh, 1796).
4. The Disruption occurred in May 1843, when over a third of the clergy and nearly half the laity left the Church of Scotland to form the Free Church of Scotland. It was the culmination of a dispute over patronage, that is the

rights of patrons, often land-owners, to present candidates for the parish ministry, as opposed to the right of congregations to choose them. For a summary of this event, its causes and effects, see *The Oxford Companion to Scottish History*, ed. M. Lynch (Oxford: Oxford University Press, 2001).

5. See John MacKenzie (Chapter 6) in this volume: 'the tradition of schism in the Scots church led to greater energy in its mission provision'.
6. For example, the Edinburgh Ladies' Association for the Advancement of Female Education in India formed in 1837, and the Glasgow Ladies' Association for Promoting Female Education in Kaffraria [South Africa] formed in 1839.
7. See L. O. Macdonald, *A Unique and Glorious Mission: Women and Presbyterianism in Scotland 1830–1930* (Edinburgh: John Donald, 2000).
8. See J. Calder, *Scotland's March Past: The Share of Scottish Churches in the London Missionary Society* (London: Livingstone Press, 1945).
9. The Missionary Association of Scottish Episcopalians, was instituted in 1846, and listed in the *New Edinburgh Almanac*, 1848. Direct support of missions began in 1871, but this is only given a passing mention in both the following: F. Goldie, *A Short History of the Episcopal Church in Scotland* (Edinburgh: The Saint Andrew Press, 1976); R. Strong, *Episcopalianism in Nineteenth-Century Scotland* (Oxford: Oxford University Press, 2002). Prior to 1871, Scottish Episcopalians sent contributions for support of missions to their sister church, the Church of England.
10. See *The White Fathers in Scotland, 1934–1984, Golden Jubilee,* 1984.
11. See, for example, A. Duff, *The Jesuits: Their Origin and Order, Morality and Practices, Suppression and Restoration* (Edinburgh: John Johnstone, 1868).
12. The *Scottish Missionary and Philanthropic Register* was first published in 1820, and continued publication till 1848. More than 20 different missionary periodicals have been identified as being published in Scotland in the nineteenth century, with a number of these being continuously published, though sometimes undergoing changes of name, from the post-Disruption period into the twentieth century. See E. Breitenbach, *Empire, Religion and National Identity: Scottish Christian Imperialism in the Nineteenth and Early 20th centuries* (Unpublished PhD thesis, Edinburgh University, 2005).
13. J. Wilson, *Memoir of Mrs Margaret Wilson* (Edinburgh: John Johnstone, and London: Whittaker and Co and J Nisbet and Co., 1838). John Wilson (1804–1875) went with the Scottish Missionary Society to Bombay in 1829, subsequently becoming a Church of Scotland, then Free Church of Scotland missionary. His wife, born Margaret Bayne (1795–1835) accompanied her husband to Bombay in 1829. She ran schools for girls in Bombay, one of the first women to do so.
14. Robert Moffat (1795–1883) served with the London Missionary Society in South Africa, where he spent 52 years. He became David Livingstone's father-in-law with the marriage of Livingstone to his daughter Mary; John Philip (1775–1861) was a minister in Aberdeenshire, before going to South Africa with the London Missionary Society in 1819. He was an active advocate of the rights of African peoples; Robert Nesbit (1803–1855), like John Wilson, went with the Scottish Missionary Society to Bombay in 1829, subsequently becoming a Church of Scotland, then Free Church of Scotland missionary; Hope Waddell (1804–1895) spent 20 years in Jamaica with the

Scottish Missionary Society, and then went in the 1840s with the United Presbyterian Church to Nigeria to establish the mission at Old Calabar, accompanied by freed slaves.

15. Alexander Duff (1806–1878) was the first Church of Scotland missionary, arriving in India in 1829. He subsequently went over to the Free Church of Scotland. He played an active role in organizing support in Scotland for foreign missions, and campaigned for education in the English language in India.
16. W. P. Livingstone, *The White Queen of Okoyong* (London: Hodder and Stoughton, *c.*1919). Mary Slessor (1848–1915) worked in the mills in Dundee, before becoming a missionary for the United Presbyterian Church in Old Calabar in 1875.
17. Robert Moffat was the subject of at least 15 biographies between 1884 and 1961, with five of these being in the period prior to 1914, and a number of them being reprinted. At least 10 biographies of Mary Slessor were published between 1915 and 2001. Alexander Duff was the subject of at least six biographies between 1879 and 1992.
18. See J. M. MacKenzie, 'David Livingstone: the Construction of the Myth' in *Sermons and Battle Hymns: Protestant Culture in Modern Scotland*, ed. G. Walker and T. Gallagher (Edinburgh: Edinburgh University Press, 1990). Only one biography is quoted here, that by W. G. Blaikie, *The Life of David Livingstone* (London: John Murray, 11th impression, 1906) [first published 1880]. David Livingstone (1813–1873) worked in a cotton mill in Lanarkshire, before becoming a missionary in South Africa in 1840. From the 1850s onwards he carried out journeys of exploration across Africa, becoming internationally famous as a consequence.
19. See, for example, MacKenzie, 'David Livingstone: the Construction of the Myth', p. 40.
20. D. A. Dow, 'Domestic Response and Reaction to the Foreign Missionary Enterprises of the Principal Scottish Presbyterian Churches, 1873–1929' (Unpublished PhD thesis, Edinburgh University, 1977).
21. The religious census of 1851 indicated that 25.6 per cent of the Scottish population were church attenders, and levels of church membership subsequently increased to reach a peak in 1905 at 50.5 per cent of the population. See C. G. Brown, *Religion and Society in Scotland since 1707* (Edinburgh: Edinburgh University Press, 1997), pp. 59, 64.
22. J. Johnston, *Dr Laws of Livingstonia* (London: Partridge and Co., 1909), p. 149.
23. In the discussion in this chapter of dominant discourses in missionary literature, all references are to writings about India and Africa. While Scots missionaries worked elsewhere, for example, the South Seas and China, the majority were located in India and Africa. In 1904, for example, 82 per cent of missionaries supported by the two main Presbyterian churches, the United Free Church of Scotland (formed in 1900 through the union of the Free Church of Scotland and the United Presbyterian Church of Scotland) and the Church of Scotland, were in India or Africa, with India retaining the bigger share, reflecting a consistent pattern over time.
24. Wilson, *Memoir of Mrs Margaret Wilson*, p. 340.
25. See T. G. Gehani, 'A Critical Review of the Work of Scottish Presbyterian Missions in India, 1878–1914' (Unpublished PhD thesis, University of Strathclyde, 1966).

26. *Home and Foreign Missionary Record of the Church of Scotland,* IX (May 1854), p. 98.
27. Blaikie, *The Life of David Livingstone,* p. 121.
28. Peter van der Veer has noted that 'Christianity, at least from Kant on, is portrayed as the rational religion of Western modernity, whereas Hinduism is mystified as Oriental wisdom or irrationality'. P. van der Veer, *Imperial Encounters: Religion and Modernity in India and Britain* (Princeton, NJ: Princeton University Press, 2001), p. 26.
29. 'From our Heathen Mission-fields', in *Church of Scotland News of Female Missions,* New Series, XII (December 1898) 94.
30. J. M. Mitchell, *In Western India: Recollections of My Early Missionary Life* (Edinburgh: David Douglas, 1899), p. 120.
31. Several Scottish missionaries, like Richard Taylor, the subject of Peter Clayworth's chapter (Chapter 11) in this volume, were 'part of an imperial network of natural history information'.
32. W. Ferguson, 'Christian Faith and Unbelief in Modern Scotland' in *Scottish Christianity in the Modern World,* ed. S. J. Brown and G. Newlands (Edinburgh: T & T Clark, 2000), p. 81.
33. J. M. MacKenzie, 'Missionaries, Science, and the Environment in Nineteenth-Century Africa', in *The Imperial Horizons of British Protestant Missions, 1880–1914,* ed. A. Porter (Grand Rapids: Wm B Eerdmans Publishing Co, 2003), p. 128.
34. Church of Scotland, *News of Female Missions,* III (July 1863), p. 66.
35. Ibid., p. 75.
36. B. Bell, in *Addresses to Medical Students* (Edinburgh: EMMS, 1855–1856), pp. 149–50.
37. For a discussion of how missionaries and other Europeans represented *sati,* see A. Major, *Pious Flames: European Encounters with Sati* (New Delhi: Oxford University Press, 2006).
38. E. Hewat, *Vision and Achievement, 1796–1956: A History of the Foreign Missions of the Churches United in the Church of Scotland* (London: Thomas Nelson and Sons, 1960), p. 75.
39. Address by Alexander Duff, in Scottish Ladies Association, *Report of Annual General Meeting,* 1839.
40. Church of Scotland, *News of Female Missions,* New Series, XI (November 1898), p. 88.
41. This issue was taken up in parliament by the Duchess of Atholl, after she attended a Church of Scotland meeting around 1929 or 1930. See Katharine, Duchess of Atholl, *Working Partnership* (London: Arthur Barker, 1958).
42. *Kikuyu News,* Church of Scotland Mission, British East Africa, XVII (March, 1910), p. 12.
43. Church of Scotland, *News of Female Missions,* New Series, X (January 1898), p. 2.
44. Ibid., p. 2.
45. Speech by Alexander Duff, Scottish Ladies Association, *Report of Annual General Meeting* (1839).
46. Speech by Rev. J. R. Macduff, Scottish Ladies Association, *Report of Annual General Meeting* (1844), p. 22.
47. Miss Read, Chamba Mission in India, in Church of Scotland, *News of Female Missions,* New Series, X (1898).

48. See *Account of the Proceedings and Debate in the General Assembly of the Church of Scotland, 27 May, 1796* (Edinburgh, 1796).
49. I. D. Maxwell, 'Civilization or Christianity? The Scottish Debate on Mission Methods, 1750–1835' in *Christian Missions and the Enlightenment,* ed. B. Stanley (Grand Rapids, MI/Cambridge: William B. Eerdmans, 2001), p. 140.
50. These competing discourses might even be articulated by the same individual. As O'Brien notes in her chapter (Chapter 9) in this volume, 'many missionaries were racist and anti-racist at the same time'.
51. See *Scottish Missionary and Philanthropic Register* (1831), pp. 43–7.
52. See *Scottish Missionary and Philanthropic Register* (December 1840), pp. 181–3.
53. *Missionary Record of the United Presbyterian Church,* XII (November 1857), p. 189.
54. Ibid., p. 190.
55. See *Missionary Record of the United Presbyterian Church,* I, no. 8 (1846), p. 114 ff.
56. Advert for the Society for Exploring and Evangelizing Central Africa, by Means of Native Agency, a London based organization, with a Scottish branch, in *Home and Foreign Missionary Record of the Church of Scotland,* IX (November 1854), p. 282. Rev. David Livingston (sic) is listed as an Honorary member.
57. See Smith, quoting from Duff's 'Chronicles' in G. Smith, *Life of Alexander Duff* (London: Hodder and Stoughton, 1908).
58. J. Wells, *Stewart of Lovedale: The Life of James Stewart* (London: Hodder and Stoughton, 1908), p. 12.
59. MacKenzie, 'Missionaries, Science, and the Environment in Nineteenth-Century Africa', p. 124.
60. In William Ferguson's view, Drummond's best-selling *Natural Law in the Spiritual World,* was a credit to neither science nor religion, being 'stuffed' with 'gnomic riddles'. See Ferguson, 'Christian Faith and Unbelief in Modern Scotland', p. 80.
61. For discussion of this theme see C. Hall, *Civilising Subjects.*
62. See the Chapter 11 by Peter Clayworth in this volume.
63. See, for example, A. Porter, 'An Overview, 1700–1914', in *Missions and Empire,* ed. N. Etherington (Oxford: Oxford University Press, 2005), pp. 40–63.
64. E. H. Gould, 'Prelude: The Christianizing of British America' in *Missions and Empire,* pp. 19–39.
65. See J. M. MacKenzie's chapter (Chapter 6) in this volume.
66. *Scottish Missionary and Philanthropic Register* (1832), p. 141.
67. See *Scottish Missionary Register* (December 1840), pp. 181–3.
68. However, as John McAleer shows in his chapter (Chapter 10) in this volume, Alison Blyth's journal reveals her husband's views as more ambivalent than those he later claimed to have espoused.
69. R. Philip, *The Elijah of South Africa, or the Character and Spirit of the late John Philip DD* (London: John Snow, 1851), p. 61.
70. G. Smith, *Life of Alexander Duff* (London: Hodder and Stoughton, 1899), p. 99.
71. *Missionary Record of the United Presbyterian Church,* XII (November 1857), p. 185.

72. Ibid., p. 186.
73. For a detailed account of how the expedition came about, see J. McCracken, *Politics and Christianity in Malawi 1875–1940: The impact of the Livingstonia Mission in the Northern Province* (Cambridge: Cambridge University Press, 1977).
74. As well as appearing in *The Scotsman*, articles appeared in papers such as the *Dundee Courier, Glasgow Herald, Daily Review, Edinburgh Courant, Aberdeen Weekly Herald and Free Press, Wick Gazette,* and also in Christian papers and missionary periodicals. See MS 7906, National Library of Scotland, which contains a collection of newspaper cuttings on the Livingstonia mission expedition to Nyasaland.
75. McCracken, *Politics and Christianity in Malawi,* p. 158; see also A. C. Ross, 'Scotland and Malawi, 1859–1964', in *Scottish Christianity in the Modern World* ed. S. J. Brown and G. Newlands (Edinburgh: T & T Clark, 2000), pp. 283–309.
76. Ross, 'Scotland and Malawi, 1859–1964', p. 289. Lord Balfour of Burleigh was Conservative politician who served as Scottish Secretary from 1895–1903 and was a prominent Church of Scotland member and promoter of presbyterian reunion.
77. McCracken, *Politics and Christianity in Malawi*, p. 157.
78. Interventions were also made in domestic politics, among other things in pursuit of religious aims, such as disestablishment, or the re-unification on the main Presbyterian churches, eventually achieved in 1929, since these entailed the enactment of legislation by parliament.
79. J. S. Moffat, *The Lives of Robert and Mary Moffat* 3rd edn (London: T Fisher Unwin, 1885), p. 248.
80. Johnston, *Dr. Laws of Livingstonia,* p. 16.
81. W. Robertson, *The Martyrs of Blantyre* (London: James Nisbet, 1892), p. 38.
82. Johnston, *Dr. Laws of Livingstonia,* p. 88.
83. See, for example, newspaper cutting dated 6 March 1877, no title given: MS 7906, National Library of Scotland; 'Scotland and Geographical Work' in *Scottish Geographical Magazine,* I (1885).
84. See MacKenzie, 'David Livingstone: Construction of the Myth'; and J. M. MacKenzie, 'Heroic Myths of Empire' in *Popular Imperialism and the Military, 1850–1950,* ed. J. M. MacKenzie (Manchester: Manchester University Press, 1992).
85. Given Livingstone's celebrity, publishers of works on his life might expect a sizeable readership. John Murray, the publisher of an early biography on Livingstone, had initially approached the popular writer Samuel Smiles to undertake this work, though this proposal did not come to fruition. William Garden Blaikie who authored the biography was, however, judged to have produced 'a Masterly Account'.
86. MacKenzie, 'David Livingstone: Construction of the Myth', p. 38.
87. Blaikie, *Life of Livingstone,* p. iii.
88. Ibid., p. iv.
89. Ibid., p. 395.
90. Smith, *Life of Duff,* p. 337.
91. W. P. Livingstone, *Laws of Livingstonia: A Narrative of Missionary Adventure and Achievement* (London: Hodder and Stoughton, 1921), p. 9.

92. Ibid., p. 153.
93. For a discussion of Slessor's view of herself as a 'pioneer of the frontier' see C. McEwan, '"The Mother of all the Peoples": Geographical Knowledge and the Empowering of Mary Slessor', in *Geography and Imperialism, 1820–1940*, ed. M. Bell, R. Butlin and M. Hefferman (Manchester: Manchester University Press, 1995), pp. 125–150.
94. See J. H. Proctor, 'Serving God and the Empire: Mary Slessor in South-Eastern Nigeria, 1876–1915', in *Journal of Religion in Africa,* XXX, no. 1 (2000), pp. 45–61.
95. *The Scotsman,* 18 January 1915.
96. Ibid.
97. 'The Livingstone Centenary', in *The Scotsman* , 20 March 1913.
98. MacKenzie has argued that this represented the re-appropriation of Livingstone by the Scottish cultural revival of the 1920s. MacKenzie, 'David Livingstone: The Construction of the Myth'.
99. Hall, *Civilising Subjects,* p. 12.
100. Thorne, *Congregational Missions and the Making of an Imperial Culture,* p. 156.
101. For an instance of this, see John McAleer's chapter (Chapter 10). Alison Blyth's private journal shows her husband taking a more equivocal stance on slavery than his own subsequent representation of his anti-slavery position.
102. See Chapter 14 in this volume by Ruth Compton Brouwer.
103. See Chapters 8 and 13 in this volume by Shurlee Swain and Fiona Bateman.
104. See, for example, S. Thorne, *Congregational Missions and the Making of an Imperial Culture,* and C. Hall, *Civilising Subjects,* on the impact of foreign missions on the construction of identities in England. Fiona Bateman, in her chapter (Chapter 13) in this volume, has also noted that the function of Irish missionary discourse was 'not so much to strengthen the functioning of their "colonial" power, but to strengthen Irish identity as Catholic and civilized'.
105. For a discussion of the importance of the church to national identity in Scotland and to Scottish civic life, see, for example, L. Paterson, *The Autonomy of Modern Scotland* (Edinburgh: Edinburgh University Press, 1994); G. Morton, *Unionist-Nationalism: Governing Urban Scotland, 1830–1860* (East Linton, Tuckwell Press, 1999); J. Hearn, *Claiming Scotland: National Identity and Liberal Culture* (Edinburgh, Polygon, 2000).

Further reading

Hargreaves, J. D. *Aberdeenshire to Africa* (Aberdeen: Aberdeen University Press, 1981).

Piggin, S. and Roxborogh, J. *The Saint Andrews' Seven: the Finest Flowering of Missionary Zeal in Scottish History* (Edinburgh: Banner of Truth Press, 1985).

Piggin, S. *Making Evangelical Missionaries 1789–1858: The Social Background, Motives and Training of British Protestant Missionaries to India* (Abingdon: Sutton Courtney Press, 1984).

Proctor, J. H. 'Serving God and the Empire: Mary Slessor in South-Eastern Nigeria, 1876–nineteen15', *Journal of Religion in Africa,* XXX 1 (2000) 45–61.

Proctor, J. H. 'The Church of Scotland and British Colonialism in Africa', *Journal of Church and State*, XXIX (1987) 475–93.
Ross, A. *David Livingstone: Mission and Empire* (London: Hambledon and London, 2002).
Ross, A. *John Philip, 1775–1851: Missions, Race and Politics in South Africa* (Aberdeen: Aberdeen University Press, 1986).
Semple, R. A. *Missionary Women: Gender, Professionalism and the Victorian Idea of Christian Mission* (Woodbridge and Rochester: The Boydell Press, 2003).

Part II

Colonies and Mission Fields

Greater Britain: Whiteness and its Limits

6

'Making Black Scotsmen and Scotswomen?' Scottish Missionaries and the Eastern Cape Colony in the Nineteenth Century

John MacKenzie

Introduction

J. G. Pocock's famous plea for a four-nation approach to the history of the British and Hibernian Isles has been followed more eagerly by historians of Britain and Ireland than by those concerned with imperial history. Indeed, Empire was supposed to be about the suppression of such separate ethnicities. The very word 'British' applied to Empire was intended to convey the allegedly joint overseas project in which the distinct ethnic communities of these isles would be dissolved in global endeavour. The Union, particularly that with Scotland, but to a certain extent that with Ireland as well, was forged as much abroad as at home. The formation and deployment of Scottish and Irish regiments; the activities of politicians seeking bipartisan causes; emigrants abandoning the distress of home countries and pursuing new opportunities; as well as churches aiming for the expansion of Christendom could use empire as a realm of conciliation. That at least was the theory. On the other hand, Seeley's *The Expansion of England* seemed to suggest that these ambitions would take place on England's terms and would ultimately represent the creation of a world-wide English polity. This was a conscious attempt at proposing English exceptionalism, an exceptional history which could first embrace the cultural overwhelming of these islands as a prelude to the global expansion which was theoretically the central defining purpose of English history. But such a programme was a chimera. Empire served to re-emphasize rather than obscure the diverse nationalities of these islands. It is surely not right to suggest that it was only with the end of Empire that the nationalities could re-emerge: they had never been suppressed; they had never been in abeyance.

This chapter is designed to demonstrate how much this was true in relation to missionary activity. The Scottish missions emphasized their Scottishness and created remarkable connections between periphery and centre. Scots missionaries had a distinctive approach to imperial opportunity. They created a particular environment for the pursuit of relationships with Africans, but were also closely bound up with warfare and violence, white settler projects and the environment, education and publishing and much else at the so-called periphery. This is also a highly gendered story and one that passes through a considerable dynamic in the course of the nineteenth century.

Scots missionaries and the Cape frontier

The histories of the Cape frontier, of Scots military figures and of missionaries are inseparably intertwined. Yet, in a significant corpus of historical writing on the frontier, these have rarely been satisfactorily combined. Moreover, until recent times the Scots missionaries have seldom been examined as a separate ethnic group with different objectives and methods, although their activities upon the frontier were important in both white and, more particularly, black history. The missionaries constituted a separate pressure group with connections to the imperial metropole and to Scottish society and its various churches. They were frontier 'pioneers' who arrived when that frontier was still 'open' – that is, an incipient zone of contact between white and black, not yet fully under colonial rule. They often attempted to establish their mission stations during the period when the frontier was 'closing', that is the time of turbulence and violence when imperial power was being imposed, sometimes aggressively, at times reluctantly. They usually withdrew when war broke out, but they also weathered the vagaries of imperial policy: successively efforts to set up buffer zones and treaty systems, the prosecution of forward policies and periods of apparent retreat and finally the pushing of the colonial border through the frontier zone. Once this had happened, the frontier had been 'closed'. Blacks were forced to adjust to the new conditions. And the missionaries began to make more headway both with their spiritual and their educational objectives. But as Lamar and Thompson have pointed out, the white takeover of the American frontier was a great deal more complete than the southern African one.[1] Despite continual and endemic violence, African societies were more resistant and, in some senses, more ready to adjust and assimilate (in the sense of a two-way assimilation) than the indigenous peoples of North America. Perhaps missionaries helped in this.

The Eastern Cape frontier was distinctive in a number of ways. African peoples were relatively densely settled, but the southern Nguni had no central political authority as the northern Nguni did. The western and southern part of this region was more or less suitable for white settlement. Such settlement, begun by Afrikaners in the last days of the Dutch company and British pioneers in the early days of British imperial rule, was overlaid by the arrival of the 1820 settlers in that and subsequent years. Scottish missionaries, in various societies, positioned themselves on this frontier and became embroiled within the processes of frontier closure. They acquired relatively large tracts of land; they established complex relationships, not always benign, with African peoples; with the colonial authorities; and also, often hostile, with settlers. The environment of the region constituted a significant underpinning of all of this activity. It seemed to offer attractive, extensively timbered and seemingly well-watered lands, beyond the relatively arid Karoo, suitable for some cultivation as well as the running of sheep and cattle. Yet its fertility was often exaggerated; it lurched from severe drought to excessive rainfall; and it sometimes experienced extremes of heat and cold. Violence was inevitably related to these environmental shifts. But its hills and river valleys also rendered it an appealing, even romantic, landscape for whites, offering some analogy with Scotland itself.

If the Eastern Cape frontier had many unique characteristics as a missionary field, we should also consider the distinctiveness of the traditions of the Scottish missionaries who settled there. It is certainly the case that the work of many Scots missionaries should be analysed in terms of interactive relations with Scotland, both from the point of view of whites and blacks. This constitutes a sort of 'third way' between the predominantly white focus of older mission history and the necessary corrective of the black perspectives of a more recent historiography. The intention here is to consider the effects upon the metropole as well as upon African peoples, for there can be no doubt that Scottish society itself was modified by the existence of the missions. As Esther Breitenbach argues in Chapter 5, in this volume, the culture of empire was spread throughout Scotland by its missionary discourses in complex and dynamic ways from the late eighteenth century. She also notes that Scotland's missionary enterprise was used 'to fashion the nationalistic claim of a specifically Scottish contribution to empire'.[2] Lesley Orr Macdonald has written of the considerable and mutually supportive role of women in Scottish missions in India, China and Africa, as well as in the support systems in Scotland.[3] This culminated with the role of female heroes of the late nineteenth century in spreading the ideology of imperialism.[4] Moreover, missionaries undoubtedly worked within the ideologies of an international movement, as Ruth Compton Brouwer has argued in respect of Canadian missions.[5]

A wave of local Scottish missionary societies, often focusing on Africa, emerged at the turn of the eighteenth and nineteenth centuries. It was not long before various Scots societies developed associated women's organizations in Scotland, helping to emphasize the ties to 'home'. Sometimes, these were framed along the lines of 'women's work for women', but often they raised funds for more general objectives, including the financing of 'native teachers'. Such organizations were a setting for both middle- and working-class female activism, in which 'home missions' and 'foreign missions' were undoubtedly connected.[6] By the end of the nineteenth century women were beginning to play an increasing role in the governing structures of the Scottish churches.[7] Their activities in the missionary field undoubtedly contributed to this. Another very different social group pulled into the missionary endeavour were some of the industrialists of the Central Belt. As major contributors to missionary projects from the 1840s, they seem to have seen little conflict between their holding down of workers' pay and conditions in Scotland while supposedly contributing to the 'raising' of the spiritual and economic situation of Africans.

At the Cape, the Scots missionaries had a highly ambivalent relationship with both imperial and African authorities, and their fortunes ebbed and flowed with the vagaries of colonial policies. The Scottish missionaries developed their spiritual campaigns in the midst of violence, initially on or beyond the effective frontier. And if their activities were conditioned by the nature of that frontier, there can also be little doubt that there was much that was distinctive about their Scottish experience and character. Even if some may doubt whether their methods were much different from those of other 'nonconformist' missions, they themselves stressed the unique nature of their contribution. They often wrote of historic Scottish frontiers and the cattle raiding that had taken place across them. Their own propaganda insisted that they understood the problems and were hardy and determined enough to overcome them. They became, in effect, landowners with 'tenants' (sometimes literally so, though by the early twentieth century they were repudiating this role)[8] who were drawn into a whole range of economic, environmental, spiritual and westernizing relationships. The Scottish estate, of which some of them had had experience in Scotland, was a model in socio-economic as well as environmental terms. After the initial open frontier phase, ministers and teachers became increasingly well-educated and were full of ambitions not just for conversions, but also for the 'modernisation' process.[9] As we have seen, education was at the centre of their mission. They were not unique in this, but they were particularly seized by the need for 'industrial' training.[10] Although they were seldom popular with settlers, they saw themselves as having an obligation both to African and to colonial society to produce black artisans and an educated black proto-bourgeoisie while supposedly avoiding unnecessary competition with whites. In other words, they had a distinctively southern African approach to the labour market. Their educational policies were also designed to be self-sustaining, by producing 'native agents' – African teachers to spread the educational word, agricultural demonstrators to create a market-orientated farming mentality and catechists for further conversion – the triple prerequisite of the religious objectives.

The arrival of Scots missionaries

Although the Scottish churches were relatively slow to join the missionary enterprise, with the Established Church initially resistant, Scots and those influenced by Scottish education were active from the early stage of frontier evangelicalism. Both the London Missionary Society (LMS), founded in 1795, and the Glasgow Missionary Society (GMS), established in 1796, were non-denominational[11] and represented the new evangelical thrust which coincided precisely with the first British conquest of the Cape. The LMS attracted many Scots who had left the Established Church and joined the Congregational movement. The Rev. John Love, minister of a Scots Presbyterian Church in London, was one of the first secretaries of the Society, and presided over the emergence of southern Africa as a major sphere for its operations. In 1799 he helped to select Dr Johannes T. van der Kemp (1747–1811) as a missionary for the Cape. The Dutch Van der Kemp had studied medicine in Edinburgh in the 1780s.[12] He set up the first mission beyond the frontier, but he was forced to withdraw almost immediately on the orders of the Governor Sir George Yonge who was fearful of Jacobinism on the frontier.[13] Instead, Van der Kemp founded missions to the Khoi within the colony, which also became highly controversial. The Scottish connection was maintained when the LMS decided to send a director to survey the situation and make recommendations for future operations. The man chosen was John Campbell (1766–1840), the son of an Edinburgh greengrocer. He had entered the ministry and soon revealed his evangelical credentials by founding the Religious Tract Society of Scotland in 1793. He produced religious books for youth and founded a Sunday School to which he hoped to bring young blacks from South Africa. He sailed for the Cape in 1812, after the death of van der Kemp and, encouraged by the Governor Sir John Cradock, he spent almost two years as an explorer and natural historian in the interior. Partly as a result of his proposals, Robert Moffat from East Lothian arrived at the Cape in 1817, within three years moving to Tswana country and founding the Kuruman mission.[14] The LMS mission on the frontier was briefly re-established between 1816 and 1818 by Joseph Williams, when a Xhosa prophet figure called Ntsikana, one of Ngqika's counsellors, decided to make a partial conversion.[15] The other influential convert was Nxele, who was a prophet attached to the chief Ndlambe.[16] Williams also built a furrow for irrigation purposes and this became a *sine qua non* of all missions, evidence of their advanced approach to agriculture. Campbell was back at the Cape between 1818 and 1821 arriving with the celebrated LMS superintendent John Philip, a Congregationalist minister who had a keen sense of social injustice. Campbell published his journals in 1815 and, in two volumes, in 1822.[17] These books kept the Cape in missionary sights, particularly in Scotland.

Meanwhile, Love, back in Scotland after 1800, had become chairman and then secretary of the GMS. The GMS, initially similar to other societies founded throughout Scotland,[18] was an 'auxiliary' movement, raising funds, holding meetings, issuing propaganda, in effect, and often sending money and recruits on to the LMS.[19] But the GMS developed greater ambitions: it did become directly involved in missionary endeavour and in the recruitment of missionaries. After considering Sierra Leone and India, Love, with his LMS experience, may well have been influential in persuading it to turn its attention to southern Africa.[20] As a result, the GMS became a celebrated force on the missionary frontier until its energies ran out with the continuing secessionist tendencies of Presbyterianism in the 1830s and 1840s.[21]

The Tory Governor Lord Charles Somerset had a highly ambivalent attitude to Scots. On the one hand, he was to become locked in a fierce battle with a group he called the 'Scotch Independents' at the Cape.

These included John Philip, the poet and settler Thomas Pringle, the printers and publishers George Greig and John Fairbairn, among others. He regarded them as dangerous dissidents and their educational establishment as a seminary of sedition. Yet Somerset was also inclined to attempt to use missionaries to pacify the frontier, even Scots if necessary. The Rev. John Brownlee, a gardener from Lanarkshire, was encouraged to take up this highly dubious role of 'government missionary' in 1820. But in 1830, missionaries themselves recognized that such an official role often conflicted with their spiritual purposes. In the course of the 1820s, a number of Scottish missionaries arrived, setting up missions in the disputed frontier areas. The Scottish character of these missions is symbolized in a number of ways: first, through their physical appearance; second, through the immediate attention to education; third, through their social organization; and fourth, of course, through the specific theology dispensed there. We should also remember that Scotland was itself a trilingual country – with ministers working in English, Scots, and Gaelic, and several seemed to have little difficulty in turning themselves into Xhosa linguists.

When the Rev. William Ritchie Thomson arrived at Gwali or Tyumie he was struck by the allegedly Caledonian beauty of the place. He insisted that the mission should be laid out afresh with cottages erected parallel to the street with gardens stretching out behind (the Scottish parallel is clear).[22] No wonder Thomson thought that, black faces apart, going to church was like dropping into 'a little Scotch village'.[23] Ross had arrived with a small printing press (he had worked in a printing office in Glasgow and regarded himself as a professional) and founded the considerable tradition of printing and publishing which became a mark of the missions of the area. The first Xhosa words, using a primitive orthography (later superseded) invented by the Rev. John Bennie, were printed on this press in 1823. A larger press arrived in 1831 and was succeeded by others as printing technology improved.[24] Indeed, wherever Scots went in the world, printing was an important part of their mission, both religious and secular.

The GMS now harboured, as it turned out, unrealistic ambitions for a string of stations into the interior of Africa. It was perhaps this vision which partly motivated David Livingstone's journeys into the interior. Nevertheless, starting in 1824 a number of missions appeared with names like Lovedale, after the recently deceased John Love, Balfour, Burnshill, Pirie and Glenthorn. Each of these missions was endowed with several outstations and schools, staffed by Africans, around which the Scots missionary was expected to make frequent itinerations. The triumph was always to build a stone church at each, invariably in an architectural style reminiscent of Scotland, often with contributions from Africans themselves. It is significant that when John Philip visited the Khoi missions in 1821, he urged the inhabitants to enhance their reputation by building in stone.[25] Stone- and brick-built architecture would indeed be a defining characteristic of the Scottish missions, physically symbolic of the shift from the open to the closed frontier. Later, from the 1850s, as the frontier closed, missions with names like Macfarlan, Henderson, Cunningham, Paterson, Duff, Main and Buchanan, later again Donhill and Stuartville, were founded further beyond the frontier, many of them attracting missionaries from Scotland, but largely staffed by Africans. This activity climaxed with the founding of the Gordon Memorial Mission in Zululand. Most of these places retain their names to this day. These naming and architectural policies seem to have been a conscious means of inscribing Scotland upon the landscape just as Scottish mores and social forms would be inscribed upon society. But at first such ambitions were heavily circumscribed by war.

For at this stage the Scots were incapable of avoiding the violence of the contact zone. And Scots were far from being sympathetic to either the colonial authorities or the white settlers. The wife of the missionary John Ross, Helen, wrote a series of letters to her sister in Scotland which has survived. She tells of the unwelcome visits of a white commando and colonial troops in 1829. The Xhosa Chief, Maqoma and his people were cleared from land which later became the Kat River Settlement of the Khoi, ensuring that the Khoi were themselves targets in subsequent wars. Her husband remonstrated with officers, some of them fellow Scots, for making war on the Sabbath.[26] Indeed, Ross established close and fruitful relations with the Chief Maqoma, elder son of Ngqika.[27] In a letter to her parents in 1830, Helen wrote about the Maqoma case, revealing the extent of the political involvement of the missionaries. Dr Philip, Mr Brownlee, Mr James Read and Mr Fairbairn, and two French missionaries had visited Lovedale to discuss Maqoma's grievances. The Chief came to dine with them and next day the party visited his place. Helen Ross describes the admiration the missionaries felt for him and the extent to which they were exercised by the grave injustice that had been done. On another occasion, her husband went over to Grahamstown to put Maqoma's case to the Governor, an action which aroused the great enmity of the settlers (although she went on to say, somewhat enigmatically, that it would have been much worse had he not been a Scottish minister).[28]

Ross showed his allegiance by opening the Pirie mission, deeper into Xhosa country, and remained there until his death in 1878.[29] And he continued to go to Grahamstown to argue against the warlike sentiments of the settlers.

Nevertheless, the missions were repeatedly swept up in war, their staffs forced to flee leaving their charges and the buildings to their fate. While the frontier remained 'open', their position was highly ambiguous: they were caught between the Scylla of identification with Colonial settlers and authorities and the Charybdis of supposed alliance with the Xhosa, placing them constantly at risk. Missions were destroyed in the wars of 1835, 1846 and 1850–1851. Some survived because of the support of local Xhosa leaders. Still, nothing could obscure the fact that the missions had so far largely failed in their objectives. The energetic founding of stations was not matched by rapid conversions and the missions suffered frequent financial problems. The Free Church nearly withdrew in 1848, when a deficit of £2,400 was discovered, but the problem was solved by funds from wealthy adherents in Scotland. By now many missionaries were beginning to recognize that the best chance of success lay with the closure of the frontier and the establishment of colonial authority. This is well represented in the manner in which many congregations failed to join African resistance during the 1877–1879 frontier war. By then, the so-called 'native agency' had become a very significant part of the activity of the Scottish missions.

Moreover, it has even been claimed that the Transkei land system was ultimately based upon the iniquities of the insecurities of the Scots tenantry.[30] The objective of this system was to break communal land tenures and create an individualized peasantry. The Rev. Henry Calderwood, a Scottish missionary who had resigned from the LMS in 1846 after a bitter dispute, became native commissioner in the district of Victoria East. He saw parallels between the 'injustice and hardship' of a bad land system in northern Scotland and the situation of Africans. He also found biblical justification in Joshua's settlement of the tribes in Palestine. Calderwood, supported by Richard Ross and Governor Sir Harry Smith, envisaged a small-scale peasantry in which each African male would have eight to ten acres of arable land, grazing rights on commonage, in return for an annual rent and security of tenure. The Mfengus had been settled with individual tenure and were the precedent. The idea was later transferred from Victoria East in the Ciskei to the Transkei and was widely favoured by Cecil Rhodes in the 1890s.

Education

Calderwood's land proposals indicate the extent to which the Scots were intent upon a complete social and cultural revolution. And the prime aid to such a revolution was Western education, the principal route to achieving their spiritual purpose, their civilizing and christianizing mission. In this they were attempting to create parallel institutions to their compatriots in India, if they subsequently diverged in notable ways. Significantly, the celebrated Dr Alexander Duff of the Scottish Mission in Calcutta (Kolkata) claimed in an address to the General Assembly of the Church of Scotland in 1837 that his work in India

> simply consists in transporting to the plains of Hindustan that very system of teaching and preaching which in the hands of our own Knoxes and Melvilles once rendered Scotland an intellectual, moral and religious garden among the nations of the earth.[31]

This notion of Scotland as having a significant educational and religious history was to be much repeated in South Africa. Duff, who represented a tradition of 'rational Calvinism', rooted in the ideas of the Enlightenment, went on extensive tours of Scotland when at home on furlough, determining that Scots and their parishes should have 'ownership' of missions. In this he followed John Inglis, the minister of Greyfriars in Edinburgh and leader of the Moderate faction, who in 1826 had addressed a public letter to the people of Scotland, to be read in every parish of the Established Church, seeking financial support for missions (the tradition of the parish letter to raise funds went back to the seventeenth century). The furlough lecture tour became a Scottish tradition, much reported in the national and local press. Duff, and his compatriot John Wilson of Bombay, founded colleges that would later become the nuclei of the Universities in those cities. Both joined the Free Church after the Disruption in 1843 leading to the establishment of further Established Church missions in those cities. The schismatic tendency of the Scots church energized missions in South Africa as well as in India.

Indeed, after the sixth frontier war, the Lovedale mission became the setting for the development from 1841 of the most important educational establishment in the region, if not – in terms of black education – South Africa as a whole. At the Lovedale Seminary, later Institution, the missionaries sought to transform Africans into workers, at various levels, who would contribute their labour and skills to the settler economy.[32] But the original conception of Lovedale was infused with the early nineteenth-century race mentality of the frontier. It was designed to be a multiracial establishment in which the children of missionaries and other settlers would be educated together with Africans. They would share all the activities of the school, although, perhaps symbolically, they would sit at separate tables in the dining room and sleep in separate dormitories. Racial co-education had its limits. Although schools were founded at other missions, Lovedale was special in a number of respects. It grew to an extraordinary extent. Its buildings, funded locally and from Scotland, were of a high standard. Its teaching staff was well-educated. It set about the teaching of girls, and it began to attract black pupils from throughout southern Africa. It valued technical education from the beginning, but this was particularly stressed during the principalship of James Stewart after 1870. It also represented the origins of tertiary education for Africans, developing a teacher training school, and the future and nearby Fort Hare University College was planned within its walls.

Lovedale developed as a complete environment. Like eighteenth-century Scottish estate owners, the missionaries planted trees, 'stately oaks and pines'. There were gardens and a farm. All pupils were expected to work on outdoor projects for 13 hours a week, on a campus that was three miles across. Pupils were taught that idleness and Christianity were incompatible, that intellectual development had to run parallel with technical, horticultural and agricultural activities. Photographs, mainly dating from the late nineteenth century, offer much evidence of this philosophy.[33] The gates to Lovedale open to an arboretum of exotica. Pupils are marshalled for outdoor work parties. They sit at benches in workshops as at pews in church. Printing and bookbinding were significant from an early stage in the history of the mission and this opened out into major publishing and journalistic ventures.[34] This helped to reflect the commercial and marketing objectives of the missionaries. The surplus produce of farm and garden were available for sale, as were the products of the workshops and the printing establishment. Departments involved with brickmaking and building accepted construction contracts from local settlers and towns. This attracted the approval of colonial government and grants for dedicated workshops were provided in the 1850s. By then, the economic ambitions of mission and colony coincided. Stewart, unlike his predecessor, the Rev. William Govan, believed in an elementary education more suited to large numbers.[35] This was allied to manual work and industrial training. Although the activities of Lovedale were interrupted by the seventh frontier war in 1845, its position compromised by occupation by the military, its activities were renewed in 1850.

Stewart, apart from periods in Central Africa and travels overseas, dominated Lovedale until his death in 1905. By then, Lovedale had become a key institution of the closed frontier.

Stewart became one of the most celebrated missionary and educational figures of the period. Partly brought up in Edinburgh and partly on a farm in Perthshire, he had developed his oft-repeated agricultural philosophy of ploughing as an act of Christian service, working within the environment with a gun in one hand and a Bible in the other. He also developed a passion for botany and other natural sciences. After undergoing what he saw as a visionary experience on his father's farm, he embarked on a lengthy education in Edinburgh – eight years in arts and divinity, followed by medical studies (interrupted by a visit to Central Africa to see Livingstone). Stewart's great social coup was to marry Mina, a daughter of the wealthy Glasgow shipbuilder Alexander Stephen. This gave him status as well as access to an opulent class of industrialists who supported and contributed to his many financial appeals. He raised considerable sums for buildings at Lovedale: for the founding of the Mfengu mission at Blythswood, for the Livingstonia Mission in northern Malawi and for the Gordon Memorial Mission on the fringes of Zululand. His wife became the rather grand chatelaine of Lovedale, the hostess of many visiting celebrities, from the Colony and elsewhere, including members of her own family.[36] The Stewarts lived in a house which cost £800 (contributed by Mrs Stewart's friends). Style had arrived on the frontier, an appropriate setting for the entertaining of notable imperial visitors such as the High Commissioner Alfred Milner in the late 1890s.[37]

Women and missions

Although Stewart desired to keep women as well as blacks within a hierarchical missionary establishment, the missions became dependent on the work of women. On the white side, women – initially as missionaries' wives, but also as unpaid workers – were very important in the organization and running of missions.[38] The earliest LMS missionaries famously formed unions with Khoi women, and Robert Moffat intriguingly wrote to his Scottish fiancée that if she did not get out to southern Africa quickly, he would be forced to do the same.[39] She went and Scots were now consistently arriving with their white, usually Scottish, wives. They too were 'pioneers' of the open frontier, embroiled in its turbulence and insecurities. The numbers on the frontier were considerable and their role as lynchpins of mission organization remarkable. They were key in maintaining contacts with Scotland through active correspondence. They often maintained Scots cultural traditions. They published accounts of their activities in missionary and church publications.[40] They were remembered at home and even had obituaries in the local Scottish press. They acted as hostesses to visiting dignities. They suffered great dangers and hardships, from war, disease and childbirth, when they were often tended by black midwives. Some of the marriages to missionaries have the appearance of being arranged, particularly after the death of a first wife. And like the men, they experienced geographical, meteorological and ethnic dislocation, which helped to compound their isolation, melancholia and depression.[41] But they busied themselves with maintaining westernization through their bodies, securing clothes from home, not just for themselves, but also for servants and converts.[42]

From the middle of the century, as the frontier came increasingly under colonial control, wives became even more influential. Moreover, single women were beginning to arrive as missionaries and teachers in their own right, underpaid, but still a vanguard for female professionalism, often more advanced in the missionary context than in home society, and consequently rebounding upon it. Nevertheless, marriage was still seen as the natural state of women, and independent female missionary workers who married were usually deemed to have resigned. The missions had a similarly paradoxical effect upon the gendered relationships of those Africans who were within the missions' orbit. The roles of women and men swiftly changed under mission influence. Men became agriculturalists and eventually migrant labourers – missionary education equipping them for such a role – while women were subjected to efforts to transform them into 'keepers of home and hearth' like their white counterparts, although some African women, as we shall see, emerged as teachers and nurses in the late nineteenth and early twentieth centuries. If the influence of African women in traditional society declined at first, it may have increased as male migrant workers left home and some women secured jobs. In any case, John Philip and many of the other missionaries advocated education for girls and promoted it strongly at the missions.

Thus the creation of a girls' department at Lovedale in the 1860s was already based on a tradition of female education.[43] It was also rooted in a long tradition of involvement of Scottish women in missionary activity. As early as 1800, the Northern Missionary Society of Inverness formed a Women's Society, raising funds and organizing meetings and sermons. But by the time of the foundation of the Glasgow Ladies'

Association for promoting female education in Kaffraria in 1839, there was a considerable tradition of such activity. In 1825, the Dunfermline Ladies' Society offered to fund an African teacher, called Robert Balfour, at the Chumie Mission. Two other black teachers seem to have been called Charles Henry and John Burns, a naming policy seemingly representing conversion in every sense.[44] In 1832, the minute books of the Presbytery of Kaffraria, indicate that another black teacher, Thomas Brown, was to be 'specially under the patronage of the ladies of Cumnock Union'.[45] But by this time, women teachers, pioneers of white female professionalism, were already arriving from Britain. The Ladies' Kaffrarian Society sent out Miss McLaren in 1839 to establish a girls' school at Igqibigha. She was followed in the 1840s by the Misses Ogilvie, Thomson, Isabella Smith, Harding, Ross, McDiarmid and Weir.[46] A girls' school was established at Mgwali in 1861, under the auspices of the still active Ladies Kaffrarian Society in Scotland, and Miss Blair later headed the Ross Industrial School for girls at Pirie. By 1868, the buildings for the Lovedale girls' school had been completed and a fresh infusion of women teachers had arrived, including the celebrated Jane Waterston, the Misses Macdonald and Marianne MacRitchie, followed in 1881 by Mrs and Miss Muirhead.[47] Photographs of the girl pupils reveal the concentration on 'domestic science' and other female activities deemed appropriate within a Western concept of the 'separation of gendered spheres'. Missionary wives and teachers, certainly in the early period, also set about creating projections of themselves. Yet by the end of the century Lovedale had contributed to the development of black female professionalism. By 1886, Lovedale had produced, for example, 158 black female teachers. Indeed, several Lovedale students, including Sana Mzimba and Martha Kwatsha, went to schools in Scotland. While in Britain, Kwatsha became a Sunday School teacher, reversing the missionary role, and later married the Rev. M. J. Mzimba. The medical mission founded in 1898 set about the training of black nurses. Two African nurses (Mina Colani – presumably named after Mrs. Stewart – and Cecilia Makiwane, former pupils in the girls' school) began their training in 1903, joining hospital assistants already under training. But this led to segregation. Up until then, Victoria Hospital had admitted white and black patients, but the matron decreed that African nurses could not possibly train with European patients, and in 1904 the European ward was closed. This was a further indication of the manner in which early multiracialism was almost at an end. European numbers of pupils declined as more white schools opened, and white participation effectively ended when the Scots Superintendent of Education of the Cape, Sir Thomas Muir, decreed that white teachers could not be trained there.[48]

African ministers and evangelists

Scots had already contributed a good deal to social change through the training of African ministers. The first of these was the celebrated Rev. Tiyo Soga (1829–1871). He was born in a Xhosa community, the son of a senior councillor of Ngqika, at the time of Maqoma's dispossession and was educated at a United Presbyterian mission school before moving on to Lovedale in 1844. Clearly talented the missionaries sent him to Glasgow at the time of the seventh frontier war in 1846. He travelled to Scotland with sons of white missionaries and a Scottish businessman provided funds for his fare and his studies. He returned to the Cape as a catechist, but in 1851 he was back in Scotland to train for the ministry. He was attached to a United Presbyterian Church in Glasgow, the congregation paying for his studies, including the acquisition of some medical knowledge.[49] (Subsequently, the church's Juvenile Missionary Society paid his £25 p.a. salary on the frontier.) In Glasgow, he met and married Janet Burnside and was involved in missionary work in the poor East End of the City. Tiyo's ordination was a dramatic affair, before a packed congregation with all the leading ministers of the UP church present. The Moderator Dr William Anderson delivered an impassioned prayer in which he indulged in 'a tirade against the colonial policy of England', directing his wrath against the Prime Minister, the Government and the Colonial Secretary whose 'blundering acts were confessed as if by his own lips'. On the other hand, he presented supplications for 'the noble Kafir chieftain, Sandilli' (Sandile).[50] Soon after their wedding in 1857 the Sogas returned to the frontier and founded the Mgwali mission. There they were joined by the Rev. John Chalmers, who had been born on the frontier, but inevitably sent to Scotland for his education. Tiyo's life was prematurely cut short by TB, but only after he had founded new missions and had provided himself with a literary reputation by translating *Pilgrim's Progress* and the four New Testament gospels into Xhosa.

Tiyo and Janet had four sons and three daughters,[51] all but two of whom studied in Scotland. Two became missionaries, one also a doctor, one a government official and another a veterinary surgeon. Three of them married Scots women. Tiyo Soga always insisted to his sons that they should think of themselves as 'Kaffirs' and they certainly maintained an interest in the history and culture of their black forebears. But they could not escape the fact that their inheritance was multiply hybrid – their genes, their education, their religion, and their work placed them on the margins of two societies, interpreting modernization to their fellow Africans. And the white part of their hybridity was unquestionably Scottish rather than British. It was said, for example, that Soga's dedication to the underdog was partly based upon his love of the Burns song, 'A Man's a Man for a' That'. But Tiyo Soga has also been seen as representing an early form of negritude. As the first westernized South African who received a University education, he has been credited with helping to develop both black consciousness and early concepts of African nationalism. He was certainly seen as a precursor by Africans, ministers, professionals and politicians who followed him.[52] More Africans were ordained soon after this death: the Revs. Elijah Makiwane and M. J. Mzimba in 1875, followed by the Revs. Candlish Koti, J. Knox Bokwe and Ndongo Matshikwe. By this time, some of the 'native agents' were even used as missionaries further into the interior of Africa: in 1876, four products of Lovedale, William Koyi, Shadrach Mngunana, Isaac Williams Wauchope and Mapassa Ntintili, were sent from the Eastern Cape to Central Africa as teachers, evangelists and artisans to the Scottish mission field in what would become Malawi.[53]

Conclusion: Scots missions and imperialism

By this time, the Scottish missions had become closely allied with imperial rule and soon with white authority, something that seems inconceivable back in the 1820s to 1850s. The South African Scottish missions, closely associated with their Indian counterparts, and later joined by those in West, Central and East Africa, became significant sources of imperial propaganda, through meetings, lectures, publications (not least the major genre of missionary biography) as the century progressed. The Rev. James Stewart entitled his first address to the General Assembly of the Free Church of Scotland 'The King of the World, or Christian Imperialism'.[54] By the end of the century missionaries had enhanced and confirmed the Scots' sense of an imperial role and had consequently contributed to those aspects of a Scottish identity that were rooted in involvement in the British Empire. Stewart, a friend of General Gordon, Cecil Rhodes, Bartle Frere and Alfred Milner, exhibited a passionate adherence to British imperial expansion.[55]

By the First World War it seemed as though the original convictions of the Scottish missionaries had been achieved, however dangerously and hesitantly at first: that Christianity was the essential concomitant of civilization; that their message was so overwhelmingly liberating that it would blow down 'barbarism' like the walls of Jericho; that in the process they would free women from drudgery and oppression, men from warfare and violence, as well as create enlightened agriculture, introduce the elevating forces of the market, and use science and medicine to free Africans of 'superstition'. The landscape would be domesticated through the use of Scottish names, the planting of trees and gardens. But southern Nguni culture would prove resistant enough to produce 'nativist' reactions, returns to the old ways particularly marked between the 1840s and 1860s. Even Tiyo Soga remained very much a man of his own culture. While they worked for what they saw as 'God's Kingdom', they had been irretrievably sucked into the conflicts of human kingdoms. Their own missions became petty principalities in which they themselves exercised significant degrees of temporal power, interacting with local chieftaincies and with the colonial authorities. But greater certainty had its dangers: the missionaries became more arrogant and high-handed. Coming from a tradition of endemic schism, they should not have been surprised when they encountered African separatism. The Rev. Mpambani J. Mzimba (1850–1911), ordained in 1875, was denied the right to allocate money he had raised in Scotland on a visit in 1893 and seceded, taking many adherents with him. Stewart was deeply embittered when Mzimba symbolically built a church on a hill overlooking Lovedale.

The frontier missions also had a considerable impact upon Scotland. It was a commonly held view that 'there is much in common between the eighteenth-century Highlander and the African'.[56] Africa somehow demonstrated the speed with which Scotland itself had entered a modern industrial world. The education and attainments of the missionaries, as played out in so many life stories, seemed to illustrate the manner in which Scots from almost any social class could aspire to religious and educational, architectural and agricultural, engineering and industrial developments on a distant frontier. Those from quite straitened social circumstances could pay their way through university by working as tutors. It is not surprising that they brought their sense of self-help and their characteristic conviction in the multifarious benefits of education to the frontier. It was this conviction that they represented a distinctive ethnic identity and a different cultural and religious tradition that led them to believe that they were in the business of creating 'black Scotsmen'.[57] The phrase resonated down the century and was still being used by the Rev. E. Ntuli, Moderator of the Presbyterian Church of South Africa in 1941.[58] By extension, female mission workers were in the business of creating black Scotswomen.[59] Moreover, as Macdonald has written, Scotswomen in the missions, and their supporters at home, were 'engaged in a wide range of tasks and responsibilities which offered able and innovative individuals levels of independence, professional development and authority which they could never have aspired to back in Scotland'.[60] It is at least possible that Scots women had a greater sense of personal agency and responsibility.[61]

By the end of the century the former radicalism of the Scottish churches seemed to have drained away. A tradition of obedience to the civil authority was established and racially separatist tendencies emerged. Congregations became racially distinct and when the Presbyterian Church of South Africa was established in 1897 it did not include the black missions, which would be overseen by a Missions Council, thereby maintaining control of policy in white hands. The Bantu Presbyterian Church only emerged in 1924.[62] Justifications about the avoidance of white dominance were bandied about, but the separation was all too convenient in the era of apartheid. When the Nationalist Government, following its Bantu Education Act of 1955, took over mission schools, including Lovedale, there was scarcely a whimper of protest. In the end, the Scots missions were locked in a paradox: they trained many future nationalists, but they themselves became segregationist and were careful to avoid confrontation with white governments. The Scots missions had moved from radical dissent to obedience to what they saw as the civil authority. This was not, of course, universally true and they strongly maintained their tradition of dissent elsewhere in Africa.[63] The responses of Scots missions within the political arena were as complex and distinctive as in so many other areas of imperial endeavour.

Notes

1. H. Lamar and L. Thompson, *The Frontier in History: North America and Southern Africa Compared* (New Haven, CT: Yale University Press, 1981), p. 35. See also M. Boucher, 'The Frontier and Religion: A Comparative Study of the USA and South Africa in the First Half of the Nineteenth Century' (MA Dissertation, University of South Africa, 1966); E. A. Walker *The Frontier Tradition in South Africa* (Oxford: OUP, 1950), and M. Legassick, 'The Frontier Tradition in South African Historiography' in *Economy and Society in Pre-industrial South Africa*, ed. S. Marks and A. Atmore (London: Longman, 1980), pp. 44–79.
2. See Chapter 5 by Esther Breitenbach in this volume. See also E. M. Breitenbach, 'Empire, Religion and National Identity: Scottish Christian Imperialism in the 19th and Early 20th Centuries' (PhD Thesis, University of Edinburgh, 2005).

3. L. O. Macdonald, *A Unique and Glorious Mission: Women and Presbyterianism in Scotland 1830–1930* (Edinburgh: John Donald, 2000), ch. 3.
4. Macdonald, *Unique and Glorious Mission*, pp. 133–44. This is a necessary corrective to an excessive concentration on male heroes. J. M. MacKenzie, 'Heroic Myths of Empire' in *Popular Imperialism and the Military, 1850–1950* ed. J. M. MacKenzie (Manchester: Manchester University Press 1992), pp. 109– 38.
5. See Chapter 14 by Ruth Compton Brouwer in this volume.
6. These connections between home and foreign missions are well described in Macdonald, *Unique and Glorious Mission*, particularly ch. 3. See also J. Marriott, *The other Empire: Metropolis, India and Progress in the Colonial Imagination* (Manchester: Manchester University Press, 2003).
7. Macdonald, *Unique and Glorious Mission*, ch. 4.
8. J. Lennox, *United Free Church of Scotland: The Story of Our Missions* (Edinburgh: United Free Church of Scotland Foreign Missions Committee, 1911), p. 54.
9. The Wesleyans, for example, did not display the educational attainments of the Scots. N. Etherington, *Preachers, Peasants and Politics in Southeast Africa, 1835–1880* (London: Royal Historical Society, 1978), p. 32.
10. According to the memoir of his grandson, John Ross visited the Moravian mission at Genadendal on his journey to the eastern frontier from the Cape. He was particularly impressed with the industrial training there.
11. In practice, the LMS was strongly Congregational and the GMS was effectively Presbyterian.
12. I. H. Enklaar, *Life and Work of Dr. J. Th. Van der Kemp 1747–1811, Missionary Pioneer and the Protagonist of Racial Equality in South Africa* (Cape Town and Rotterdam: Balkema, 1988), pp. 15–18.
13. Van der Kemp and Read insisted that the Khoi should be free, live on a basis of equality with whites, and be released from all forms of compulsion. R. Elphick and H. Giliomee, *The Shaping of South African Society, 1652–1820* (London: Longman, 1979) p. 380.
14. R. Moffat, *Missionary Labours and Scenes in Southern Africa* (London: John Snow, 1846 – first 1842). By this edition, this book had already sold 18,000 copies, demonstrating the immense popularity of missionary works in this period.
15. Ntsikana even wrote a hymn which was said to be a great favourite with African congregations.
16. Nxele's Christianity was much more apocalyptic than Ntsikana's. In 1818 he led the forces of Ndlambe against Ngqika at the battle of Amalinde. He was captured and exiled to Robben Island. Natasha Erlank, 'Re-examining initial encounters between Christian missionaries and the Xhosa, 1820–1850: the Scottish case', unpublished paper. This is based on a chapter of Erlank's PhD thesis, 'Gender and Christianity among Africans attached to Scottish Mission Stations in Xhosaland in the Nineteenth Century' (Cambridge 1998). I am grateful to Dr. Erlank for supplying me with a copy of this paper.
17. Rev. J. Campbell, *Travels in South Africa, Undertaken at the Request of the London Missionary Society, Being a Narrative of a Second Journey in the Interior of That Country* (London: Religious Tract Society, 1822).
18. Similar societies to the GMS and the EMS were founded in Stirling, Greenock, Paisley, Kelso, Perth, Dundee, Aberdeen, Inverness, Elgin and Nairn. These

were all 'praying societies' interested in missionary activity, although sometimes the concern was primarily with 'home missions' in the growing industrial cities. There is a paradox here: on the one hand, difficulties in travel ensured that such bodies were founded on a local basis; yet there was clearly a national fashion being replicated around the country.

19. P. Hinchliff, 'Whatever Happened to the Glasgow Missionary Society?' *Historiae Studia Ecclesiasticae* [Church History Society of Southern Africa] XVIII, no. 2 (1992) 104–20. Hinchliff discovered that because of the nature of its organization, the manner in which its activities were handed over to the Free Church of Scotland and the United Presbyterian Church of Scotland, and the subsequent lack of interest in its archive – apart from legal documents – its papers had largely disappeared. Some of the missionary records of the Free Church are to be found in the National Library of Scotland.
20. 'Notes On the Early History of the Glasgow Missionary Society' compiled by R. H. W. Shepherd, handwritten, Cory Library Grahamstown (CLG), MS 8722. This seems to be partly based on the 'Report of a Special Meeting of the Members of the Glasgow Missionary Society held in the Trades Hall, Glassford Street, Glasgow, Monday 4th March 1837'. The GMS was originally constituted by 22 ministers and 13 laymen. The fissiparous nature of the Scottish churches was well reflected in the fact that they were representatives of the 'Relief', 'Old Light Burgher', 'New Light Burgher', 'Antiburgher', 'Reformed Presbyterian' and 'Establishment' connections of the Presbyterian Church.
21. In 1837, the GMS split (over the issue of affiliation to the Establishment or adherence to independence). One part gave itself the unwieldy name of the Glasgow Missionary Society according to the Principles of the Church of Scotland, the other the snappier Glasgow African Missionary Society. Bennie, Ross, Laing, McDiarmid, Weir and the missions at Lovedale, Burnshill and Pirie adhered to the former, while Chalmers and Niven, the missions at Tyumie and Igqibigha went with the latter. In 1844 the GMSPCS, slightly paradoxically, handed over its assets and operations to the Foreign Missions Committee of the Free Church, while the two stations of the GAMS went to the United Presbyterian Church three years later.
22. Donovan Williams, *When Races Meet: The Life and Times of William Ritchie Thomson* (Johannesburg: A.P.B. Publishers, 1967), p. 44.
23. N. Erlank, 'Kilts for Loincloths: Scottishness in the Eastern Cape in the Nineteenth Century', unpublished Paper in Africana Library, University of Cape Town, p. 3. I am grateful to Dr Erlank for permission to cite this paper.
24. Some of these presses later found their way into the South African missionary museum in King William's Town. A Murray McGregor, 'Notes on Missionary Establishments and Sites Visited by Members of the Historical Society, 15–16 May, 1976', *Looking Back*, XVI, no. 3 (September 1976), 67–74.
25. A. Ross, *John Philip: Missions, Race and Politics in South Africa* (Aberdeen, Aberdeen University Press, 1986), pp. 98–9.
26. Helen Ross to her sister, 13 October 1829, CLG MS 2638 (1827–49). Other letters are in MS 2637 (1823–27).

27. Maqoma and his people had been cleared eastwards from frontier land: Dr. John Philip in April 1830 saw Boer farmers heading for Fort Beaufort to claim farms on his former lands. W. M. Macmillan, *Bantu, Boer and Briton* (Oxford 1963 – first published in 1929), p. 41.
28. Helen Ross to her parents, 17 April 1830, CLG MS 2638.
29. Like so many other missionaries, Ross and his wife Helen established a dynasty. Bryce Ross (1825–1899) was born at Lovedale and educated at both Edinburgh University and the Andersonian Institution in Glasgow (for medicine) from 1846. He returned in 1851, worked at Lovedale for ten years, wrote textbooks and a hymnary in Xhosa and served on the 'Board of Revision of the Kaffir Bible'. His younger brother Richard Ross (1828–1902) was also educated in Scotland and married a daughter of John Brownlee. His son, Rev. Brownlee John Ross (1865–1944), returned from studies in Edinburgh to work at Lovedale and Cunningham. He was interested in Scots and clan history and impressed upon Africans the social similarities between Scotland and Africa! He frequently wore the kilt. A biographical record of the Ross family is in the CLG MS 9167. Brownlee Ross wrote memoirs of his grandfather, father and uncle for the *Daily Despatch,* republished in *Brownlee J. Ross: His Ancestry and Some Writings with an Introduction* by R. H. W. Shepherd (Lovedale: Lovedale Press, 1948).
30. Erlank, 'Kilts for Loincloths', p. 6. See also *Brownlee J. Ross: His Ancestry and Some Writings*, pp. 28–9.
31. Quoted in I. Maxwell, 'Enlightenment and Mission: Alexander Duff and the Early years of the General Assembly's Institution in Calcutta', North Atlantic Missiology Project, position paper no. 2, n.d., p. 7. Duff visited the Eastern Cape missions in 1864. A. A. Millar, *Alexander Duff of India* (Edinburgh: Canongate, 1992), pp. 189–90.
32. For Lovedale, see R. H. W. Shepherd, *Lovedale, South Africa, 1824–1955* (Lovedale: Lovedale Press, 1956); J. Wells, *Stewart of Lovedale* (London: Hodder and Stoughton, 1909), and various pamphlets, including R. H. W. Shepherd, 'Lovedale South Africa' (Lovedale: Lovedale Press, n.d., probably 1931); 'Short Biographies of Galla Rescued Slaves now at Lovedale' (Lovedale: Lovedale Press, 1891); 'The Jubilee of Lovedale Missionary Institution, July 1891' (Lovedale, Lovedale Press, 1891). See also James Stewart, *Lovedale Missionary Institution* (Edinburgh and Glasgow, 1894) and Glasgow Missionary Society, 'Quarterly Intelligence: The Opening of the Lovedale Seminary', South African Library, Cape Town.
33. These photographs are to be found in Wells, *Stewart,* passim. Others are to be found in works listed in note 32 above.
34. For accounts of Lovedale printing, see J. Atkinson, 'The Lovedale Press, A Review of Recent Activities' (Lovedale: Lovedale Press, 1931), S. G. V. Crawford, 'Lovedale Press 1823–75', *The Coelacanth,* XIV, no. 1 (1976) 28–37 and the various histories of Lovedale by Shepherd. Robert Moffat's mission at Kuruman also developed a notable press. See F. R. Bradlow, 'Printing for Africa, the Story of Robert Moffat and the Kuruman Press' (Kuruman: Kuruman Press, 1987). This press, also active from the 1820s, published religious works in several languages, periodicals, school texts, advice on alcohol and the ubiquitous *Pilgrim's Progress*.

35. Stewart abolished the teaching of Latin and Greek and some African students objected to this as discriminatory!
36. Her brother and his wife, major benefactors of Lovedale, paid a visit to the Eastern Cape and to Lovedale in 1886 and were received almost as visiting royalty. An account of this journey can be found in a transcript of the original diary kept by Mrs Stephen, to be found in the South African Library in Cape Town.
37. Wells, *Stewart*, pp. 342–9.
38. For a more general treatment of a variety of denominations in the later period, see D. Gaitskell, 'Re-thinking Gender Roles: the Field Experience of Women Missionaries in South Africa' in *The Imperial Horizons of British Protestant Missions, 1880–1914*, ed. A. Porter (Grand Rapids, MI: William B. Eerdmans, 2003), pp. 131–57. This was also published as a North Atlantic Missiology Project position paper.
39. W. Woodward, 'The Petticoat and the Kaross: Missionary Bodies and the Feminine in the London Missionary Society 1816–1828', *Kronos*, XXIII (November 1996), 91–2. Woodward points out that missionaries tended to sexualize indigenous women, but not European ones.
40. For a more general survey, see J. Rowbotham, '"This is No Romantic Story": Reporting the Work of British Female Missionaries, *c.* 1850–1910', North Atlantic Missiology Project, position paper no. 4.
41. Williams, *When Races Meet*, pp. 42–3.
42. For a more extensive treatment, see J. M. MacKenzie, *The Scots in South Africa: Ethnicity, Identity, Gender and Race* (Manchester: Manchester University Press, 2007), ch. 4.
43. Girls were said to be more amenable to education in the early days of missions, partly because they were more settled and therefore 'available' than the men.
44. This supposed abandonment by 'native agents' of African ways in favour of missionary ones did not necessarily work. Balfour went over to the Xhosa in the 1835 war. It was also said that these early agents were of indifferent quality. Clearly, a much stronger educational infrastructure was required, as subsequently provided by the Lovedale Institution. Williams, *When Races Meet*, pp. 55–66, 69.
45. Presbytery Minute Books, CLG MS 9037. There are five of these minute books covering the years from 1825 (when the Presbytery was established) to 1875. They are a mine of information and have been used rarely.
46. Lennox, *Story of Our Missions*, pp. 28, 70.
47. The 1868 and 1881 names are derived from *Lovedale Past and Present.* Significantly, the first names of the women are usually not mentioned and generally we know less about them than about the men.
48. Shepherd, *Lovedale* (1956), p. 15.
49. The Church paid £202 for his University education and raised over £132 to kit him out as a missionary. John A. Chalmers, *Tiyo Soga: A Page of South African Mission Work*, 2nd edn (Edinburgh, London, Glasgow and Grahamstown, 1878), p. 99. This work was immensely popular and went through several editions. Chalmers had access to a number of sources on Soga that were subsequently lost, as well as collecting a large quantity of oral memoirs of him.
50. Chalmers, *Tiyo Soga*, p. 89. Sandile was the Xhosa chief in this period.

51. One daughter entered missionary work in the Transkei. Another went to Glasgow and never returned; she became a singing teacher there.
52. D. Williams, *Umfundisi: a Biography of Tiyo Soga 1829–71* (Lovedale: Lovedale Press, 1978), pp. xix and 118–27.
53. T. J. Thompson, 'Xhosa Evangelists in Late Nineteenth-Century Malawi: Black Strangers or Fellow Countrymen?', North Atlantic Missiology Project, Position Paper no. 34, n.d.
54. James Wells, described Stewart as subscribing and contributing, together with the Rev. John MacKenzie, to the expansion of the British Empire. Wells, *Stewart*, pp. 304 and 330. See also James Stewart, *The Assembly Addresses of the Rev. James Stewart* (Lovedale: Lovedale Press, 1899).
55. We need a closer consideration of the role of missionaries in imperial expansion in the Eastern Cape and Natal in the nineteenth century. J. H. Procter's article, 'The Church of Scotland and British Colonialism in Africa', *Journal of Church and State*, XXIX (1987) 475–93, briefly alludes to the late nineteenth century in Malawi.
56. Ross, *Ancestry and Some Writings*, p. 6.
57. Robert Niven first used this phrase in 1845. Williams, *When Races Meet*, p. 156. In 1848 Niven also suggested that Africans should be clothed in kilts. Erlank, 'Kilts for Loincloths' p. 3.
58. 'Lovedale's Centenary: A record of Celebrations July 19–21, 1941' (Lovedale: Lovedale Press, 1941), p. 15. These celebrations made much of the distinctive Scots contribution to education, while the Bishop of Grahamstown insisted that even the Anglican Church in South African had been mainly influenced by Scotland. Tshekedi Khama, paramount chief of the Tswana and a former student of Lovedale, was present at this centennial celebration and also made a speech.
59. On the missionary education of girls, see D. Gaitskell, 'At Home with Hegemony? Coercion and Consent in the Education of African Girls for Domesticity in South Africa Before 1910' in *Contesting Colonial Hegemony: State and Society in Africa and India,* ed. D. Engels and S. Marks (London: British Academic, 1994), pp. 110–28.
60. Macdonald, *Unique and Glorious Mission,* p. 128.
61. Ibid., p. 132.
62. This later became the Reformed Presbyterian Church of South Africa. The two branches only came together again in 1999. G. A. Duncan, 'Presbyterian expressions in Southern Africa: In the Context of 350 Years of the Reformed Tradition', *Nedgeref Teleogiese Tydskrip,* XLII, no. 2 (2002) 423–31. See also Graham Alexander. A. Duncan, 'Scottish Presbyterian Church Mission Policy in South Africa, 1898–1923', MTh, University of South Africa, 1997.
63. T. Jack, Thompson, 'Presbyterianism and Politics in Malawi: A Century of Interaction', *Round Table,* XCIV, no. 382 (2005) 575–87.

Further reading

Breitenbach, E. M. 'Empire, Religion and National Identity: Scottish Christian Imperialism in the 19th and Early 20th Centuries' (PhD Thesis, University of Edinburgh, 2005).

Elphick, R. and H. Giliomee. *The Shaping of South African Society, 1652–1820* (London: Longman, 1979).

Etherington, N. *Preachers, Peasants and Politics in Southeast Africa, 1835–1880* (London: Royal Historical Society, 1978).

Gaitskell, D. 'Re-thinking Gender Roles: The Field Experience of Women Missionaries in South Africa'. In *The Imperial Horizons of British Protestant Missions, 1880–1914*, ed. A. Porter (Grand Rapids, MI: William B. Eerdmans, 2003), pp. 131–57.

Macdonald, L. O. *A Unique and Glorious Mission: Women and Presbyterianism in Scotland 1830–1930* (Edinburgh: John Donald, 2000).

MacKenzie, J. M. *The Scots in South Africa: Ethnicity, Identity, Gender and Race* (Manchester: Manchester University Press, 2007).

Procter, J. H. 'The Church of Scotland and British Colonialism in Africa', *Journal of Church and State,* XXIX (1987) 475–93.

Walker, E. A. *The Frontier Tradition in South Africa* (Oxford: Oxford University Press, 1950).

7
Archbishop Vaughan and the Empires of Religion in Colonial New South Wales

Peter Cunich

When Roger Bede Vaughan, second archbishop of Sydney (1877–1883), died in August 1883 on his way to Rome, the 'Benedictine phase' of the Australian Catholic Church's early development came to an abrupt but not unexpected end.[1] The involvement of English Benedictine monks with the Australian mission had begun approximately fifty years earlier, with the appointment in 1832 of William Bernard Ullathorne as Vicar-General of New Holland. It was, however, the figure of John Bede Polding, Vicar Apostolic of New Holland and Van Diemen's Land from 1834 and archbishop of Sydney from 1842 until his death in 1877 which dominated the formative years of the Australian Catholic Church. This period represents approximately one-quarter of the entire history of Australian Catholicism but it is a phase of development whose Benedictine characteristics are sometimes overlooked by historians intent on emphasizing the Irish nature of the colonial church. The years immediately after the death of Archbishop Vaughan marked such a triumph for what his successor Cardinal Patrick Moran called the 'Irish spiritual empire' in Australia that the earlier Benedictine contribution was almost totally overshadowed by the extraordinary progress achieved after 1884 by Moran and his fellow Irish-born bishops.[2] Indeed, Vaughan's own demise was so much coincident with the final death knell of English Benedictine rule in Sydney that any consideration of his achievements as metropolitan of New South Wales demands an assessment of the Benedictine legacy as a whole.

Evaluating Vaughan's episcopal career and the Benedictine legacy in New South Wales up to 1883 is fraught with difficulty. Historians have long disagreed as to how the history of the Catholic Church in colonial

New South Wales should be interpreted and the historiographical battle that has raged for more than a century over the respective merits of English Benedictines and secular Irish clergy in the founding epoch of the Australian church shows no signs of abating. Cardinal Moran himself fired the first salvo in this battle with the publication in 1896 of his *History of the Catholic Church in Australasia*, an interpretation which was quickly challenged by the English Benedictine Henry Norbert Birt in his *Benedictine Pioneers in Australia* (1911).[3] Nearly one hundred years later, the embers of this long-contested history of empire are still smouldering and, despite attempts by numerous scholars in the intervening years to tone down the partisan nature of the debate, books continue to appear intent on perpetuating old arguments with new evidence, new twists in the tale. Recent studies have therefore argued that Archbishop Vaughan's achievements during his short reign in Sydney have been overstated and that he should be held responsible for 'the lack of effective leadership that emanated from the Sydney archdiocese' until the advent of Cardinal Moran.[4] Vaughan is lambasted not only as a traitor to his own monastic order but also as someone who resorted to 'underhand and deceitful' methods to achieve his political goals.[5] Not only is he held responsible for the isolation of Sydney's Catholic community from the wider colonial society after the great education debate, but he is also presented as a petulant and duplicitous martinet in his dealings with the priests of his diocese.[6] Such interpretations rest uneasily alongside the more measured assessment by historians such as Patrick O'Farrell, who considered Vaughan to be 'a cleric of extraordinary abilities', a bishop who governed 'easily and well' and was able to forge a strong sense of unity among the divided Catholics of New South Wales.[7]

On what basis, then, is Archbishop Vaughan's episcopate to be assessed? It is certainly true that in his years as coadjutor to Polding (1873–1877) he and the Irish suffragans fell into a deep conflict which was often highly divisive of the Catholic community in Sydney, and it cannot be denied that the education policy which he championed from 1879 led to a 'sectarian convulsion' which perpetuated the entrenched conflict between Catholics and protestants in New South Wales until well into the twentieth century. Such interpretations take a decidedly political view of relations within the Catholic Church in the colony, and imply that Vaughan's struggle to impose his will over the Irish suffragans as their future metropolitan in some way did irreparable damage to the church. Likewise, the isolation of Catholics after the 1880 Public Instruction Act is interpreted as a tearing of the as yet unstable fabric of a multi-confessional and secularizing society in colonial Australia. These interpretations no doubt have much merit in analysing the Catholic Church in its political and social relations both internally and within the wider society. They may be a barrier, however, when attempting to assess the effectiveness of an archbishop in his role as an ecclesiastical superior appointed by Rome to assert papal authority in one of the most distant corners of the British Empire.

Australia in the nineteenth century was, in fact, just one of the many colonial territories of several European colonial powers where the centralizing imperial authority of the Roman church came into conflict with different notions of imperial spiritual authority. The idea that two different 'empires of religion' might have been operating within particular colonies in the nineteenth century is a reminder that the manifestation of political or religious identity is seldom a simple matter determined by a single influence. An 'identity' is more often made up of a multiplicity of identity fragments and these different components interact with each other to produce a complex identity mix. In the case of the Catholic Church in early New South Wales, I would like to suggest that the complex and distinct identity of the Catholic Church can be analysed with reference to at least four, but perhaps even five, different empires of religion which were vying for primacy in the colony at that time. Further, we should consider the interactions among all these empires of religion when evaluating the success of the clerics appointed by Rome to impose papal authority over Catholics in British missionary territories.

The British Empire and religion

It is widely recognized that religion can play a central role in the formulation of national identity and a sense of political coherence. The founding precept of the modern English state was the idea that the English crown was sovereign or 'imperial' in nature. This vision was enunciated in Henry VIII's legislative programme of the 1530s and created a theoretical English 'empire' free from all external political and religious interference. The boast that 'this realm of England is an empire ... governed by one supreme head and king' laid the foundations of a protestant state religion which gave the later British Empire some of its most enduring religious features. An Oath of Supremacy was imposed as a test of loyalty and the English government thus entered a long period of Catholic persecution designed to establish the crown's authority in matters of religion as well as politics. This long period from the middle of the sixteenth century until the Catholic Emancipation Act in 1829 was marked by the imprisonment and execution of Catholics, particularly Catholic priests, and a raft of other punitive legislation. By the late seventeenth century the Church of England was firmly established as the state religion of the kingdom and this relationship between church and state was enshrined in various acts of parliament.[8] From the time of the Elizabethan settlement in religion, therefore, the formal structures of the Catholic Church ceased to exist in England and were progressively replaced by an Anglican hierarchy. An attempt was also made to suppress the Catholic Church in Ireland, but this project was less successful, as has been demonstrated by Tadhg Ó hAnnracháin elsewhere in this volume. More important, perhaps, as the British Empire spread across the Atlantic and Indian Oceans, attempts were made to impose Anglican religious uniformity on the new colonies. But these attempts were hampered by a general lack of evangelical zeal within the established church during the seventeenth century. Consequently, by the end of the eighteenth century the 'aggressive Anglicanism' of some British colonists and settlers had failed to achieve an empire united by a single religion, and nowhere outside of the British Isles did the Church of England enjoy the status of an established church.[9]

Indeed, the Church of England outside the British Isles lacked any formalized structure until relatively late in the eighteenth century.[10] The bishop of London had been given spiritual authority over 'all British subjects overseas' by King Charles I, but there was very little metropolitan supervision of church affairs in the British colonies until 1786 when an act of parliament authorized the archbishops of Canterbury and York to consecrate bishops for America. By 1793, five bishops had been consecrated for the former colonies in the now-independent United States and two sees had been created in Canada.[11] From this time colonial and foreign sees of the Church of England were erected by royal letters patent, but it was not until another Act of 1813 that a bishop and three archdeacons were provided for India, with the bishop based in Calcutta.[12] It was the bishop of Calcutta who exercised jurisdiction over the penal colonies in Australia from 1814, but when two further Indian dioceses were formed under an Act of 1833, the colonies of New South Wales and Van Diemen's Land (Tasmania) were 'dissevered' from Calcutta's jurisdiction. After a brief reversion to supervision by the bishop of London, the Australian colonies were given their own bishop in 1836 with the appointment of William Grant Broughton as Bishop of Australia.[13]

By the beginning of the nineteenth century a strong evangelical spirit had arisen within the Anglican Church at home and this movement was reflected in the colonies. A number of the colonial chaplains in Sydney were noted for their zealous protestantism, especially Samuel Marsden and Archdeacon Scott. It was these men who were initially responsible for imposing within the Australian colonies the spiritual authority of the Church of England, a task which became all the more pressing once large numbers of Irish Catholic convicts came pouring into New South Wales after 1800. For the first 30 years in Australia there was no question of allowing any other spiritual authority in the colonies except that of the British imperial power, and for most of this time Catholic priests were banned from entering Australia. Only slowly were these restrictions lifted but the number of Catholic clergy was kept small until the 1840s. It was not until 1848 that a metropolitan hierarchy was established for the Anglican Church in Australia, several years after the Catholic Church had established its own hierarchy under the metropolitan jurisdiction of Archbishop Polding.[14]

The British state was never able to exterminate Catholicism or secure complete authority for the Church of England within its dominions, either within the British Isles or across the widening overseas empire, but religious persecution did have the effect of reducing the Catholics of England to a very small religious minority. Outside of Ireland, therefore, the British Empire initially included relatively few Catholics across its scattered colonial possessions and trading outposts, and most Britons had little or no acquaintance with Catholics or Catholicism. This did not, however, mean that the British state had no experience of governing Catholics or of the formal structures of the Catholic Church. Rome had appointed a vicar apostolic to England in 1622, and from 1688 the country was divided into four missionary districts under their own vicars apostolic who held titular episcopal sees.[15] Priests from various religious orders together with a large number of secular priests were sent to the English mission during the seventeenth and eighteenth centuries, and it was during the eighteenth century that many English gentlemen gained a first-hand if rather impressionistic experience of the Catholic Church during their visits to Italy on the 'grand tour'. Englishmen certainly learned something about Catholicism from their contacts with Ireland where penal laws had also been enacted to exclude Catholics from public life. Nevertheless, Catholicism remained as the religion of the majority of Irish. Although the Catholics in Ireland were predominantly peasants, a sizeable Irish Catholic gentry had survived the English attempts to impose religious uniformity and these landowners and merchants continued to provide their sons for leadership roles in the Irish Catholic Church. Rome continued to appoint bishops to the ancient Irish sees and overseas seminaries kept up a regular supply of priests. The establishment of Maynooth College in 1795 meant that priests could be trained in Ireland instead of having to travel to the Continent. Attempts to extend the authority of the protestant Church of Ireland were therefore thwarted by the continuing existence of a Catholic hierarchy and priesthood who repudiated all efforts at conversion and colonization by the English.[16] The last major attempt to enforce conformity to the established church in Ireland through a 'new reformation' aimed at the mass conversion of Catholics between 1801 and 1829 was a dismal failure.[17] From the British imperial point of view, the difficulties of keeping the Irish Catholic Church under control were magnified during the veto conflict in the early years of the nineteenth century as the British government tried to play a larger role in the appointment of Catholic bishops in Ireland.[18] Much of the British experience of dealing with Catholicism and the Catholic Church in Ireland was therefore negative up to and throughout the nineteenth century.

It was not just in England and Ireland that the British governing class became acquainted with the structures, personnel and modes of operation within the Catholic Church. During the eighteenth century, the British were extending their colonial empire outside the British Isles, and this empire included enormous tracts of territory in North America, India and Australasia. In both hemispheres the British came into contact with Catholic bishops and vicars apostolic, most of whom had been appointed either by the Congregation for the Propagation of the Faith in Rome or under the Portuguese *padrado*. In Canada, the British inherited the Catholic diocese of Quebec after the Seven Years' War (1756–1763), but there was little attempt at formally establishing the Church of England in any part of Canada at that time.[19] The British perhaps realized that toleration was the only practical policy that could be adopted in the predominantly French territories which it had acquired, so the Catholics in Lower Canada were given the right to have their own French bishop, but under British 'supervision'.[20] A little later the British government allowed the erection of a prefecture apostolic in the predominantly Catholic enclave of Newfoundland, and Irish Franciscan bishops were appointed to undertake missionary activities there.[21] Vicariates apostolic were later created in Nova Scotia and Upper Canada and placed under Irish and Scottish secular bishops.[22] By the early 1800s, then, a pattern of toleration of Catholics and their formal church structures had been established in Britain's remaining North American colonies.[23]

In India the situation was even more complicated. As the British extended their trading interests and territorial conquests they came into contact with the old Portuguese dioceses of Goa (1533) and Cochin (1558), and the mostly French vicariates apostolic of Fort St George/Madras (1642), Malabar/Verapoly (1659) and Bombay (1720). In the decades after the Seven Years' War, British India eventually swallowed up most of the territories in these dioceses and mission fields, but the British made little attempt to impose English bishops on these long-established Catholic jurisdictions. The Portuguese, Italian and French missionary societies therefore continued to dominate the sees. It was not until the 1830s that Pope Gregory XVI began to establish new vicariates in India, and at this time Britain insisted that any new bishops should be British subjects.[24] Likewise, in Malaya, Britain did not initially interfere with the French bishops of the diocese of Malacca (1558). A hardening of policy came immediately after the Napoleonic Wars, however, when Britain was faced with the immediate problem of ministering to the almost entirely Catholic population of Mauritius, a French colony in the Indian Ocean which had surrendered in 1810. The British were once again, as in the case of Canada, forced to conclude that toleration was the only practical policy to pursue with regard to religion in Mauritius. But in this instance the British government exerted its influence through Bishop William Poynter, vicar apostolic of the London District, to ensure that an Englishman was appointed as vicar apostolic of the Cape of Good Hope.[25] The appointee was an English Benedictine monk, Edward Bede Slater (Vicar Apostolic, 1818–1832), and it was from this moment that the English Benedictines began to extend their missionary role beyond the confines of the British Isles. Slater's mission field covered not only Mauritius, but also most of southern Africa, and would eventually be extended to include the whole of the Indian Ocean and the new colonies in Australia. This arrangement set a new pattern in imperial relations between Britain and the Holy See: if Catholic mission fields were to be established within the British Empire, it would in future be desirous that British subjects (and preferably Englishmen) were found to act as bishops.

It was in this expanding British Empire that the Australian Catholic Church was founded in the early decades of the nineteenth century. This new church, planted in the soil of a distant and exotic southern continent, represented an important development in relations between London and Rome, for it was an audacious experiment which challenged the very basis of British imperial rule. While the erection of the vicariate of New Holland under the leadership of Bishop Polding in 1834 was a development which the British had dealt with elsewhere in their empire and to which they responded by appointing an Anglican 'bishop of Australia', the establishment in 1842 of an Australian Catholic hierarchy with Polding as metropolitan over the two suffragan sees of Hobart and Adelaide was another matter altogether. This was the first new Catholic hierarchy to be erected by Rome in a British territory since the Reformation and it was a sensitive matter both between Rome and London and within the colony of New South Wales itself. Even though the Catholic Emancipation Act had been passed in 1829, the 1830s were still a time of undiminished Protestant suspicions about the inherent subversiveness of Catholicism within the British Empire, and the influx of Catholic priests and Irish Catholic convicts into Australia during that decade exacerbated these latent fears. In New South Wales, suspicion that the British empire of religion was being threatened by Roman spiritual imperialism was imagined by English colonists from the last years of the 1790s and throughout the nineteenth century, even though Catholics tended to demand that only toleration and equality be granted to them. The two imperial jurisdictions of London and Rome were therefore set on a collision course in the battle for spiritual authority within the Australian colonies after 1842.

The governors and senior administrators in charge of the penal colony in New South Wales had very little experience of the arrangements which had been put in place in other parts of the British Empire with regard to accommodating Catholic clerical activity and a natural desire among Catholic colonists for separate worship. In the early years of the settlement in Sydney there seemed to be little need for a Catholic chaplain as the majority of convicts were English and nominally Protestant, but after the arrival of increasing numbers of Irishmen in the early 1800s, some of whom were political prisoners after the failed 1798 rebellion, the need for Catholic priests steadily grew. But it was not only the Irishness of the Catholic population of Sydney in the early 1800s which frightened the Protestant civil and religious authorities in the colony. They were haunted as much by a deeply conditioned fear of all Catholic priests who had been branded as traitors to the crown in the penal days as they were by claims of 'Irish Catholic sedition' which were the results of 'popery and priestcraft'.[26] Catholic priests were therefore kept well away from the colony. Governors King, Macquarie, Darling and Brisbane were perhaps the most anti-Catholic of the early governors and all made efforts to use their Anglican chaplains as promoters of social order and civil obedience. Catholics were forced to attend Protestant services in an effort to promote a greater respect for English civil authority, and Samuel Marsden believed that if ever Catholic services were approved in Sydney 'the colony would be lost to the British Empire in less than one year'.[27] Governor Macquarie was open to securing the services of a few priests to help calm the rowdiness and immorality of the Catholic population, but echoed the opinion of his superiors in London when he suggested that if ever Catholic priests were to be allowed into the colony 'they should be Englishmen of liberal Education and Sound constitutional principles'.[28] This policy was accepted and reinforced by later governors, especially Sir Ralph Darling and Sir Richard Bourke.[29]

However, not even the advent of reliable middle and upper-class English priests such as Ullathorne and Polding (and later Davis and Vaughan) was enough to satisfy a powerful group within the protestant establishment of Sydney that the privileges of Anglicanism as the religion of empire were being properly protected by the colonial government. Polding did not help matters with his love of lavish ceremony and his habit of wearing Roman ecclesiastical dress in public, something that was still uncommon in England at that time. Some members of Sydney society were outraged when Polding arrived at a Government House levee in full Catholic episcopal attire in May 1537 and Bishop Broughton immediately insisted that the governor enforce the 'Dublin protocols' which had previously regulated the attire of clerics on formal occasions.[30] The Anglican establishment became even more alarmed when Governor Bourke's Church Act of 1836 effectively disestablished the Church of England, encouraged the building of Catholic chapels and gave parity to the four largest Christian denominations (Anglicans, Catholics, Presbyterians and Methodists). This legislation seemed to question the essential link between church and state which was the bedrock of the British Empire.[31] English colonists had already been disturbed by the Catholic Emancipation Act of 1829, a liberal concession by the metropolitan government which was nevertheless decried by the Sydney establishment.[32] Earlier calls by Archdeacon Scott for the formal establishment of the Church of England in New South Wales were taken up after 1836 by Bishop Broughton, and were heard time and again over the next decade. Broughton perceived the growth of the Catholic Church in Sydney as a flagrant act of aggression against the liberal principles of British rule and took a prominent lead in challenging the pretensions of Roman imperial incursion within a British colony.[33] Broughton had earlier warned the Colonial Office that he would fiercely 'oppose the enemies of the Reformation' in Sydney and once he had arrived in his new see he set about building a protestant 'citadel' in response to Polding's activities in the colony.[34] Broughton's citadel was to have a physical as well as a spiritual presence in the colony. Already Governor Darling had been shocked by the excessive size and grandeur of St Mary's chapel, fearing that it might act as a symbol of Catholic ascendancy in the colony.[35] Broughton immediately set about raising funds for an Anglican cathedral, believing that 'without such a stronghold of faith we cannot keep our position', and in May 1837 the foundation stone of St Andrew's cathedral was laid.[36]

These examples serve to indicate that the early colonial establishment in New South Wales attempted to use all means at its disposal to reinforce the authority of the Church of England as the principal religion of empire within Australia. This authority was challenged after 1836, however, and the Church Act was something of a crisis for the Anglican ascendancy. Clear and unquestioned primacy of place had always been given to the Anglican colonial chaplains before 1836, but after the passing of the Church Act the position of the Church of England had to be bolstered by the erection of a colonial diocese centred on Sydney. As the years went by, attempts were made to reinforce the central role of Anglicanism in stabilising the social foundations of Britain's furthest imperial possession, but repeated calls to establish the church failed, as they did elsewhere in the Empire, and the vociferous support of a small but powerful section of colonial society was not enough to stem the rising tide of institutionalized Catholic activity in New South Wales. When Polding was elevated as archbishop of Sydney in 1842, Bishop Broughton saw it as 'an act of invasion and intrusion', and an example of 'direct and purposed hostility towards us ... contrary to the laws of God, and the canonical order of the Church'.[37] The influx of Irish immigrants in the 1830s and 1840s was likewise seen as a 'menace to the future development of a civilized British society in Australia' because of their supposed disloyalty to the British crown. In the 1860s, Henry Parkes was able to raise the rallying cry of 'no popery' in response to the Fenian scare of 1868 when an Irishman attempted to assassinate the Duke of Edinburgh during his visit to Sydney.[38] Accompanied by the revival of orange lodges, the formation of the Protestant Political Association, and the establishment of the extremist *Australian Protestant Banner* newspaper in 1868, the sectarian divisions in New South Wales were only made worse by the promulgation of Papal Infallibility in 1870.[39] Bishop Barker of Sydney was by 1876 declaring that Catholic schools would produce citizens who were 'aliens, enemies of the English crown, of English laws', and after the Catholic bishops of New South Wales issued their Joint Pastoral on education in 1879 Henry Parkes accused Archbishop Vaughan of disloyalty and sedition against the British crown.[40]

Rome and the colony of New South Wales

It is against this background of chauvinistic British protestantism in the service of empire that we must try to measure the success of the Benedictine mission to New South Wales. If there was an 'other' empire of religion in New South Wales during the nineteenth century it was clearly that age-old enemy of the British state – Rome. Catholicism had long been feared by English protestants, but Rome and its missionary priests had ceased to be an immediate concern to the British government during the late eighteenth century as the papacy took a battering first from Joseph II of Austria and later at the hands of Napoleon. By the 1820s, however, the situation in Europe was changing rapidly. The previously disordered central bureaucracy of the Roman church was recovering from the Napoleonic ravages and the Congregation for the Propagation of the Faith was gradually strengthening itself as the church's central missionary institution. In the 1820s and during the pontificate of the former Benedictine monk Gregory XVI (1831–1846) a large number of new missionary dioceses and vicariates were established. As already noted, these included many in the major territories of the British Empire: Canada, India and Australia.[41] The bishops who were appointed to these new sees were responsible directly to the Holy See and so a very noticeable process of centralization of church authority occurred from the early 1830s until the end of the long pontificate of Pius IX (1846–1878).

The Benedictine monk-bishops sent out from England to lead the Catholic Church in Australia were sympathetic to this Roman centralization of ecclesiastical authority: they had been trained in the monastic school of obedience and demanded it of their subjects, both clerical and lay. Soon after his arrival in Sydney as vicar-general in 1833, William Bernard Ullathorne perceived that the most important need in the fledgling church was the imposition of Roman discipline, and that the divided house of colonial Catholicism would never prosper until full Roman authority was established firmly and formally. When Bishop Polding arrived in 1835, he was quick to impose this Roman authority, warning the clergy that only faculties issued by him as vicar apostolic would be recognized by the Holy See.[42] He had already taken care to ensure that Propaganda had revoked the faculties of all priests working in the colony so that his episcopal authority would not be undermined.[43] He was particularly determined to force one of the more troublesome priests, Fr John Therry, into submission.[44] Time and again as bishop he reiterated the message that 'Before everything else we are Catholics', and that all the faithful must submit to proper ecclesiastical authority.[45] This message was taken seriously by most of the clergy but only by some of the laity in the colony. W. A. Duncan affirmed in *The Chronicle* that 'Our religion is neither English nor Irish, but Catholic', but the majority followed the lead of a small but determined group of laymen who supported what Patrick O'Farrell has described as 'an aggressive Irish-Australian Catholicism of an explicitly political kind'.[46] Polding's repeated attempts to assert his episcopal authority were therefore largely undermined by resistance from an increasingly vocal pressure group within the Catholic laity of Sydney.

It was Polding's successor to the see of Sydney, Archbishop Roger Bede Vaughan, who best enunciated the theme of imperial Roman authority over the Australian church. Like Polding, he continually emphasised that his authority issued from Rome rather than London, and his first words upon landing in Sydney in 1873 included: 'You may call me an Englishman if you will...but I am a Catholic first.'[47] Vaughan made his position even clearer in his first pastoral letter, *Pius IX and the Revolution*, published in 1877. The pastoral was a robust defence of universal papal sovereignty against secular power. In it he drew the attention of the Catholics of New South Wales to the plight of Pius IX and the church in Italy, comparing the pontiff to 'a perpendicular rock of granite' withstanding the attacks of revolution, libertinism and infidelity.[48] Vaughan stressed that the bishops of the universal church owed 'implicit obedience' to the pope's 'spiritual sovereignty': 'they follow his lead, and shape their policy according to his example' and 'In his formal teachings as Universal Pastor in questions of Faith and Morals he cannot lead astray'.[49] It was only a few years later during the Catholic education campaign that he called on all Australian Catholics to recognise and submit to the authority of Rome: 'The spiritual empire, of which we are soldiers, by its very history, stirs up the fires of charity and zeal in our hearts. Ours is one of the very few causes in the world worth living for, and dying for, too'.[50] He insisted that Australian Catholics had to 'stand firm on the adamantine rock of the Catholic faith'.[51] 'If we take the Papal Chair as a centre, and cast our eyes around the world, we shall find that the Catholic Church is engaging in almost every country in a heavy conflict with her enemies.'

> Let us, then, often meditate ... on the great Spiritual Empire to which we belong; encourage in our minds a profound sense of thankfulness that we are members of so glorious a society; and think of how we can do our part towards strengthening its hold and perpetuating its power in this land of our adoption.[52]

It was not only the English bishops of Sydney who emphasized the Roman spiritual empire as the primary purpose of their mission. The Irish suffragans believed themselves to be 'true sons of Rome', and Archbishop Moran when he arrived in the colony firmly reasserted the Roman *imperium* over the Australian church.[53] All the bishops in New South Wales, whether English or Irish or Italian, were therefore keen to demonstrate their loyalty to the Holy See and its universal spiritual authority.

The English Benedictines in New South Wales

It was not always completely apparent to the laity of the Sydney diocese, however, that the universal and unitary force of the Roman spiritual empire as enunciated by Vaughan and his fellow bishops was as well manifested in the Catholic Church in Australia as it was supposed to be. Catholics in the colony of New South Wales had long sensed that there were in fact two Catholic empires of religion competing for primacy within the Catholic community in Sydney, and both of these have been written about extensively. The first of these empires was an 'English Benedictine empire' which was very much Polding's brainchild. Polding's so-called 'Benedictine dream' took on a distinctive form between 1843 and the early 1860s, but ultimately failed to gain traction in the colonial church and was all but abandoned after the appointment of Irish suffragan bishops in 1859 and 1865. The Benedictines were throughout these two decades competing with an incipient Irish spiritual imperialism which came to full bloom in the 1860s and 1870s. What is most significant about these competing spiritual empires within the broader Roman *imperium* is that while they for a time consumed much emotional energy within the Catholic population of New South Wales, ultimately they were reconciled under Vaughan's leadership in the late 1870s and early 1880s. By that time it was clear that there was no place for national in-fighting among Catholics when there were far graver issues challenging the Catholic community as a whole. The competing Benedictine and Irish spiritual empires did nevertheless have a very real existence for some time within the colony.

As has already been mentioned, the English Benedictines became part of the British Empire's global expansion with the appointment of Bede Slater as vicar apostolic of the Cape of Good Hope in 1818. Slater, a monk of Ampleforth, was allowed by his Benedictine superiors to take two fellow monks with him to Mauritius in 1819 and thus began a role of missionary outreach beyond the British Isles by the English monks of St Benedict. The English Benedictines were open to the idea of overseas missionary activity because of the peculiarity of their own congregational beginnings. From the moment of its erection as an independent congregation in 1619, the *raison d'etre* and main apostolic work of the English Benedictine Congregation was the 'mission' to Catholics in England.[54] The four houses of monks which were founded on the Continent were therefore focused on the work of the English Mission throughout the penal times. The Constitutions of the Congregation established a governance structure which gave the needs of the mission (divided into two provinces under powerful provincials) primacy in all questions regarding the deployment of monks and the distribution of resources. The three semi-autonomous priories in France and the one abbey of the congregation at Lamspringe in Germany were often regarded as little more than training colleges for the monastic novices who would one day enter the mission field. A separate mission code had been approved alongside the formal congregational Constitutions and every monk received a copy of it at the time of profession.[55] More importantly, each monk at his profession took the Missionary Oath together with his other monastic vows. The Missionary Oath obliged each monk of the congregation to go to the English Mission when asked to do so by the President.[56] When the Downside and Ampleforth communities settled in England after the French Revolution expelled them from France, they and the third surviving priory at Douai continued to think of the 'mission' as the main work of the congregation, despite the increasing size of their fashionable public schools and the attendant need for larger communities of monks based in the monasteries.[57]

This was also the Romantic Age and in the cloister at Downside there seems to have developed in the 1820s and 1830s a very strong commitment to the idea of mission in its British imperial context as opposed to its narrower English national manifestation of earlier centuries. John Bede Polding was a central character in this development. As novice master at Downside from 1824 to 1834 he guided the monastic formation of a large number of young men. He was the principal influence over William Bernard Ullathorne and was playfully referred to by his confreres as the 'Bishop of Botany Bay' because of his desire to be sent out to help the Catholic convicts in New South Wales.[58] Polding's desire for the overseas mission became well known to John Augustine Birdsall, President of the Congregation, to whom Polding acted as personal secretary after the general chapter of 1826. Birdsall recommended Polding to Rome for elevation as bishop, first in 1829 as a replacement for Bede Slater in Mauritius, and then in 1832 for the new vicariate apostolic of Madras.[59] Birdsall certainly seems to have shared Polding's imperial ambitions for the English Benedictines, and saw Britain's overseas trade routes as a means of spreading Catholicism throughout the Empire.[60] During his presidency (1826–1837) he attempted to provide monks for the new missions in Africa and Australia, and he became one of Polding's strongest supporters in the attempt to set up a third province of the English Benedictine Congregation in New South Wales. He firmly believed that the missionary opportunities in Australia represented for the English congregation 'the greatest quest... since the Reformation'.[61] Birdsall's imperial plans did not find support among the other Benedictine superiors whose missionary work within the British Isles was already restricted because of the short supply of vocations to the monastic life. A serious shortage of subjects who could be spared for missionary work outside England meant that Birdsall's imperial vision could not be fully implemented.[62] He was nevertheless able to secure the appointment of William Placid Morris to replace Slater in Mauritius in 1832, and in the same year Ullathorne was placed under Morris as vicar-general in New South Wales. It was not until the appointment of Polding as vicar apostolic of New Holland, however, that the prospect of a real Benedictine empire of religion appeared to come within reach.

Polding's romantic plan for a great missionary diocese in New South Wales which would be wholly staffed by Benedictine monks working out of a central monastery attached to St Mary's Cathedral contained elements reminiscent of the ancient cathedral priories of medieval England. The underlying principles of the 'Benedictine dream' were therefore completely consistent with the traditional missionary impetus of the English Benedictine Congregation. Despite his transparent humility, Polding was a man of no mean ambitions when it came to spreading the fame of his order. As sub-prior he had been closely involved in building the ambitious neo-gothic chapel and monastic wing at Downside in the 1820s, and upon his consecration as bishop in 1834 he told the general chapter that he had undertaken the job in New South Wales 'in the hope of extending the usefulness of our Congregation'.[63] His plan was to create a Benedictine province in Australia which would be acknowledged as 'no inconsiderable or uninteresting part of our Holy Institute'.[64] He failed to convince the general chapter, however, and the Australian province of the English Benedictine Congregation was never formed. Nor was he able to secure a reliable supply of English monks for his mission, either in 1834 or at the quadrennial chapters of 1838 and 1842.[65] Instead, Polding set about establishing a new Benedictine congregation with recruits from Ireland and Australia, attempting to build a monastic community attached to St Mary's Cathedral which would be based as much as possible on the practice of the English congregation. Polding saw himself as a modern-day St Augustine sent out to Australia by a new Benedictine Pope Gregory, and convinced himself that monk-missioners were the only priests suitable for the task of converting the new land. Initially his abbey-diocese appeared to be a great success and by 1848 the St Mary's community numbered 40.[66] His building work at St Mary's aimed at transforming the large but rather ungainly chapel into a cathedral worthy of the missionary work which was being guided from its precincts, and his other churches and ecclesiastical buildings were constructed on a similar scale. Monks in habits became a not uncommon sight in the streets of Sydney, and before long Polding had established both contemplative and active congregations of Benedictine nuns to undertake educational and relief work within the diocese.[67]

To Polding's great disappointment the early promise of his Benedictine plan for Sydney could not be sustained in the 1850s, and by the end of that decade the whole edifice of his Benedictine empire had collapsed. Defections of monks, scandals within and outside the cloister, poor leadership and Polding's lack of attention to formation all played their role in the unwinding of the Benedictine congregation in Sydney.[68] Despite this failure, Polding would remain a Benedictine at heart throughout the rest of his episcopal career, attempting to ensure the appointment of a Benedictine successor and the continuation of his much reduced monastic community at Lyndhurst Academy in Glebe. In his later years he became increasingly downcast about the ruin of his abbey-diocese, even though he had much to be proud of in the growth of the Catholic Church which his episcopate had seen.[69] He certainly remained true to his assurances to friends at home that in spite of absence and distance he would continue to love the Benedictine order 'as my Mother, to which I am proud to acknowledge my obligations'.[70] The 'Benedictine dream' for New South Wales therefore ultimately came to a disastrous end in the late 1850s, although some of the institutional remnants and Benedictine personnel remained in place until the 1870s.

By the time Vaughan succeeded Polding, however, there was little chance for the dream to be revived, even though some of the Irish suffragans and laity feared that this might be the case. Vaughan had in fact been commissioned by Propaganda to dismantle the remaining structures of Polding's Benedictine diocese and it was perhaps due in part to this implicit repudiation of his predecessor's Benedictine scheme that Vaughan was able to secure a considerable measure of trust and support from the Irish bishops.[71] The Benedictine empire of religion was therefore reduced to nothing more than a chimera during Vaughan's episcopate, a monster which did not really exist but which continued to be imagined by the Irish bishops until Patrick Francis Moran brought an end to their concerns by being appointed to the empty metropolitan see in 1884.

Archbishop Polding's dogged attempts to see his Benedictine dream realized was therefore something of a personal crusade which was not supported with any level of enthusiasm by his confreres in England. The lack of support given by the general chapter in the 1830s and 1842 can be blamed on manpower shortages, but it is also notable that several members of the chapter had perceived problems developing in Sydney between English ecclesiastical superiors and their predominantly Irish subjects. By 1842, Polding's old friend at Downside, Thomas Joseph Brown, thought that Irish bishops would eventually be necessary for Australia because Irish priests would have difficulties working under English superiors.[72] Even earlier, in 1838, Ullathorne had realized that the presence of so many Irish among the Catholic population of Sydney made it unlikely that an English Benedictine mission would ever succeed, noting pessimistically that 'To do anything Benedictine in the colony is now out of the question.'[73] The later problems which Polding and Vaughan experienced with Irish priests and laity have correctly been interpreted as more political or nationalist in nature than religious. Yet breakdowns in ecclesiastical authority were a serious matter in the centralizing Catholic Church of the mid-nineteenth century and it is tempting to suggest that we see evidence here of an Irish empire of religion at work.[74] In reality, however, it is difficult to discern any widespread recognition within the colonial church of what Cardinal Moran later referred to as an 'Irish spiritual empire'. This is a term which became popular among Irish political leaders after 1922 but it seems to have had very little real existence as a policy of the Irish church or churchmen in the nineteenth century.[75]

An 'Irish' or an 'Australian' Catholic Church?

How, then, might Irish spiritual imperialism be considered as one of the empires of religion operating in New South Wales from the 1840s? It has been argued that the refashioning of the Irish church along fiercely ultramontane lines by Cardinal Paul Cullen from 1849 until his death in 1878 engendered a triumphalism and a new feeling of confidence which was the basis of 'a Hibernian spiritual empire to surpass Britain's money-grubbing imperialism'.[76] Cullen certainly exerted considerable control over episcopal appointments in Ireland and was able to place his own kinsmen within the Australian hierarchy because of his influence in Rome, but it is difficult to discern any explicit 'imperial' policy in this use of his power. It is true that he disliked Englishmen, especially English Catholics, but his concerns for the Australian church seem to have been motivated by the same uncompromising ultramontanism which marked his policies in Ireland. He was, therefore, a 'Roman ecclesiastical imperialist' first and foremost.[77] Rather than attempting to create a spiritual empire with Irish characteristics by providing bishops and priests for the Australian and American missions, Cullen expected his episcopal clients to show absolute devotion and obedience to the Holy See. These were sovereign virtues which Cullen had inculcated among his students during his time as rector of the Irish College in Rome and, in general, his acolytes demonstrated that Cullen's trust had not been misplaced. James Murray and the Quinn brothers were utterly devoted to Cullen, and the stamp which they sought to put on the Australian church was ultramontane rather than Irish.[78] Matthew Quinn therefore spoke as a true Cullenite when he said of his brother Irish bishops that 'We are the true sons of Rome.'[79] Cardinal Moran, a nephew of Cullen and his secretary for a time, was perhaps the most enthusiastically ultramontane of all the Irish bishops appointed to Australia, for it was he who completed the work of subjecting the Australian church to Roman authority.[80] In this he was simply following the policy of his two English Benedictine predecessors in Sydney and so it is doubtful that the Irish empire of religion about which he spoke was anything more than a way of praising the pioneering work of Irish clergy in Australia. The elements of Irish piety and episcopal authoritarianism which are so often alluded to as distinguishing marks of the Irish empire of religion in Australia are therefore perhaps more correctly described as manifestations of a certain type of Romanizing reform within the church which the Irish adopted with particular fervour under Cardinal Cullen. The notion of a Hibernicized church in Australia therefore has serious limitations.[81]

Where does this leave our evaluation of Archbishop Vaughan's brief episcopacy and the more general legacy of the Benedictine pioneers in New South Wales? I would contend that of the four empires of religion which have been identified as competing forces in colonial Sydney, two stand out as being of prime importance. It was the protestant imperialism of the British Empire and the Catholic imperialism of Rome which were the major competitors for spiritual authority in the colony of New South Wales. It might further be argued that this contested authority in the spiritual realm was a problem which was negotiated to a greater or lesser extent in all British colonies, whether or not the main protagonists were English or Irish.[82] The peculiar situation which developed in the Catholic Church in New South Wales was one which in turn created a particular political and national context which had an unfortunate impact on a church government which was divided between English and Irish clergy. It is therefore unfortunate that historians have focused so much attention on the failure of Polding's 'Benedictine dream' and the resulting internecine conflict between English and Irish bishops who should have united in a more harmonious way to achieve the work which they had been sent to do by Rome. The focus on national conflict between the English and Irish is understandable, of course, considering the wider currents of Irish-Australian history in the years leading up to federation and nationhood. The particular historiographical context within which the religious conflicts of the Benedictine phase in the Australian Catholic Church's development have been interpreted has also cast a long shadow. Yet there is a large literature which questions too simple a national interpretation of this history and insists that it was the triumph of Rome rather than Dublin which lies at the heart of the Catholic story in colonial New South Wales.[83]

Archbishop Vaughan's episcopal career should therefore be judged on the basis of his performance in advancing the imperial goals of the Holy See. In this endeavour he was remarkably successful, for by the time he left for Rome in 1883 he had reasserted the metropolitan authority of Sydney which had waned in Polding's last years. Moreover he was able to bind the New South Wales bishops together as a solid bulwark against the secularizing policies of the government of Sir Henry Parkes. That he was respected and admired in Sydney is abundantly clear from the farewell he received, the reports of his death, and his obsequies in St Mary's Cathedral.[84] His efforts in rebuilding St Mary's on a grand scale, doubling the number of Catholic Churches and trebling the number of Catholic schools in Sydney and in presiding over an extraordinary growth in the number of religious orders in the colony are ample testimony of his achievements.[85] Perhaps his greatest achievement, however, was the stabilization of a church which was seriously divided when he arrived from England in 1873. When his successor arrived from Ireland in 1884 he inherited a church in which the authority of the metropolitan see was not simply a matter of at last having an Irishman sitting on a Benedictine throne but a testament also to Vaughan's success in imposing Roman discipline over all his subjects. At the same time, however, Vaughan and Polding had laid down sturdy foundations of an essentially Australian church which was neither Roman nor Irish. This new manifestation of Catholicism in colonial New South Wales was one which Cardinal Moran wisely accepted with its unfamiliar features rather than trying to remould the Sydney church into something which was more recognisably Irish. The Sydney diocese therefore stood out in the late nineteenth century as being far more 'Australian' in character than its several suffragan sees which had long been under the rule of Irish bishops. Vaughan, Polding and the Sydney Benedictine experiment therefore played a much larger role in establishing a distinct Australian form of Catholicism than is generally recognized.

Notes

1. I am grateful to Dr John Carroll and the Right Rev. Dom Aidan Bellenger for their perceptive comments on an earlier version of this chapter.
2. P. J. O'Farrell, *The Catholic Church and Community: An Australian History*, 3rd edn (Sydney: New South Wales University Press, 1992), p. 195.
3. P. F. Moran, *History of the Catholic Church in Australasia* (Sydney: Oceanic Publishing, n.d. [1896]); H. N. Birt, *Benedictine Pioneers in Australia*, 2 vols (London: Herbert & Daniel, 1911). For Cardinal Moran, see P. Ayres, *Prince of the Church: Patrick Francis Moran, 1830–1911* (Melbourne: Miegunyah Press, 2007).
4. A. Cunningham, *The Rome Connection: Australia, Ireland and the Empire, 1865–1885* (Sydney: Crossing Press, 2002), p. 225.
5. Ibid., pp. 90–2.
6. Ibid., p. xv; R. Lehane, *Forever Carnival: A Story of Priests, Professors and Politics in 19th Century Sydney* (Canberra: Ginninderra Press, 2004), pp. 296–7.
7. O'Farrell, *Catholic Church and Community*, pp. 175–6, 185.
8. The Act quoted is from 1533 (24 Henry VIII, *c*.12), 'An Act in Restraint of Appeals to Rome', the text of which appears in *The Tudor Constitution: Documents and Commentary*, ed. G. R. Elton, 2nd edn (Cambridge: Cambridge University Press, 1982), pp. 353–8.
9. B. S. Schlenther, 'Religious Faith and Commercial Empire', in *The Oxford History of the British Empire, Vol. 2, The Eighteenth Century*, ed. P. J. Marshall (Oxford: Oxford University Press, 1998), p. 128.
10. R. Strong, *Anglicanism and the British Empire c.1700–1850* (Oxford: Oxford University Press, 2007), pp. 196–7, 210.

11. R. Border, *Church and State in Australia: A Constitutional Study of the Church of England in Australia* (London: SPCK, 1962), pp. 2–3. The two sees established in Canada were for Nova Scotia (1787) and Quebec (1793); see *Church and State in Canada, 1627–1867: Basic Documents*, ed. J. S. Moir (Toronto: McClelland and Stewart, 1967), p. 111.
12. S. Neill, *A History of Christianity in India, 1707–1858* (Cambridge: Cambridge University Press, 1985), pp. 261–75.
13. Border, *Church and State in Australia*, pp. 4–5.
14. Ibid., p. 8.
15. These districts were erected on 30 January 1688 as the London District (originally erected in 1622 as the Vicariate Apostolic of England), the Western District, the Midland District and the Northern District. These four districts were subdivided in 1840 to create the new Yorkshire, Lancashire, Eastern and Wales districts. Another five vicariates were later created but it was not until 29 September 1850 that the English hierarchy was finally re-established consisting of 13 dioceses.
16. R. Foster, 'Ascendancy and Union', in *The Oxford Illustrated History of Ireland* (Oxford: Oxford University Press, 1989), p. 186.
17. S. J. Brown, 'The New Reformation Movement in the Church of Ireland, 1801–1829', in *Piety and Power in Ireland, 1760–1960: Essays in Honour of Emmet Larkin* (Belfast: Institute of Irish Studies, 2000), pp. 12–13, 180–208.
18. O'Farrell, *Catholic Church and Community*, p. 12.
19. The diocese of Quebec was originally erected as the vicariate-apostolic of New France on 11 April 1658 and elevated to a diocese on 1 October 1674. Details of the creation of vicariates apostolic and missionary dioceses in Canada and other British colonies are to be found in *Catholic Encyclopedia: An international work of Reference on the Constitution, Doctrine, Discipline and History of the Catholic Church*, ed. C. G. Herbermann et al., 15 vols (New York: Robert Appleton, 1907–1912). For Canada, see Moir, *Church and State in Canada*, pp. 72–110.
20. Schlenther, 'Religious Faith and Commercial Empire', 147–8.
21. The prefecture apostolic of Newfoundland was erected on 30 May 1774.
22. The vicariate-apostolic of Nova Scotia was erected on 4 July 1817 and that of Upper Canada on 12 January 1819.
23. It is interesting to note, however, that a prefecture apostolic was not established in the United States until after the colonies had broken away from British control, the Jesuit John Carroll being appointed in 1784 as prefect apostolic for the United States(later bishop of Baltimore from 1789).
24. S. Neill, *Christianity in India*, pp. 279–83.
25. A. Bellenger, 'The English Benedictines and the British Empire', in *Victorian Churches and Churchmen: Essays presented to Vincent Alan McClelland*, ed. S. Gilley (Woodbridge, UK: Boydell and Brewer, 2005), 97–100. See also A. Bellenger, 'Revolution and Emancipation', in *Monks of England: The Benedictines in England from Augustine to the Present Day*, ed. D. Rees (London: SPCK, 1997), pp. 206–8.
26. O'Farrell, *Catholic Church and Community*, pp. 4–5.
27. S. Marsden, 'A Few Observations on the Toleration of the Catholic Religion in N. South Wales, *c.* 1806–7', in *Documents in Australian Catholic History*,

ed. P. J. O'Farrell and D. O'Farrell, 2 vols (London: Geoffrey Chapman, 1969), vol. 1, p. 73.
28. Governor Lachlan Macquarie to Earl Bathurst (18 May 1818), in *Documents*, vol. 1, p. 53.
29. F. O'Donoghue, *The Bishop of Botany Bay: The Life of John Bede Polding, Australia's First Catholic Archbishop* (Sydney: Angus & Robertson, 1982), pp. 32–3.
30. G. P. Shaw, *Patriarch and Patriot: William Grant Broughton 1788–1853, Colonial Statesman and Ecclesiastic* (Melbourne: Melbourne University Press, 1978), p. 117.
31. O'Donoghue, *The Bishop of Botany Bay*, pp. 66–7.
32. O'Farrell, *Catholic Church and Community*, p. 56.
33. Ibid., pp. 49–51.
34. Shaw, *Patriarch and Patriot*, pp. 97, 108.
35. O'Farrell, *Catholic Church and Community*, p. 27.
36. Shaw, *Patriarch and Patriot*, p. 116.
37. Ibid., 164; *Documents*, vol. 1, pp. 260–1.
38. O'Farrell, *Catholic Church and Community*, pp. 86–7.
39. Ibid., pp. 157–60.
40. Ibid., pp. 163, 188.
41. Bellenger, 'English Benedictines and the British Empire', p. 98. The following dioceses and vicariates were established in the various British colonial territories during Gregory XVI's pontificate: Madras (1832), Bengal (1834), Ceylon (1834), New Holland and Van Diemen's Land (1834), Western Oceania (1835), Montreal (1836), Toronto (1841), Hong Kong (1841), Malacca-Singapore (1841), Halifax (1842), New Brunswick (1842), Sydney (1842), Hobart (1842), Adelaide (1842), New Zealand (1842), the North-West (Canada, 1844), Perth (1845), Maitland (1846) and Vancouver Island (1846).
42. Polding to his clergy, 15 October 1835, in *The Letters of John Bede Polding O. S. B.*, ed. M. X. Compton et al., 3 vols (Sydney: Sisters of the Good Samaritan, 1994–1998), vol. 1, p. 54.
43. Polding to Propaganda, 2 October 1834, ibid., p. 43.
44. Father Therry's submission, 1 October 1836, ibid., p. 9.
45. Polding's Lenten Pastoral for 1856, *Documents*, vol. 1, p. 155.
46. O'Farrell, *Catholic Church and Community*, pp. 68, 95.
47. *The Freeman's Journal*, XXIV, no. 1552 (20 December 1873) 9.
48. R. B. Vaughan, *Pius IX and the Revolution: A Pastoral Letter to the Clergy and Laity of the Diocese of Sydney* (Sydney: Edward F. Flanagan, 1877), p. 6.
49. Ibid., pp. 9–10.
50. O'Farrell, *Catholic Church and Community*, p. 189.
51. Ibid., p. 190.
52. Ibid., p. 192.
53. Ibid., p. 213.
54. Geoffrey Scott, *Gothic Rage Undone: English Monks in the Age of Enlightenment* (Downside, UK: Downside Abbey, 1992), p. 10.
55. D. Lunn, *The English Benedictines 1540–1688: From Reformation to Revolution* (London: Burns & Oates, 1980), p. 148.
56. Ibid., pp. 160–1; Scott, *Gothic Rage Undone*, p. 42.

57. The Abbey of Saints Denis and Adrian at Lamspringe was suppressed in 1803 and was not re-established in England like the other priories.
58. O'Donoghue, *The Bishop of Botany Bay*, pp. 4–11.
59. Ibid., pp. 12–15.
60. Bellenger, 'English Benedictines and the British Empire', p. 101.
61. O'Donoghue, *The Bishop of Botany Bay*, pp. 18–19.
62. Athanasius Allanson, *Biography of the English Benedictines* (York, UK: Ampleforth Abbey, 1999), pp. 432–3.
63. Polding to the General Chapter, 15 July 1834, in *Polding Letters*, vol. 1, p. 33.
64. Polding to Birdsall, 29 June 1834, ibid., p. 32.
65. O'Donoghue, *The Bishop of Botany Bay*, p. 63.
66. Polding to R. R. Madden, 26 June 1848, in *Polding Letters*, vol. 2, p. 108.
67. For the Good Samaritans see M. Walsh, *The Good Sams: Sister of the Good Samaritan 1857–1969* (Melbourne: John Garratt Publishing, 2001).
68. M. Shanahan, *Out of Time, Out of Place: Henry Gregory and the Benedictine Order in Colonial Australia* (Canberra: Australian National University Press, 1970), especially ch. 9.
69. O'Donoghue, *The Bishop of Botany Bay*, pp. 166–7.
70. Polding to Heptonstall, 15 January 1839, in *Polding Letters*, vol. 1, p. 126.
71. T. Kavenagh, 'Vaughan and the Monks of Sydney', *Tjurunga*, XXV (1983) 151.
72. O'Donoghue, *The Bishop of Botany Bay*, p. 64; Shanahan, *Out of Time, Out of Place*, p. 46.
73. Birt, *Benedictine Pioneers*, vol. 1, p. 371.
74. T. L. Suttor, *Hierarchy and Democracy in Australia 1788–1870: The Formation of Australian Catholicism* (Melbourne: Melbourne University Press, 1965), p. 4.
75. M. E. Daly, *The Slow Failure: Population Decline and Independent Ireland, 1922–1973* (Madison, WI: University of Wisconsin Press, 2006), pp. 16–17. I am grateful to Professor Hilary Carey for alerting me to this reference.
76. K. T. Hoppen, *Ireland Since 1800: Conflict and Conformity* (London: Longman, 1989), p. 145.
77. D. Bowen, *Paul Cardinal Cullen and the Shaping of Modern Irish Catholicism* (Dublin: Gill & Macmillan, 1983), p. 293.
78. Ibid., pp. 28–9.
79. O'Farrell, *Catholic Church and Community*, p. 213.
80. Bowen, *Paul Cardinal Cullen*, p. 28.
81. Cunningham, *The Rome Connection*, p. xiv.
82. In Hong Kong, for example, the conflict was between English missionaries of the Church Missionary Society and Italian missionaries appointed as vicars-apostolic by Pius IX.
83. See especially Suttor, *Hierarchy and Democracy*, p. 5; J. N. Molony, *The Roman Mould of the Australian Catholic Church* (Melbourne: Melbourne University Press, 1969), pp. 1–7, 166–9.
84. See especially J. T. Donovan, *The Most Rev. Roger Bede Vaughan, D.D., O.S.B., Archbishop of Sydney: Life and Labours* (Sydney: F. Cunninghame, 1883) and the report by Fr Timothy McCarthy recently published as 'A Letter to Propaganda Fide in 1878', ed. Mary Xavier Compton, in *Tjurunga*, LXX (2006), 83–96.

85. For Vaughan's work in completing the first stage of St Mary's Cathedral, see A. E. Cahill, 'Archbishop Vaughan and Cardinal Moran as Cathedral Builders', in *St Mary's Cathedral Sydney, 1821–1971*, ed. P. J. O'Farrell (Sydney: Devonshire Press, 1971), pp. 127–43.

Further reading

Ayres, P. *Prince of the Church, Patrick Francis Moran* (Melbourne: Miegunyah Press, 2007).

Bellenger, A. 'The English Benedictines and the British Empire'. In *Victorian Churches and Churchmen: Essays Presented to Vincent Alan McClelland*, ed. Sheridan Gilley (Woodbridge, UK: Boydell and Brewer, 2005), pp. 97–100.

Birt, H. N. *Benedictine Pioneers in Australia*, 2 vols (London: Herbert & Daniel, 1911).

Border, R. *Church and State in Australia: A Constitutional Study of the Church of England in Australia* (London: SPCK, 1962).

Cunningham, A. *The Rome Connection: Australia, Ireland and the Empire, 1865–1885* (Sydney: Crossing Press, 2002).

Lunn, D. *The English Benedictines 1540–1688: From Reformation to Revolution* (London: Burns & Oates, 1980).

Molony, J. N. *The Roman Mould of the Australian Catholic Church* (Melbourne: Melbourne University Press, 1969).

O'Farrell, P. *The Catholic Church and Community: An Australian History*, 3rd edn (Sydney: New South Wales University Press, 1992).

Scott, G. *Gothic Rage Undone: English Monks in the Age of Enlightenment* (Downside, UK: Downside Abbey, 1992).

Shanahan, M. *Out of Time, Out of Place: Henry Gregory and the Benedictine Order in Colonial Australia* (Canberra: Australian National University Press, 1970).

Shaw, G. P. *Patriarch and Patriot: William Grant Broughton 1788–1853, Colonial Statesman and Ecclesiastic* (Melbourne: Melbourne University Press, 1978).

8
'Brighter Britain': Images of Empire in the International Child Rescue Movement, 1850–1915

Shurlee Swain

The literature produced by the child rescue movement, disseminated from England in the second half of the nineteenth century, was replete with images of Empire. Organizations such as Dr Barnardo's, the Church of England Waifs and Strays Society, the National Children's Homes and the National Society for the Prevention of Cruelty to Children all produced magazines for both adult and child supporters from the earliest years of their operation. Although the circulation of such magazines is unknown, the publication of correspondence and subscription lists show that they were read widely throughout Britain and the colonies, positioning the work of child rescue organizations, and their charismatic male founders, within the larger context of the 'civilizing mission'. The term, the 'civilizing mission', was used primarily to describe the attempts by Evangelical Christians to inscribe a benevolent purpose within British imperial expansionism, providing the justification for much of the missionary activity described elsewhere in this volume.[1] However, in this chapter, the focus is on the civilizing mission at home, a process by which insights gained from missionary endeavour abroad were used to evangelize the great unchurched in the slums and rookeries of England's cities.

The use, or misuse, of photography in Barnardo's publications has attracted some interest from historians alert to its persuasive misrepresentations and its sexual suggestiveness.[2] This chapter, however, moves beyond the image to the text, arguing that the whiteness implicit in the child rescuers' construct of a 'Christian childhood' led to their becoming complicit in the imperial project of dispossession in the colonies to which their ideology (and often also the children they claimed to have rescued) were transported. It is concerned not with the disparity between publicity and practice which has preoccupied many later critics of such organizations,[3] but focuses instead on the nature of the discourse which, through its influence on public opinion, was to have a significant influence on the shaping of child welfare policy across the Empire.

Darkest England as a site of mission

Child rescue literature was a subset of that much larger body of literature which sought to represent the 'poor' to middle- and upper-class readers, a literature which, Childers has argued, promised to bridge the gap between the 'two nations' but simultaneously functioned as a *cordon sanitaire,* built upon a series of 'misrepresentations and illusions'.[4] The self-proclaimed child rescuers, who provide the focus for this chapter, constructed a series of 'artistic fictions'[5] which functioned to reconstitute the everyday phenomenon of the street child as an object of pity, a victim of vice and neglect, simultaneously a threat to, and the embodiment of, the future of nation, race and Empire. Thomas Barnardo (1854–1905), founder of Dr Barnardo's Homes (1868), Thomas Bowman Stephenson (1838–1912), founder of the Wesleyan National Children's Homes (1869), Edward de Montjoie Rudolf (1852–1933), founder of the Anglican Waifs and Strays Society (1881) and Benjamin Waugh (1839–1908), secretary of the National Society for the Prevention of Cruelty to Children (1889), constituted a distinct subset of the much larger group of 'slummers' identified by Seth Koven.[6] They were supported in their literary effort by an array of writers, artists and poets who incorporated child rescue themes into their work. Included amongst these supporters were such famous authors as R. M. Ballantyne[7] (1825–1894), Sir Arthur Conan Doyle[8] (1859–1930), the Rev. Dr W. Fitchett[9] (1841–1928), Rider Haggard[10] (1856–1925), W. T. Stead[11] (1849–1912) and Hesba Stretton (1832–1911).[12] Their stories and poems were published alongside the regular contributions of staff writers and contributors whose poignant tales filled the pages of the journals month after month. Designed to alter national sensibilities while attracting ongoing financial support, Barnardo's *Night and Day* and *Young Helpers' League Magazine,* Stephenson's *Children's Advocate, Highways and Hedges* and *Our Boys and Girls,* Rudolf's *Our Waifs and Strays* and *Brothers and Sisters* and Waugh's *The Child's Guardian* and the *Children's League of Pity Paper,* provide a rich archive for scholars interested in the interplay of empire and mission. Like the publications from overseas missions, discussed elsewhere in this volume by Bateman and Breitenbach (Chapters 13 and 5), these magazines constructed images of the other for adults and juveniles alike, images which were heavily embedded in an existing imperialist discourse.

Central to such literature, as Catherine Hall's work has shown, was the mutually constitutive discursive relationship between centre and periphery, or metropole and colony, a relationship which had missionaries at its core. Race, she has argued, was foundational to this discourse, 'a space in which the English configured their relation to themselves and others'.[13] Missionaries, who moved freely between the self and the other, provided a crucial conduit for this developing discourse, having an influence 'at home' equally significant to that which they had in the countries to which they were sent.[14] The child rescue movement was a subset of this larger missionary enterprise; its leaders were active participants in 'the competition for bodies and souls' both at home and abroad.[15] Despite a primary concern with conversion and salvation, their literary output also functioned to reflect and refract developing discourses around race. Missionary endeavour filled the minds of the readers with images of a foreign other which were readily applied to the children of the poor.[16]

In the classical mission narrative, Cox argues, 'male clerical heroes... move from the Christian heartland into a kind of global religious vacuum peopled by non-Christians who are sometimes portrayed as noble, sometimes as vicious, but always as ignorant of the benefits of the Christian gospel'.[17] The child rescuers found their mission field much closer to home. In a poem written in the voice of a boy rescued and sent to Canada, Charles Barker positions Barnardo as a missionary to the English:

> I allers prays for the Doctor, and means to till I die,
> And I bless the Lord wot told him in the courts and lanes to go,
> And rescue the outcast children from want and sin and woe.
> It's jist sich a work as Jesus likes on the earth to see;-
> Look at His words on the walls, sir- 'Let the little ones come
> unto Me'.[18]

The work was always seen as Christ's work, its message that these children too were God's children, 'named by Jesus in prayer'.[19] By contributing to the child rescue cause supporters were assured that they were participating in the Lord's work and would, accordingly, receive their reward in heaven.[20]

Although the children in need of rescue were resident in English cities, they were simultaneously 'other' to the people on whom they called for help, an otherness depicted in terms derived from the mission experience. Living in 'dusky dens' in a state of 'practical heathendom',[21] 'Fresh from the mud of river-bank or street/Rude as the heathen of benighted lands',[22] their need was presented as more pressing than that of 'the far heathen' whose calls were believed to be far more readily heard.[23] The need for rescue, Mrs E. S. Craven Green argued, was closer to home:

> Wild ocean isles ye need not seek, nor need your footsteps roam
> In heathen lands, for peril paths–they lie around your home.[24]

The term heathen implied a possibility of conversion, realized most tellingly in John Longley's story of the boy, rescued because of his attachment to his mother's Bible, who grew up to become a preacher.[25] The problem, in such cases, was ignorance. Presented with the truth of the gospel, combined with the 'loving care' of the rescuer, the heathen could be transformed.[26] However, as Frank Horner pointed out in his classic reversal story, 'The little savages of Nodlon', the 'home heathen',[27] had an added advantage for these 'little heathens ... will *wash white*'.[28] Thus transformed, the threat of the 'heathen at our very gates' could be allayed.[29] 'As veritable a heathen as could be found in the Cannibal Isles', miraculously could become an asset to the nation.[30]

The term 'pagan' was put to a similar use. It not only had a more ancient provenance, with references back to Greece and Rome,[31] but also implied a 'savagery' from which 'civilized England' should be free.[32] As such, it was used most powerfully by the National Society for the Prevention of Cruelty to Children in its campaigns for law reform,[33] but it also provided a valued point of contrast for other organizations seeking to emphasize the value of the work which they were undertaking.[34] Its strength derived from the implied contrast with a 'so-called Christian England' where such evil or primitivity should not be allowed to exist.[35] 'The story would better befit a barbarous period centuries ago, and of an uncivilised country', the author of 'Benny's wrongs and rescue' warned his readers, 'but it must be told ... as having occurred in Christian England, the favoured country wherein, as a rule, childhood is specially dear and sacred, and whose child-life, in its pure pleasures and happiness, is perhaps unequalled in any other country.'[36] It was a message repeated by Adelaide Proctor in her poem 'Homeless':

My dogs sleep warm in their baskets,
Safe from the darkness and wind;
All the beasts in our Christian England
Find pity wherever they go;
Those are only the homeless children
Who are wandering to and fro.[37]

The term 'pagan' could also function as a source of shame in cases where non-Christians had proved more compassionate than Englishmen and women.[38] The 'great cities' of 'Christian England', Rider Haggard declared, had produced 'abominations not known amongst savage tribes'.[39]

A racialized discourse

It was in the contrast drawn between a Christian England and the savage tribes that this religious discourse became embroiled in race. As Catherine Hall has argued, the use of a similar language to describe working-class Britons and Aboriginal, African or other indigenous peoples did not mean that they were thought of as the same. The 'common language of race ... employed to map these peoples' assumed that Anglo-Saxons were marked out by their blood, and racial instincts, 'as a people who would conquer and propagate, in the name of their superior civilization and Protestant religion'.[40] The child rescue literature consistently argued that the outcast child's claim to salvation lay in this shared heritage. Because of their inherent whiteness they alone could be 'washed clean'. Whiteness, Ruth Frankenberg argues, has three elements: 'a location of structural advantage, of race privilege ... a place from which white people look at ourselves, at others, and at society ... [and] a set of cultural practices that are usually unmarked and unnamed'.[41] Most mission literature encoded such whiteness practice, designating non-European societies as 'primitive' to buttress the superiority of the colonizing nation.[42] While comparisons drawn between the urban poor and the 'tribes' and 'savages' of foreign lands, may appear to point to a form of 'class racism', the implicit advantages of whiteness remain.[43] It is the threat to white privilege, not the denial of the privilege that such comparisons signify.

Representations of the crowded areas of the inner city as an alien land were standard fare in child rescue literature throughout the second half of the nineteenth century. Here race, class and tribe were intertwined, embellished with such negative, even threatening descriptors as 'feeble and famished', 'ragged' and 'predatory'.[44] Implicit in such descriptions was the notion that such 'tribes' or 'races' were 'rising' or 'breeding',[45] creating a 'domain of barbarism' that would need to be 'invaded' in the interests of creating a 'better race'.[46] Invoking fears of atavism, Bowman Stephenson warned his readers that all Britons were descended from 'savages not dissimilar to the Maori' and that savagery was returning amongst those who are not loved and cared for today.[47] By the end of the century this discourse had taken on an additional eugenic strain, arguing that in order for degeneracy and deterioration to be avoided efforts needed to be made to 'save the ... British breed'.[48] The child savers who went amongst the 'chilling slush of our streets and gutters' in the interests of the 'future of the ... Empire', were, a young Winston Churchill declared, the 'champions of our race'.[49]

When non-whites are incorporated into this narrative, their role is to buttress the assumed superiority of the British which the images of urban decay threatened to undermine. Whereas in the first half of the nineteenth century, Susan Thorne argues, the symmetry of representations of poor and colonized peoples was used as a justification of class and racial subordination, the threat to the privileges of whiteness was clearly uppermost for child rescuers 50 years later.[50] Like Mayhew, they understood it as pertinent to convince their readers that poor residents of London and other urban centres 'were *of* English society though separate from it, *related* to the middle class but a "race" apart from it, fellow inhabitants of the same city but members of a different "tribe"'.[51] The argument that, like indigenous peoples, the poor of the inner city were unable to love and care for their children was intended to shame the latter, or at least those who allowed such neglect to continue.[52] Newly incorporated into the Empire, the people of New Guinea were presented to readers of the *Child's Guardian* as having a lesson for the people of England. Confronted with stories of child abuse in England, a missionary replied: 'If anybody dealt so with a child in New Guinea, the natives would spear him, but such treatment of a child is impossible to New Guineaers.' 'We live in a land in which Christianity has come to mean mawkish mercy to child-murderers', the article continued. 'Better days will come when New Guineaers teach us righteousness.'[53]

But even such faint praise was rare. As Breitenbach argues, in Chapter 5, in this volume, even missionaries committed to the equality of all humanity spoke in a patronizing tone. Travel, to the child rescuer, served only to confirm their convictions of European superiority with the non-white people they encountered almost uniformly depicted as objects of pity, lacking affection for their children, essentially primitive and inferior and destined to disappear.[54] Recounting his visit to a Mohawk reservation in Canada, Stephenson wrote about his teaching the children to play, an essential element of childhood in 'civilisation'.[55] In a later instalment he expressed his surprise at encountering an 'Indian baby... [who] crowed, and smiled, and kicked just as heartily as though his face had been white.' The baby's father was described as 'a nobleman, though only an Indian'.[56] From South Africa Stephenson wrote: 'I can't help feeling, as I see all these strange, heathen people around me, that English boys and girls ought to be thankful ... for their birth in a Christian land.'[57] This theme continued when he arrived in Australia. Marvelling at Melbourne's progress in its 50 years of existence he compared the city which he had visited with the 'huts' which had previously occupied the site:

> I need hardly tell you that the people who used to lie under such miserable huts were not white people. They were black, and they were amongst the very lowest and most degraded races that have ever been found.[58]

A final stop in Tasmania allowed him to reflect on the 'sad' but inevitable decline of the 'great tribes and nations of red men and black men... fading away before the white races' but readers were assured: 'we cannot stop this; but we can at least have the pleasure of helping the missionary societies, which are endeavouring to save the remnants of these people from destruction'.[59]

Whitening the Empire

The theme of 'inevitable decline' ran through most such accounts, simultaneously justifying and excusing white incursion. The 'march of white man' was 'irresistible', wrote missionary journalist, the Rev. Egerton R. Young,[60] before concluding, 'perhaps it is right that a superior race should be paramount in that great Continent, yet the treatment of the Indian might have been much more kindly and Christian'.[61] For most writers, the solution to this problem lay, not in kind treatment, but in immigration.[62] Canada, it was argued, was an 'Englishman's birthright' and should not be left to the 'alien and inferior races'.[63] The colonies of settlement constituted a 'brighter Britain' which residents of 'Darkest England' could embrace and redeem.[64]

The child rescuers were not the first to use poor children to build the Empire. Workhouse children were apprenticed to planters in the Virginia settlement in Britain's first imperial adventure, a practice continued, following the loss of the American colonies by Annie MacPherson and Maria Rye who negotiated with Poor Law officials in overcrowded unions to take parties of children to Canada. In the second half of the nineteenth century, the new child rescue organizations built upon such foundations, but distinguished their operations from those of their predecessors by emphasizing the training and aftercare extended to the children to fit them for their imperial mission.[65] The preference for white children in the colonial labour market was assumed. Why, Maria Rye asked, would Indian children need to be trained to become servants if there were sufficient child migrants available?[66] Deploring the survival of slavery, the Rev. S. Coley expressed surprise that 'a black fellow was worth buying' when, in England, 'there were thousands of white men for whom it was said the best thing would be to send them out of the country'.[67] In a darker tone, Vanoc warned that the benefits of child migration would only be completely understood at some time in the future when in colonies 'sparsely peopled with white emigrants ... white men may be the hunted'.[68]

Central to the argument for English superiority advanced in child rescue publications was the assertion that Evangelical Christianity had developed a new notion of childhood.[69] However, unlike the Mothers' Union leaders discussed by Prevost in Chapter 12, in this volume, child rescuers, both at home and abroad, seemed unable to bring children of colour within this new definition. The photographs included in the various magazines make it clear that, amongst the 'rescued' children, there were increasing numbers of those who could not be washed clean. Both Seth Koven and Caroline Bressey have discussed the anxieties which the presence of these children aroused. Most striking is the photograph of three naked black children, the only children photographed naked in the entire Barnardo archive.[70] To Koven, this is indicative of Barnardo keying into wider social anxieties around miscegenation.[71] Bressey would agree, pointing to the falsification of the racial identity of the children's father, an Englishman in the admissions book who becomes 'coloured' in the *Night and Day* story, as evidence of Barnardo's unwillingness, to embrace the notion of a multiracial Britain. Rather, she concludes, he looked for degrees of 'whiteness' in deciding who was eligible to be saved.[72]

The imperative to rescue white children from savagery was always dominant. The 'little English girls and boys' rescued by Barnardo from the Turkish theatre troupe to which they had been sold by their parents, was celebrated as a victory over 'heathenism'.[73] Similarly the people (of many races) in the West Indian port, who subscribed to an appeal to rescue a 'golden-haired' child from her depraved mother, were widely applauded.[74] In a later article, the Waifs and Strays' Society chose to link the story of a boy, brought back to England after having been rescued from imprisonment amongst other races, with that of another, perceived as white, who was kept from his mother when it became apparent that she was a West Indian planning to take her son back to her homeland.[75] When the major cities in colonies of settlement began to reproduce the worse features of cities 'at home', evangelicals established locally based child rescue societies. Central to their mission was the need to rescue white children from 'contamination', maintaining and uplifting the 'prestige of the white race'. Colonial child welfare legislation, often enacted in response to child rescuers' concerns, was similarly blind: the child whose rights and innocence were to be protected was inherently white.[76]

Through this means the child rescue movement became complicit in the imperial project. Having borrowed so heavily from missionary imaginings in describing its initial endeavours, it now projected those imaginings back to their source to support imperial claims. In the colonies, whiteness brought an identity and a set of accompanying privileges to settler children, justifying the oppression and dispossession of indigenous peoples, adult and children alike.[77] Child rescue literature, like the missionary literature of which it was a part, circulated in both metropole and colony, shaping and sustaining 'imperial mythologies'.[78] 'I would like every English child to think about Australia, not as a wild and savage and half-heathen country, but as the home of great and intelligent and prosperous English communities' wrote Stephenson in 1883.[79] But, if that transformation was to be maintained, the hierarchy between the races had to be sustained, with rescue work amongst settler children of particular importance.[80] They could not be allowed to display behaviour condemned in England as 'heathen', 'savage' or 'wild',[81] for, in the colonies, white children stood as exemplars of the racial and religious superiority which had delivered the land and its people to the colonizers.[82]

In April 1898, the National Children's Homes magazine, *Highways and Hedges*, included a tribute to a former resident who, rescued as a child, had chosen in adulthood to go as a missionary to Canada, where she worked with her husband at an orphanage. This article combined the classic 'before and after story' with the older Evangelical narrative of the 'good death' for, having returned to England in late pregnancy,

Mrs Joseph William Butler died in childbirth. Readers were assured that here was a woman who was doubly saved.[83] The colonies were routinely represented as a place of opportunity for English children. Here they could take up land or further their education and enter the professions, opportunities that were never offered to the graduates of the Homes who remained in England. The 'heathens who could be washed clean' could capitalize on their whiteness to advantage themselves, but their duty to the indigenous peoples whom they were displacing rarely extended beyond the obligation to rule or to aid. As the rescued children grew to maturity, the magazines began to feature their letters home, most of which, predictably, celebrated their success in promoting the imperial mission.[84] Amongst this mass of self-congratulatory correspondence there appears but one dissenting voice. 'We often see black people in town', wrote a young man placed in Bunbury, Western Australia. 'These people are called Aboriginals. Australia really belongs to this race of people, they live in the bush, and some of them are civilised.'[85] But they were, of course, black, and as such, could feature as exotica in child rescue literature but had no place in the new world for children which child rescuers claimed to be creating.

In its intertwining of child, race and nation, the imperial child rescue movement both drew on and contributed to the mutually constitutive relationship between periphery and centre, colony and nation and mission and empire which was so central to discourses around Britishness and whiteness in the latter half of the nineteenth century. Using notions of childhood innocence, transcended as they were being constructed both at home and abroad, child rescuers reinforced notions of the rightness of Empire and the superiority of the white Britons to whom God had entrusted large portions of the world's landmass. But that endowment was both fragile and conditional. Without continuing attention to the fate of childhood at home and in the settler colonies, the Empire, and ultimately the nation would be lost, and God's bounty squandered. What these discursive formations ignored was the fate of those excluded from such racially exclusive notions of childhood. Denied the protection extended by law to children identified as the future of both nation and Empire, indigenous children were destined to disappear, or at best, to be retrained to serve white interests.

Notes

1. See for example the Chapters 5 and 11 by Esther Breitenbach and Peter Clayworth respectively.

2. S. Koven, 'Dr Barnardo's "Artistic Fictions": Photography, Sexuality, and the Ragged Child in Victorian London', *Radical History Review*, LXIX (1997) 6–45; S. Koven, *Slumming: Sexual and Social Politics in Victorian London* (Princeton, NJ: Princeton University Press, 2004); A. McHoul, 'Taking the Children: Some Reflections at a Distance on the Camera and Dr Barnardo', *Continuum: The Australian Journal of Media & Culture,* V, no. 1 (1991).
3. Many scholars who have had access to case records have made this observation. See, for example, B. Bellingham, 'Waifs and Strays: Child Abandonment, Foster Care, and Families in Mid-Nineteenth-Century New York', in *The Uses of Charity: The Poor on Relief in the Nineteenth-Century Metropolis*, ed. P. Mandler (Philadelphia, PA: University of Pennsylvania Press, 1990), pp. 133ff; J. Parr, *Labouring Children: British Immigrant Apprentices to Canada, 1879–1924* (London: Croom Helm, 1980), ch. 4.
4. J. W. Childers, 'Observation and Representation: Mr. Chadwick Writes the Poor', *Victorian Studies*, XXXVII, no. 3 (1994) 408.
5. Koven, *Slumming*, p. 88.
6. For biographies of these child rescuers see M. Weddell, *Child Care Pioneers* (London: The Epworth Press, 1958).
7. R. M. Ballantyne, 'A Northern Waif', *Our Waifs and Strays,* I, no. 13 (1885) 1–2; no. 14 (1885) 5.
8. Sir A. Conan Doyle, 'The Seed and the Tree', *Night and Day,* XXXI, no. 247 (1908) 77–8.
9. Rev. Dr. Fitchett, 'Children's Home Conference Meeting', *Highways and Hedges,* XII, no. 141 (1899) 205–7.
10. H. Rider Haggard, 'A Soldier's Child', *Children's League of Pity Paper,* III, no. 11 (1896) 95–6; 'Mr H. Rider Haggard on the Society', *The Child's Guardian,* XVIII, no. 8 (1904) 85–6; 'The Real Wealth of England', *Night and Day,* XXX, no. 243 (1907) 74–6.
11. W. T. Stead, 'For All Those Who Love Their Fellow Men: A Supplement for Christmas Time', *Review of Reviews* (July–December 1901) 670–90.
12. H. Stretton, 'London Society for the Prevention of Cruelty to Children', *The Children's Advocate,* VI, no. 65 (1885) 89–90.
13. C. Hall, *Civilising Subjects: Metropole and Colony in the English Imagination 1830–1867* (Cambridge: Polity, 2002), pp. 12–13.
14. I. Hofmeyr, 'Inventing the World: Transnationalism, Transmission, and Christian Textualities', in *Mixed Messages: Materiality, Textuality, Missions,* ed. Jamie S. Scott and Gareth Griffiths (New York: Palgrave Macmillan, 2005), p. 20.
15. F. Prochaska, *Christianity and Social Service in Modern Britain: The Disinherited Spirit* (Oxford: Oxford University Press, 2006), p. 13.
16. C. Hall, 'Going a-Trolloping: Imperial Man Travels the Empire', in *Gender and Imperialism,* ed. C. Midgley (Manchester: Manchester University Press, 1998); Hugh), p. 180; H. Cunningham, *The Children of the Poor: Representations of Childhood since the Seventeenth Century* (Oxford: Blackwell, 1991), p. 6.
17. J. Cox, 'Narratives of Imperial Missions', in *Mixed Messages: Materiality, Textuality, Missions,* ed. Jamie .S. Scott and Gareth. Griffiths (New York: Palgrave Macmillan, 2005), p. 6.
18. C. W. Barker, 'Told in Canada', *Night and Day,* XI, no. 120 (1887) 74–5.
19. H. E. W., 'My Boys', *Night and Day,* XIII, nos. 137 and 138 (1889) 138.

20. Miss Charlotte Murray, 'The League Hymn – We'll All Help', *Young Helpers' League*, I (1892) 19; 'Leagued in Love', *Young Helpers' League*, I (1892) 188.
21. 'Scenes and Sights in London. No. II. Golden Lane and Whitecross Street, St. Luke's', *Ragged School Union Magazine*, I, no. 8 (1849) 141–6. See also: Rev. W. Wade, 'The Enemies of Childhood. No. 3. – Overcrowding (Cont)', *Our Waifs and Strays*, VIII, no. 216 (1902) 273–5.
22. J. P., 'The Ragged School', *Ragged School Union Magazine*, I, no. 1 (1849) 13–14.
23. Mrs E. S. Craven Green, 'The Claims of the Needy', *Ragged School Union Magazine*, I, no. 2 (1849) 80. For similar use of the term 'heathen' see: T. J. Barnardo, *My First Arab: or How I Began My Life-Work* (London: Dr Barnardo's Homes; reprint, Serialized in Young Helpers League 1894; Revised pamphlet version, 1917); J. B. Orris, 'The Increase of Juvenile Crime', *Reformatory and Refuge Journal*, no. XXXIV (1867) 14–17; 'The Lay Mission', *The Children's Advocate and Christian at Work*, IV, no. 21 (1873) 131–3; 'Stories of Our Work: Big Joe', *The Children's Advocate and Christian at Work*, V, no. 33 (1874) 129–30; 'A Little English Heathen', *The Children's Advocate*, VI, no. 4 (1876) 31; F. Herschell, 'Annual Meeting of the Reformatory and Refuge Union', *Reformatory and Refuge Journal*, VI, no. LXXII (1876) 79–84; J. Pendlebury, 'A Story from the Children's Home: How an Outcast Bairn Was Gathered In', *The Children's Advocate*, I, no. 8 (1880) 113–19; T. J. Barnardo, 'The Bitter Cry of the Outcast Children', *Night and Day*, VII, no. 79–80 (1883) 140–3; T. J. Barnardo, 'The Continued Necessity for Voluntary Agencies in Reclaiming Children of the Streets', *Night and Day*, V, no. 54 and 55 (1881) 202–8; A. W. Mager, 'Stories of Our Work VI: Which Is His Head?', *The Children's Advocate*, V, no. 57 (1884) 167–9; 'Honora's Effort', *Our Waifs and Strays*, I, no. 3 (1884) 5–7; 'The Children's Charter', *Highways and Hedges*, II, no. 24 (1889) 225–6; T. J. Barnardo, 'The Economics of Child Rescue', *Night and Day*, XIX, no. 182 (1895) 1–3; T. Bowman Stephenson, 'Reject?', *Highways and Hedges*, IX, no. 97 (1896) 11–13; M. E. Lester, 'Cameos from Life. No.8. – Feeling for the Child', *Our Waifs and Strays*, VI, no. 173 (1898) 357–8; Ballard, 'The Outcast Lives', 151–2; M. E. Lester, 'Cameos from Life. No 14. – Quartettes', *Our Waifs and Strays*, VII, no. 195 (1900) 325–7. Given the sectarianism rife within the child rescue movement it is not surprising that the term had a particular resonance in relation to Catholic children, described as 'practically heathen', neglected by their church until child rescuers became interested in them. T. J. Barnardo, 'The Roman Catholics and My Children', *Night and Day*, XIII, nos. 137–8 (1889) 122–6.
24. Mrs E. S. Craven Green, 'Unto This Work Ye Are Called', *Ragged School Union Magazine*, I, no. 5 (1849) 92. See also: B. M. H., 'Why Not Both?', *Our Waifs and Strays*, XII, no. 310 (1910) 462–3; J. J. B., 'By Highway and Byway. XV – the Bubble on the Surface', *Our Waifs and Strays*, XIV, no. 342 (1914) 284–5.
25. J. Longley, 'My Mother's Bible', *The Children's Advocate* (May 1871) 4–5.
26. E. Hopkins, 'Little Mary', *Night and Day*, V, nos. 46 and 47 (1881) 24–30.
27. A. W. M, 'Our Home Heathen', *The Children's Advocate -Quarterly Budget*, VI, no. 3 (1876) 18–20; Rev. S. Naish, 'Grace and Glory: Being an Account of the Life and Death of a Former Inmate of the Children's Home', *The Children's Advocate*, III, no. 33 (1882) 129–33.

28. F. Horner, 'The Little Savages of Nodlon', *The Children's Advocate and Christian at Work* (January 1873) 2–5.
29. T. J. Barnardo, 'Ever-Open Doors', *Night and Day*, XVI, no. 162 (1892) 33–5. See also Archdeacon Farrar, 'Archdeacon Farrar on the Children's Home', *Highways and Hedges*, VI, no. 66 (1893) 114–16; Rev. Canon Barker, 'A Plea for the Homes', *National Waifs' Magazine*, XXV, no. 217 (1902) 73.
30. Sister Grace, 'Waste or Worth?', *Highways and Hedges*, VII, no. 78 (1894) 108–10.
31. 'Crown and Sword', *The Child's Guardian*, III, no. 25 (1889) 132.
32. Rev. A. P. Stanley, 'A Festival Address', *The Children's Advocate* 1, I, no. 2 (1880) 17–20.
33. 'The Cry of the Children', *Christian Australian World*, May 16 1889; 'Cruelty from Sire to Son', *The Child's Guardian*, IV (1890) 54; 'The Angel of the Little Ones, or the National Society for the Prevention of Cruelty to Children', *Review of Reviews* (July–December 1891) 521–30; 'Not Charity to Children but Justice', *The Child's Guardian*, VI, no. 8 (1892) 101; 'King Dirt's Allies', *The Child's Guardian*, VII, no. 9 (1893) 122. 'Our Eleventh New Year's Day: A Review', *The Child's Guardian*, IX, no. 1 (1895) 6; 'Baby-Farming', *The Child's Guardian*, X, no. 5 (1896) 65–6.
34. 'All Sorts and Conditions of Children', *Highways and Hedges* (1906)177–82; W. Baker, H. Williams and G. Code, 'For the Empire', *Night and Day*, XXXII, no. 250 (1909) 51–3.
35. 'Honora's Effort', 5–7; 'Our New Year', *The Child's Guardian*, II, no. 13 (1888) 1–2; 'Baby-Farming', *Australian Christian World*, 17 July 1890; 'The Daily Romance of Child Life', *Highways and Hedges*, III, no. 34 (1890) 213–15; T. J. Barnardo, 'Personal Notes', *Night and Day*, XIV, no. 148 (1890) 197–200; 'Insured; a Boy's Story', *Highways and Hedges*, IV, no. 41 (1891) 92–3; 'Uncrowned King Brute', *The Child's Guardian*, V, no. 5 (1891) 60; 'The Children's Home and Orphanage VI: The Alverstoke Branch', *Our Boys and Girls* (1900) 47–8.
36. J. Pendlebury, 'Lights and Shadows of Child-Life: Benny's Wrongs and Rescue', *Highways and Hedges*, I, no. 1 (1888) 3–5.
37. A. Proctor, 'Homeless', *The Children's Advocate*, II, no. 23 (1881) 176.
38. Rev. G. C. Lorimer, 'A Lesson in Christian Charity', *Night and Day*, II, no. 2 (1878) 24–5.
39. Rider Haggard, 'Mr H. Rider Haggard on the Society', pp. 85–6.
40. C. Hall, 'Of Gender and Empire: Reflections on the Nineteenth Century', in *Gender and Empire*, ed. Philippa Levine (Oxford: Oxford University Press, 2004), pp. 46–7, 52.
41. R. Frankenberg, *White Women, Race Matters: The Social Construction of Whiteness* (Minneapolis, MN: University of Minnesota Press, 1993), p. 1.
42. Hofmeyr, 'Inventing the World: Transnationalism, Transmission, and Christian Textualities', p. 33.
43. L. Mahood and B. Littlewood, 'The "Vicious" Girl and The "Street-Corner" Boy: Sexuality and the Gendered Delinquent in the Scottish Child-Saving Movement, 1850–1940', *Journal of the History of Sexuality*, IV, no. 4 (1994) 552.
44. A., 'The Ragged School Union: Its Objects and Claims', *Ragged School Union Magazine* I, no. 1 (1849) 4–9; A., 'The Late Earl Shaftesbury, K.G.', *Night and Day*, IX, nos. 98–103 (1885) 153–7.

45. J. P., 'The Ragged School Teacher's Appeal to All Classes', *Ragged School Union Magazine,* I, no. 3 (1849) 52; 'Emigration: Industrial Schools', *Australasian* (April 1851) 385–95. F. Peek, 'The Boarding-out of Pauper Children', *Examiner* (16 March 1872) 283–4; 'To Our Readers', *The Child's Guardian,* V, no. 1 (1891) 1; T. J. Barnardo, 'Personal Notes', *Night and Day,* XVII, no. 176 (1893) 89.
46. M. Carpenter, 'Day Industrial Schools for Neglected and Destitute Children', *Reformatory and Refuge Journal,* no. LV (1872) 269–76; Lord Brabazon, 'State-Directed Emigration: Its Necessity', *Night and Day,* VIII, no. 92 (1884) 178–80; Rev. Lord S. G. Osborne, 'The Rev. Lord Sydney Godolphin's Letters to The 'Times'', *Night and Day,* XII, no. 129 (1888) 114–15.
47. T. B. Stephenson, 'To My Little Friends', *The Children's Advocate,* VII, no. 75 (1886) 69–70.
48. Rev. W. Wade, 'Physical Deterioration', *Our Waifs and Strays,* IX, no. 246 (1904) 379–80; 'The Children's Friend: After a World-Wide Tour', *The Child's Guardian,* XVIII, no. 9 (1904) 102; Rev. Canon Fleming, 'The Arms of God's Mercy and Love', *National Waifs' Magazine,* XXVIII, no. 232 (1905) 10–11; 'The Cry of the Children', *Our Waifs and Strays,* XI, no. 274 (1907) 81; Kingsley Fairbridge, 'Juvenile Emigration and Farm School System', *The Times,* 24 May 1910; Owen Seaman, 'The Social Aspects of Child-Saving', *Night and Day,* XXXI, no. 245 (1908) 33–4; J. J. B., 'By Highway and Byway. IV. – Down South', *Our Waifs and Strays,* XIII, no. 324 (1912) 322–3.
49. Winston Churchill, 'The Friendless Young', *Night and Day,* XXXI, no. 247 (1908) 68–70.
50. S. Thorne, '"The Conversion of Englishmen and the Conversion of the World Inseparable": Missionary Imperialism and the Language of Class in Early Industrial Britain', in *Tensions of Empire: Colonial Cultures in a Bourgeois World,* ed. A. L. Stoler and F. Cooper (Berkeley, CA: University of California Press, 1997), p. 248.
51. D. E. Nord, 'The Social Explorer as Anthropologist: Victorian Travellers among the Urban Poor', *Visions of the Modern City: Essays in History, Art and Literature* ed. W. Sharpe and L. Wallock (Baltimore, MD: John Hopkins University Press, 1987), pp. 132–3.
52. A. L. Stoler, 'Tense and Tender Ties: The Politics of Comparison in North American History and (Post) Colonial Studies', *Journal of American History,* LXXXVIII, no. 3 (2001) 852.
53. 'Her Majesty, by the Grace of God', *The Child's Guardian,* III, no. 29 (1889) 75.
54. 'Victorian Notes', *Victorian Independent,* 4 June 1870, 34; Ballantyne, 'A Northern Waif', p. 5.
55. T. Bowman Stephenson, 'Letters to My Little Friends, No. II', *The Children's Advocate,* III, no. 26 (1882) 21–2.
56. T. Bowman Stephenson, 'Letters to My Little Friends, No. III', *The Children's Advocate,* III, no. 27 (1882) 35–6.
57. T. Bowman Stephenson, 'Letters to My Little Friends, No IX', *The Children's Advocate,* III, no. 35 (1882) 161–3.
58. T. Bowman Stephenson, 'Letters to My Little Friends XI', *The Children's Advocate* IV, no. 38 (1883) 19–21.
59. T. Bowman Stephenson, 'Letters to My Little Friends, No XIII', *The Children's Advocate,* IV, no. 41 (1883) 66–8. Barnardo makes a similar trip in 1885. For

his first impressions of Canadian indigenous peoples see T. J. Barnardo, 'Personal Notes', *Night and Day,* IX, no. 93 and 94 (1885) 24–5.

60. For a further discussion of Young's importance as a missionary 'image maker' see D. Francis, *The Imaginary Indian: The Image of the Indian in Canadian Culture* (Vancouver: Arsenal Pulp Press, 1995), p. 49.
61. Rev. E. R. Young, 'Missionary Life among the Red Indians of the Dominion of Canada', *Highways and Hedges,* II, no. 16 (1889) 71–3.
62. T. J. Barnardo, 'Personal Notes', *Night and Day,* XVIII, no. 178 (1894) 7–11; Rev. G. S. Rowe, 'Our Critics', *The Children's Advocate,* V, no. 59 (1884) 216–18.
63. D. Crane, 'In the Path of the Sun: Canadian Homes for British Boys', *Highways and Hedges,* XXVI (1913) 156–7. The notion of Christian providentialism as both a motivation and justification for empire was well established by this time. For a recent discussion of this concept see J. Gascoigne 'The Expanding Historiography of British Imperialism', *The Historical Journal,* XLIX, no. 2 (2006) 586.
64. Rev. W. J. Mayers, 'My Tour In 'Brighter Britain'', *Night and Day,* XVI, no. 165–8 (1892).This contrast had a long history amongst advocates of emigration. See R. D. Grant, *Representations of British Emigration, Colonisation and Settlement: Imagining Empire, 1800–1860* (Houndsmills: Palgrave Macmillan, 2005), ch. 6.
65. For a fuller discussion of child migration see G. Wagner, *Children of the Empire* (London: Weidenfeld and Nicolson, 1982).
66. M. Rye, 'Emigration', *Our Waifs and Strays,* VII, no. 197 (1900) 362. Maria Rye was one of the founders of the nineteenth-century child emigration movement. For further details of her work see Parr, *Labouring Children,* ch. 2.
67. 'Our Annual Meeting', *The Children's Advocate,* April 1872, 5–6.
68. Vanoc, 'Our Handbook: The New Emigration: The Oxford Movement', *Referee,* VIII May 1910.
69. Rev. J. Wells, 'The Child under Paganism', *Night and Day,* XXXII, no. 248 (1909) 18–19.
70. C. Bressey, 'Forgotten Histories: Three Stories of Black Girls from Barnardo's Victorian Archive', *Women's History Review,* XI, no. 3 (2002) 359.
71. Koven, 'Dr Barnardo's "Artistic Fictions"', 33.
72. Bressey, 'Forgotten Histories', 359, 369.
73. T. J. Barnardo, 'The Beni-Zou-Zougs', *Night and Day,* V, no. 56 (1881) 215–21. This article stands in stark contrast to the attitudes expressed in a later article dealing with children of colour: T. J. Barnardo, 'Leaves from My Notebook. 1. Three Woolly Black Heads', *Night and Day,* XXI, no. 202 (1898) 10–11.
74. T. J. Barnardo, 'A Rescue across the Sea', *Night and Day,* XVI, no. 163 (1892) 62.
75. 'Thy Going out and Thy Coming In', *Our Waifs and Strays,* XIV, no. 345 (1914) 359–61.
76. L. Chisholm, 'Class, Colour and Gender in Child Welfare in South Africa, 1902–1918', *South African Historical Journal,* XXIII (1990) 110–11.
77. Hall, 'Of Gender and Empire', 48–9.
78. G. Griffiths, 'Popular Imperial Adventure Fiction and the Discourse of Missionary Texts', in *Mixed Messages: Materiality, Textuality, Missions,* ed. Jamie S. Scott and Gareth Griffiths (New York: Palgrave Macmillan, 2005),

p. 51. See also Hofmeyr, 'Inventing the World: Transnationalism, Transmission, and Christian Textualities', p. 23.

79. T. Bowman Stephenson, 'Letters to My Little Friends, No. XV', *The Children's Advocate*, IV, no. 44 (1883) 115–17.
80. 'Children and Religion', *Australian Christian World* (4 February 1887) 716–17.
81. 'A Successful Experiment in Home-Mission Work', *Highways and Hedges*, IV, no. 37 (1891) 15–17. 'Sydney Larrikinism', *Australian Christian World* (9 July 1891) 4.
82. 'From National to Imperial: An Appeal to British Colonies, Paper 1', *The Child's Guardian*, XII, no. 10 (1898) 117–18; M. West, 'Work at Grahamstown', *Our Waifs and Strays*, I, no. 15 (1885) 6.
83. A. W. Mager, 'Mrs Joseph Wm. Butler (Deceased): A Sketch of Missionary Devotion and Enterprise', *Highways and Hedges*, XI, no. 124 (1898) 75–6.
84. The letters home are, of course, not representative of the experiences of the typical child migrant. See Parr, *Labouring Children*, ch. 7.
85. 'The Old Boys' Corner', *Our Waifs and Strays*, XIII, no. 311 (1911) 13.

Further reading

Bressey, C. 'Forgotten Histories: Three Stories of Black Girls from Barnardo's Victorian Archive', *Women's History Review*, XI, no. 3 (2002).

Cunningham, H. *The Children of the Poor: Representations of Childhood since the Seventeenth Century* (Oxford: Blackwell, 1991).

Frankenberg, R. *White Women, Race Matters: The Social Construction of Whiteness* (Minneapolis, MN: University of Minnesota Press, 1993).

Koven, S. *Slumming: Sexual and Social Politics in Victorian London* (Princeton, NJ: Princeton University Press, 2004).

Murdoch, L. *Imagined Orphans: Poor Families, Child Welfare, and Contested Citizenship in London* (New Brunswick, NJ: Rutgers University Press, 2006).

Parr, J. *Labouring Children: British Immigrant Apprentices to Canada, 1879–1924* (London: Croom Helm, 1980).

Scott J. S. and G. Griffiths, eds. *Mixed Messages: Materiality, Textuality, Missions* (New York: Palgrave Macmillan, 2005).

Weddell, M. *Child Care Pioneers* (London: The Epworth Press, 1958).

9
Saving the 'Empty North': Religion and Empire in Australia

Anne O'Brien

The task of understanding the relationships between religion and empire in Australia is compounded by the fact that the historiography of each of these fields is not highly developed. While transnationally, histories of modernity have generally been conceived as representing 'the triumph of secularism', in Australia the marginalization of religious history is arguably more pronounced than elsewhere.[1] Anti-clericalism in its convict origins and in the symbols of national identity – the bushman and the Anzac – have left their mark on the writing of Australian history, as has a long sectarian strand that has seen denominational defensiveness shadow religious history.[2] There are gaps in the historiography – sustained study of the Australian missionary movement one of the most serious – that other national historiographies do not share.[3] Nor has Australian history been much concerned with questions of Empire. The struggle for legitimacy of an Australian national history in the past 40 years, with its associated rejection of the 'cultural cringe', pushed to one side its Imperial connections.[4] All this is changing: feminism, 'the cultural turn' and the ascendancy of trans-national histories have increased Australian historians' sensitivity to both religion and empire and the relationships between them.[5]

What can an Australian perspective contribute to understandings of religion and empire? It can reveal the complex entanglements that arose when Indigenous cultures more than 40,000 years old encountered the satellite outpost of a large and wealthy imperial metropolitan culture. The physical context of those entanglements was similar to that of other settler societies: to paraphrase Andrew Porter – continual emigration, natural population increase, land hunger, greed, mutual fear and misunderstanding resulted in 'unmitigated cruelty and warfare'.[6] Unlike most other imperial settings, however, the violence and destruction of colonization in Australia became submerged during the first half of the twentieth century in what the anthropologist W. H. Stanner called 'the Great Australian Silence'.[7] Among the chief supports of this silence were two enduring legends of national identity, the bushman and the pioneer – each defined by slightly different characteristics but neither allegedly capable of 'unmitigated cruelty' to the traditional owners of the land.[8] Much scholarship in the past 40 years has been devoted to putting an end to that silence. The field of Indigenous history is one of the most fertile and complex in Australian historiography.[9] In the past few years, following the publication of Keith Windschuttle's *The Fabrication of Australian History* (2002), it has become embroiled in public contestation. Windschuttle's reading of Tasmanian history, widely disseminated by conservative journalists and supported by John Howard, Prime Minister 1996–2007, disparages what has become known as the 'black-armband' view of history, which he sees as exaggerating frontier violence for political purposes.[10] 'History wars' have been waged in other Western countries over various challenges to cherished understandings of national identity in the past ten years or so. In Australia they draw strength from the fact that the struggle for possession of the country has been buried beneath understandings of the bush and pioneer legends as essentially benign.

To examine the role that religion played in the colonization of Australia, I want to focus on the life histories of two men who played an important part in saving the empty north of Australia in the years of high imperialism at the turn of the twentieth century: Gilbert White (1859–1933) founding Bishop of Carpentaria and James Noble (1876?–1941), the first Indigenous deacon in the Anglican Church.[11] Focusing on how frontier Christianity shaped the identity of each man shows how race inflected 'the tensions between hierarchy and inclusion' that confronted Christian missionaries wherever they went.[12] As Catherine Hall has argued 'the time of empire was the time when anatomies of difference were being elaborated'.[13] In the north of Australia, anatomies of race difference shaped immigration, welfare, medical and Aboriginal policy; they shaped the way differences of gender and class were experienced and they shaped religious subjectivities. But while the church was embedded in an imperial culture that articulated difference it also aspired to an inclusiveness that derived from belief in the equality of all in the sight of God. The final aim of Christianity was to abolish difference – to make us 'all one in Jesus Christ'. Noble and White were both men of considerable stature, admired in their own time and since. White was one of the most influential and outspoken critics of northern colonialism, Noble a committed Christian leader of Indigenous people.[14] Focusing on the life histories of these two men enables investigation at close hand of the tensions between churchmen's belief in the equality of souls and their immersion in the politics of difference.

The north of Australia at the turn of the century provides a fertile context to examine these tensions. In many senses it was 'the last frontier' of the missionary world. The pattern of encounter, resistance, pacification and dispossession was similar to that which had characterized the south, but there were differences. Distance from dense population centres made it arguably more lawless. Timing played a part: in Queensland, the push north occurred after responsible government was granted in 1859, so local administrators were unrestrained by the Colonial Office, and the native police became an institutional force of destruction. In Western Australia, Aboriginal affairs were, until 1898, a responsibility of the Colonial Office which resisted establishing a native police force. While this may have meant that murders were less common, the dependence of settlers on Indigenous labour gave rise to brutal exploitation and kidnapping.[15] In both Western Australia and Queensland demographic factors increased the level of tension among the encroaching white population – whites were outnumbered by Indigenous people and terror on both sides was easily evoked.[16] Further, its location heightened invasion anxieties at a time when France and Germany were also colonising the Pacific; so did the fact that in large population centres such as Darwin, Broome and Thursday Island Asians made up more than half the population.[17] The trope of 'the empty north' was fed by invasion fears, but built on experience of Indigenous resistance. It sat uneasily on the north, where the real and imagined presence of 'the other' was so overwhelming that it drove immigration and native policy. Racial diversity was more threatening than a vacuum for, at a time when scientific racism was gaining intellectual credibility, it posed bio-political problems centred on miscegenation.[18]

In a more fundamental sense, the trope of the Empty North jars as a descriptor. For over 40,000 years, Northern Australia had been 'nourishing terrain' to Indigenous people whose ancestor spirits had travelled across the country, encountering myriad specific sites, imbuing them with deep meaning. And it had been a site of contact for centuries between Indigenous Australians, Torres Strait Islanders and Macassarese fishermen.[19] The trope of the empty north not only entailed denial of the white intruders' fears of a racially diverse population but also denial of pre-existing occupation and enduring layers of memory. Northern colonization, then, had its own tenor, its own momentum and bore considerable threat in the eyes of southern politicians. The title of a recent collection of essays in a Queensland-based journal – *Up North: Myths, Threats and Enchantment* – is indicative of an ongoing perception of difference surrounding Australia's north.[20]

What role did the church play in this context? By the turn of the century, when Gilbert White was appointed to the newly established Diocese of Carpentaria the church was becoming mobilized into missionary work by the actions of the state. Up until then the church had seen its main duty as to the white population.[21] Despite the fact that communities of Indigenous Christians were developing on late nineteenth century reserves and missions in Victoria, the image of failure that dogged the early missions in New South Wales – though not unusual in early colonial encounters – reinforced the notion that Indigenous people were, in the words of the Baptist missionary William Carey 'poor barbarous naked pagans'.[22] Aborigines' resistance to early missionary endeavours demonstrated not the original condition of humanity – fallen though it was in biblical terms – but humanity that had degenerated to 'the ultimate depths of human degradation'.[23]

By the end of the century, after 40 years of advancing white settlement and reports detailing abusive labour conditions, laws were passed in Queensland (1897), WA (1905) and the NT (1911) not only to try to protect Aborigines from exploitation but also to control them – to move them off their traditional lands, put an end to their resistance and to re-socialize their children. In the wake of these laws, the state invited the church to run missions and over the next 40 years 18 new mission stations were established by the main Protestant missionary societies and by the religious orders in the Catholic Church in Queensland, WA, the NT and SA.[24] Intended to become self-supporting, these missions had minimal funding from either the government or the church.They were not all the same and they are not remembered in the same ways. However, in most food and shelter were poor, children were separated from their parents and they operated according to a didactic regime that assumed Indigenous people to be a 'child race'.[25] In the wake of the 1997 Human Rights and Equal Opportunity Commission report, *Bringing Them Home*, the destructive consequences of this process have become widely known in Australia, but at the turn of the century missions were seen by both progressive churchmen and humanitarian politicians as the only way of protecting and saving the 'remnant' of Indigenous population that had survived in the north of Australia.[26]

Gilbert White and James Noble were among their most vocal and active supporters.

How did religion and empire shape the identities of these two men? Unsurprisingly, we know much more about Gilbert White than James Noble. Like many missionary prelates, White wrote a great deal – letters, diaries, sermons, memoirs – but he was more prolific than most. According to his sister and companion Lucy 'he wrote because he must sometimes even when travelling by train or otherwise'.[27] And he wrote outside the usual clerical mode – poems, travel books, an autobiography, lectures and political tracts. A powerful orator, he took every opportunity to deliver addresses on social and political issues – on missions, on White Australia, on race relations. James Noble, on the other hand, would seem to have written very little. However, as an ordained deacon of the Anglican Church he addressed large gatherings where his words were recorded; it is easier to hear his voice than that of his wife, Angelina. She too worked on the missions and she was a skilled linguist – acting as translator for police court hearings and at the Royal Commission on the Forrest River massacre in 1927. There is a poignant irony in the fact that we have almost no account of the words of a woman renowned for her linguistic skills.

Gilbert White and the limits of speaking out

White's writings reveal important elements shaping his identity and how these influenced his ideas on race, empire and religion. He was an energetic 'missionary bishop' – Carpentaria was one of the three missionary dioceses created at this time: British New Guinea was established in 1898 and North Western Australia in 1909.[28] With nearly three times as many non-whites as whites, it was a huge and unwieldy mass, encompassing the Torres Strait Islands, the Cape York Peninsula north of Cairns, the Gulf country and the whole of the Northern Territory.[29] White made 'marathon tours' of it regularly. He assisted the foundation of new missions at Roper River and the Mitchell River and became an outspoken defender of the rights of Aborigines and Pacific Islanders, one of very few churchmen to openly condemn the nature of northern colonization.[30] He told a Melbourne audience in 1907 that Indigenous people had had their land taken 'without process of law, and without any compensation given'; they were moved on to make way for cattle 'and moving on had too often meant death'.[31] With George Frodsham, Bishop of North Queensland, he was responsible for an Australian Board of Mission statement in 1903 which recognised Aborigines as the 'original owners and inhabitants of the country in which we dwell, whose land we have taken and whose means of subsistence we are daily diminishing.'[32] In sheeting home the blame for Aboriginal demise to white colonists White rejected a blanket application of Social Darwinism – 'the gathering force' in intellectual life.[33] He denied that Aborigines were 'one of the lowest human types' and disputed that they were a dying race. Evidence from missions such as Yarrabah, he insisted, showed that they were capable of 'advancement' and that far from dying out, they were increasing.[34] White also condemned the racism inherent in Australia's immigration restriction policy. In 1905, he moved a motion in General Synod that 'it is unreasonable to assume that the white man is, necessarily and inherently, superior to every race of another colour' and he also opposed the deportation of Pacific Islanders under the Pacific Islanders Act of 1905.[35]

While White saw himself as a defender of Aborigines and Pacific Islanders and an opponent of race superiority, he did not deny underlying assumptions of race difference. He disputed that Aborigines were on the lowest rung of the race hierarchy, but in keeping with the evolutionary thinking of his time, did not dispute the validity of such a hierarchy. In *Thirty Years in Tropical Australia* he wrote that the Aborigines 'do not fall far below the average standard of uncivilised humanity'.[36] And he differentiated the 'quality' of Indigenous groups across the continent. Those near Pine Creek, were 'finer made' than those in Queensland; those who lived near the Gulf of Carpentaria were 'probably the least intelligent tribe in Australia', while Torres Strait Islanders were 'physically, mentally, and spiritually... altogether of a higher type than the mainland aborigines'.[37] Similarly, while he thought the race superiority of the Immigration Restriction Act was repellent he gave considerably more space in his speeches to the fact that the Empty North was a 'standing danger' to Australia and that it had to be 'occupied in some way or another' if Australia was to retain it. His solution was to import more indentured labour from one of 'Britain's eastern possessions'. In combination with a mission system that would transform Aborigines into a self-sufficient and self-determining workforce, these workers would populate the north, protect its coastline and make the land productive – all of which would be far less expensive than subsidizing white settlement.[38] In some respects, his vision for these communities was ahead of its time: he acknowledged the importance of land to Aborigines – they would know that 'the territory on which they lived was their own' and they would not be moved from it; and the very fact that he saw these stations as eventually self-determining shows that he regarded Aborigines as 'potentially equal citizens'.[39] But it was grounded in a Christian paternalist position that assumed whiteness to be superior. 'The less advanced races' should not be excluded from the polity for it was 'a Christian duty to afford them such protection and education as is due from the elder to the younger brethren in the great human family of God.'[40]

An Anglo-Catholic in a region where the Anglican Dioceses had evolved in an Anglo-Catholic direction, White's outspokenness on Indigenous issues reflected the incarnational theology that distinguished his style of churchmanship.[41] Just as Anglo-Catholics felt impelled to work among the poor in the slums of England, White's priority was the dispossessed in north Australia.[42] His sympathy for Aborigines was also likely to have been influenced by his sexuality. His writing shows an attraction to Indigenous masculine beauty suggestive of a homoerotic sensibility, with which Anglo-Catholicism in England was also associated.[43] I have explored elsewhere the complexities of these elements of White's character.[44] I would like to explore here how White's views also reflected his identity as a man from the centre of Empire.

One of a long line of clergymen stretching back to the well known naturalist Gilbert White of Selbourne, he was born in South Africa, grew up in England and moved to Australia for the sake of his health in 1885, aged 26. He lived in Australia until his death in 1933 and he found much to admire in his 'adopted country'.[45] Much of his autobiography, *Thirty Years in Tropical Australia*, is travel writing, marked by careful observations of the changing countryside and delighted descriptions of highly coloured flowers. He found spiritual enrichment in the landscape: 'How wonderfully real the Psalms seem when read out in the Australian bush' he wrote after a long description of the country 200 miles west of Townsville.[46] But England remained the yardstick of all things civilized, its history 'an amazing record' of the spread of Christianity.[47] His poetry, in particular, was critical of Australian materialism and greed.[48] It is not surprising, then that the theme of colonial degeneracy – common in missionary writing – surfaced in White's critiques of Australian race relations.[49] In an article published in 1910 in an English missionary journal he argued that Australia's record of ill treatment of Aborigines undermined Britain's 'remonstrances on the long continued and ghastly atrocities in the Congo'. How could England 'obtain redress for the unhappy victims of red rubber' when Congo officials could point to newspaper reports of ill treatment of Aborigines in Western Australia?[50]

White's 'Englishness' strengthened his authority in addressing colonials.[51] Like many middle class English-born clergy, White's identity derived in large part from his dedication to work.[52] And this colonial context enabled White to counter the possibility of effeminacy not only through work but also through adventure. He pushed himself to keep moving around his huge Diocese, travelling 'rough', and regularly sailing the Torres Strait in bad weather, dodging the treacherous hidden coral, bearing the miseries of sea-sickness. But rather than showing the overblown athleticism that historians have identified as symptomatic of underlying shifts in Imperial masculinity in this period, White's style was quietly heroic, austere and self-denying.[53] He lived very frugally and carried not an ounce of Episcopal fat. One of the roughest sea passages was to Friday Island, which he coursed regularly to visit the coloured leprosy sufferers who were banished there by the Queensland government after 1892 – one of very few churchmen to do so.[54] After his death the new Bishop at Carpentaria wanted to erect a tower costing £5000 on the Cathedral on Thursday Island as a memorial but his family objected that such expenditure would have been repellent to him as his primary concern was 'the welfare of the native races'.[55]

White was a man of Empire but he saw himself as primarily building the Kingdom of God on earth, a Kingdom that would, at the end of time, eradicate all difference. This was the fundamental source of his commitment to hard work, to the 'welfare of the native races' and to 'speaking out'. But there were limits to White's outspokenness. Delineating those limits allows us to trace the tensions inherent in his negotiation of the conflicting imperatives of God's Kingdom and those of the British Empire. The incident which illustrates this most clearly centres on White's involvement in the discovery of murdered Aborigines at Moreton Telegraph Station in 1902, described in *Thirty Years in Tropical Australia*. He had been visiting the mission at Mapoon with Dr Walter Roth, the northern Protector of Aboriginals, when they got word of the murder of some Aborigines a few weeks earlier.[56] He and Roth were taken by Aboriginal guides to the remains of a big fire 'evidently made by a white man, as the natives never make a fire of big logs lighted in the centre'. There they found several knee-caps, other human bones and two skulls. 'Under one of the skulls was a little lump of lead of the exact weight of the bullets which had been supplied, as we knew, to the assailants, a large and unusual size.' They had enough evidence to induce William Parry-Okeden, the Commissioner of Police, to come from Brisbane to investigate and a police inquiry was held. No one was convicted.[57] A white policeman resigned and two black troopers were 'sent away'. It was a common pattern. A study by Jonathan Richards and Mark Finnane of coronial inquiries in Queensland between 1859 and 1900 shows that attacks on whites usually resulted in an official inquiry and serious charges, but that violent deaths of Aborigines or Torres Strait Islanders were either not investigated at all or, as in this case, no charges were laid.[58]

What is important here is how White wrote about this incident in his autobiography. He leaves the reader with no doubt that murder was committed and that whites were involved. But he omits the details of the outcome, assuring his readers instead that 'such justice *as was possible* was ... done' (my italics). This is a telling phrasing, but instead of explaining what happened he moves on to a lengthy general comment about frontier violence, the main thrust of which is to exonerate the Queensland government from any charge of neglecting its duty. He writes that he had always found the government 'most anxious on all occasions to do impartial justice to the natives'. He acknowledges that undoubtedly Aborigines had suffered 'cruel wrongs' but these were owing to vast distances, the impossibility of getting evidence and, worst of all, the jury system: 'it was almost impossible to get a white jury to convict a white man of any outrage, however flagrant, on an aboriginal.' What stands out from this analysis is that it does not apply in this case. There was evidence. And White himself gave a sworn statement to the inquiry of what he had seen and of how it tallied with what the Aboriginal informants had told him.[59] Further, there was no jury in this case. The inquiry had been convened by William Parry-Okeden, Commissioner of Police and Chief Protector of Aborigines for the state of Queensland. It was he who found not enough evidence to prosecute, not a jury.

It is difficult to know precisely why White went to such trouble to exonerate the government. He certainly knew that the government had to be onside for missionary work to succeed – so he didn't want to give offence.[60] But why initiate such a lengthy defence of 'the government'? Perhaps he was thinking, not of Parry-Okeden but of his friend Walter Roth, the northern protector with whom he had been travelling. Roth was a 'rigid enforcer' of the 1897 Protection Act, hated by the squatting interests in Queensland parliament for interfering with their labour supply and in this case he had issued a report naming the police as murderers. The memory of Roth's persistent and tenacious opposition to the exploitation of Aboriginal labour and his efforts to have their assailants brought to justice – until he was driven from Queensland in 1906 – may have underpinned White's blanket exoneration of 'the Queensland government'.[61]

Whatever the complexities of his motivation, White's decision to acquit the government of blame and his inability to articulate the injustice of the outcome is one small but telling instance of how an outspoken critic of colonialism came to comply with its ends. His was not a 'polite sliding over the cruel and gruesome acts' of colonization. Indeed it is important to acknowledge his unusual courage in naming those acts. But it was a 'polite sliding over' the inadequacy of the rule of law in this outpost of Empire.[62] The respect he commanded as an Anglican Bishop, one not afraid to condemn frontier violence, gave his endorsement of the rule of law even greater authority. That endorsement, when the law had so clearly failed, is a potent example of 'white noise' – in Tom Griffith's words, 'an obscuring and overlaying din of history making'. If it was only half-conscious, it was more deeply embedded for that.[63] Both noisy and silent, the long history of denial at the core of Australian history is still fought out in the pages of the conservative Australian press.

James Noble and the hope of equality

The only clergyman in this context who loudly demanded that the rule of law be applied against whites was Ernest Gribble. Described by his biographer Christine Halse as 'a terribly wild man' he was erratic, self-righteous and authoritarian and was sacked by the Anglican Board of Mission (ABM) from both Yarrabah and Forrest River Missions. His edginess barred him from high office but gave him the drive to ensure, for example, that a Royal Commission was held into the massacre at Forrest River in 1927.[64] There is no space here to do justice to his complex psychology but it is significant that it was Gribble who was the advocate and mentor of the Indigenous deacon James Noble. Gilbert White admired Noble and their lives intersected at a number of points, but Noble lived on Gribble's missions for most of his adult life, following him to Forrest River in 1914 and to Palm Island in 1933. It was Gribble who most consistently represented to him the face of Christianity. Gribble knew the pain of exclusion as Noble did and Noble's attachment to Gribble was strong. James Noble's life history provides insights into how religion and empires shaped the experience of an Indigenous man.

Little is known about James Noble's early life. He was born near Normanton, the largest town in the Gulf country, in the mid-1870s – a time when settlers were establishing cattle stations there. A recent study by Tony Roberts has meticulously researched the violence and its cover-up that accompanied this process.[65] Presumably Noble had seen or knew of white men at their worst. But he worked on cattle stations in Queensland and New South Wales for good employers. Later he was to recall that he had been introduced to Christianity by 'one young fellow I was droving for' who was 'a good Christian man...he gave me a Catechism and taught me the Creed'.[66] The impression is of a comradely relationship between young men, one that Noble valued. It is hardly surprising that in a treacherous colonial environment, Noble was receptive to the religious ideas of one who offered him work and education and treated him fairly.

While working for the Doyle brothers Noble was baptized and confirmed in the Church of England. Churchmen intervened in his life at various points after this. He worked for Rev. Edwards, at Hughenden. Then, in the mid-1890s, he 'drifted' into the hands of 'some sporting gentlemen' who discovered he was a fine sprinter. Around this time, according to his son John, he had a vision: 'an old man came up to him and asked him if he could help Revd Gribble to open Mitchell River... Dad, he thinks he was God, you know ... He gave up everything, smoking, swearing, all those ways'[67] The Bishop of North Queensland arranged for him to go to Yarrabah mission, the sort of intervention that was not unusual on the eve of the 'protection' era. At Yarrabah he met Angelina who was brought there by police after being abducted as a child by a horse dealer. They lived on the mission until 1910, when they helped establish the mission at Roper River and in 1914 they joined Gribble in founding the Forrest River mission in WA, where they lived until 1932. They had six children. In both Roper River and Forrest River their liasing with the local tribes, as well as their manual work, were crucial for the foundation of the missions.[68]

It is difficult to get a close understanding of how the particular dynamics of Noble's ancestral religion might have oriented him towards Christianity. Nor is it clear to what extent his Christianity existed with traditional beliefs. Such issues of religious change have been considered by historians and anthropologists in many parts of Australia.[69] But Noble's recorded words are few. They suggest that in a fractured colonial environment, Noble saw the missionaries as 'kind' to black people, providing them with 'a good home'.[70] To him, their message that God loved all equally offered comfort. As Noble put it 'Black people begin to sing to the same God and love the same God.' The fact that God was all-powerful was particularly attractive. God would make converts, Noble said, 'not in missionaries time, but in His own time'.[71] The rituals and ceremonies of High Church Anglican Christianity were tangible celebrations of oneness. They would seem to have spoken to him of beauty as well as solace. He described the Easter ceremonies at Forrest River to Rev. Gribble in 1932:

> On the first evensong of Easter we had a lovely service after we have finished the prayed we had a little procession round the church and we sang the hymn 'The churches one Foundation. and we into the church again and formed the Holy Cross in the middle of the church floor And on Saturday the girls brought home armfuls of lilies. I decorated the church very lovely we had a very happy Easter. [sic][72]

Ancestral religions valued particular places. Given the disruptions of colonialism, a God who transcended place must have had considerable appeal. And Noble made a number of moves over the course of his life. When he left Yarrabah for the Roper River he attested that he was impelled by the transcendent Christian God to take his message to those who had not heard it. The *Aboriginal News* reported the 'grand services' held before they left where James 'spoke of his love for Yarrabah.... He knew that God had called him to leave his beautiful home – the happiest life he had ever had – to go and help those blacks who were wild, wild as he himself had been for many years'. In Christian missionary thinking the imperative to leave one's home was compensated by belief in 'another and better home' as Ernest Gribble put it, 'where we might be reunited when we have done our work for God'.[73] For an Indigenous missionary, whose deep connections to home and country had been disrupted, such hopes must have assumed great significance.

At Yarrabah and later at Forrest River, Noble had considerable authority even though Ernest Gribble ruled both missions with an erratically authoritarian hand. Many Aborigines from Forrest River considered Noble the 'boss' of the mission.[74] Tall and strong, he taught the men how to make mud bricks and cut mangrove poles to build the houses and church; he trained the stockmen, acted as barber, dentist and doctor, assisted Gribble at Sunday services.[75] He saw his authority as biblical, telling the North Queensland synod in 1911 that after his lessons with the Doyle family 'I never got the Creed out of my head.... The words of the Bible were always in my mind, "Go and teach others"'.[76] As a lay reader, Noble was entitled to read the lessons and to preach. By the time of his ordination to the diaconate in 1925, he was conscious of the vital role he had in representing the Aborigines of Forrest River to the white world. 'They have no black people to stand and speak for them when I die.'[77]

There were, however, clear limits to his authority. Despite Ernest Gribble's attempts to have him ordained to the priesthood, he was never ordained.[78] And Gribble's inclusiveness had its limits. He wrote in 1929 that Noble had been 'as a son to me for over thirty years' – but he was not a son.[79] When Jack Gribble, Ernest's son, joined him at Forrest River after the war, he took over most of the management tasks that had been Noble's.[80] And when Gribble was desperate for someone to take over Forrest River, he was convinced it could only be a member of his family; he did not even consider Noble.[81] More basically, James and Angelina Noble endured the poverty of the missions they lived on. Angelina did all the domestic work on the missions, no teaching, despite her manifest gifts.

It is hard to know how difficult it was for Noble to live in two worlds. He remained a missionary to his people all his life and he would seem to have seen himself having made a clear separation from traditional customs. 'When I left my native habits and went to school' he told the Royal Commission into the Forrest River massacre in 1927, 'I was 20 years old.'[82] But there were times when he wanted to return. Once when he crossed swords with Gribble at Forrest River he threatened to 'go back to his own country'.[83] On another occasion he said that only God had made him 'stick' to the mission, implying that it had not been easy to do so.[84]

James and Angelina were not the only Indigenous Christian leaders at Yarrabah but they are the ones who appear most consistently in the missionary records and they came to symbolize to white Australians the best that missions could achieve. By the 1920s James in particular was admired in Anglican Church circles. He spent most of the year of his ordination on a missionary tour of Australia where he spoke in 'numerous churches, several cathedrals' and at 'a large missionary meeting' in Melbourne, where 'as he came on the platform, the whole audience, the bishops, and speakers on the platform, rose as one man in homage'.[85] Tall and handsome, his demeanour was imposing but he posed no threat. *The Daily News* commented approvingly on his 'shy reserve' and 'entire absence of swagger'.[86] At a time when Indigenous people in Melbourne and Sydney were joining together to demand land, the end to the removal of their children and the right to welfare benefits, James Noble seemed to demand nothing.[87] Indeed within the mission he participated in the punitive and coercive practices that were intrinsic to it.

At Forrest River, he rode out after runaways and, in the words of historian Neville Green, 'when the mission court found a prisoner guilty, it usually fell to James to wield the strap'.[88] Cast in this light, he is like an agent of the white missionaries – less brutal, but fulfilling a function within the imperial project not unlike that of the native police.

But Noble's views were not identical to those of the white missionaries and he was not entirely uncomplaining. His address to the Australian Church Congress in 1925 is the best insight that we have into his views; it betrays deep ambivalence. Its overarching idea is of Aboriginal demise. It begins with the words: 'I am very lone. I am the only black people left. When I speak like this you know what I mean.' [sic] This situation, he insists, is 'not your fault' but the logic of his perspective is one of reproach. He wishes that they could go back two or three hundred years when there were plenty of blacks 'and we might have saved them': but it is now too late. Of the idea that black people should be left alone, he says, 'Why didn't they say that long ago when they were all over this country.' To the idea of letting black people rule over themselves he responds 'Too late, now!' for there are only a few left, but then he adds 'I hope they can rule over themselves.' Black people, he said, wanted both more land and more churches. They needed missions, 'because they have no country of their own'. They are thankful to the government who gave the land at Forrest River – 'the government is good, the government does its best' – but black people want more land, 'to run about in, in the bush.' It is a speech that fluctuates between resignation and request, between gratitude and reproach.

Noble's address also shows that he felt a solidarity with the mission and its missionaries but that this was felt as strongly against the rest of the church as against society generally. Presumably thinking of Ernest Gribble, he saw the missionaries identifying with Indigenous people – he was unequivocal about their goodness. The missionaries, he said, were kind to black people: 'A missionary is a great picture to black people; the way he lives; the character he has; and always ready to face hardship.' They did not have good food to eat. 'At my good breakfast this morning with big bishops and men, I thought of missionary at home with only cup of tea sometimes.' In his view, the essence of conversion lay in missionaries modelling right behaviour: the missionaries 'teach the same as they work'; 'by loving him the missionary teaches people to love God, too.' But his respect for the missionaries did not mean that his vision of what the mission should be was identical with theirs. Black people, he said, want more land 'to run about in, in the bush.' His concept of mission was one where there was mobility, access to land and freedom. And they only needed the missions because 'they have no country of their own'.

Conclusion

In his recent study of missionaries in the Punjab, *Imperial Faultlines,* Jeffrey Cox constructs a typology of meta-narratives underlying missionary history: in addition to the established tradition of imperial history that marginalizes missionaries, newer work influenced by Edward Said seeks to unmask missionaries as always and everywhere utterly complicit in the ambitions of the imperial power while the school of mission studies, though characterized by critical analysis and drawn to ideas of non-Western agency, nevertheless relies on a 'providentialist master narrative of progress towards a multiracial Christian community' which has difficulty addressing the Saidian accusation of complicity 'head on'. Cox argues that rather than trying to make a quantitative assessment of the extent to which racism directed missionary activity, it is more productive to accept that many missionaries were racist and anti-racist at the same time.[89] Gilbert White can best be understood in this way. One of the most influential churchmen to defend Indigenous rights, White denied that 'the white man was superior to every other race of another colour' but was untroubled by the existence of a hierarchy of difference that allowed such judgements to be made. His belief in the equality of souls led him to denounce the cruelties of colonialism and to express respect for its victims, but not to denounce the failures of the rule of law: the anonymous 'colonist' or 'jurist' was a relatively easy target. To name government officials as negligent could well have jeopardized the missionary venture that relied on government funding – however paltry – to survive. And White believed in missions. He saw them as providing physical protection and Christian conversion – he did not see them as entrenching Indigenous poverty or severing ties of kinship. And it would seem he saw no connection between the need for 'protective' missions and the failure of the rule of law. The fact that missions 'advanced the cause of empire' by containing Indigenous resistance and re-socializing children may not have been his primary concern but it was a fortunate by-product, confirming his belief that English history was 'an amazing record' of the spread of Christianity.

It is far more difficult to know how things looked from James Noble's perspective. But the fullest surviving testimony we have suggests that he believed in missions too. For him they had saved Aboriginal people.

They introduced them to a Christian God who transcended place, loved all equally and was all-powerful. And he believed in white missionaries: he felt a strong affinity with them and saw a clear distinction between them and the rest of the church and society. But Noble's vision of the ideal mission was not identical to that of white missionaries. He wanted land and freedom and his very articulation of that vision stemmed from his conviction of the spiritual equality he shared with them. It would be misleading to read Noble's words as representing an archetypal Indigenous Christian voice, but his is the one that remains most clearly articulated. James Noble's words suggest that hope for equality was not extinguished. Empires rule by affective as well as material conditions – and they can never entirely control the affective.

Notes

1. T. Ballantyne, Religion, 'Difference, and the Limits of British Imperial History', *Victorian Studies*, XLVII, no. 3 (Spring 2005) 431; for discussions of the historiography of Australian religion see J. D. Bollen et al., 'Australian Religious History', *Journal of Religious History*, XI (1980) 8–44, H. M. Carey et al., 'Australian Religion Review', *Journal of Religious History*, XXIV (2000) 296–313 and XXV (2001) 56–82.
2. M. Hogan, *The Sectarian Strand: Religion in Australian History* (Ringwood: Penguin, 1987).
3. For a discussion of Australian missionary historiography see, Carey, 'Australian Religion Review', pp. 307–10.
4. A. Curthoys, 'We've just Started Making National Histories, and You Want Us to Stop Already?', in *After the Imperial Turn: Thinking with and through the Nation*, ed A. Burton (Durham and London: Duke University Press, 2003), pp. 70–89.
5. 'Australian Religion Review', pp. 302–6.
6. A. N Porter, *Religion versus Empire? British Protestant Missionaries and Overseas Expansion, 1770–1914* (Manchester and New York: Manchester University Press, 2004), p. 24.
7. W. E. H. Stanner, *After the Dreaming; Black and White Australians – An Anthropologist's View* (Sydney: Australian Broadcasting Commission, 1969).
8. J. Carroll, *Intruders in the Bush: The Australian Quest for Identity* (Melbourne: Oxford University Press, 1982).
9. For an introduction to the history and politics of writing indigenous history see *Contested Ground: Australian Aborigines under the British Crown*, ed. A. McGrath (Sydney: Allen & Unwin, 1995), ch. 10, pp. 168–207.
10. K. Windschuttle, *The Fabrication of Aboriginal History* (Sydney: Macleay Press, 2002); for an account of this conflict see S. Macintyre and A. Clark, *The History Wars* (Melbourne: Melbourne University Press, 2003).
11. R. Teale, 'White, Gilbert', *Australian Dictionary of Biography*, Vol. 12, (Melbourne: Melbourne University Press, 1990), p. 466; J. Kociumbas, 'Noble, James', *Australian Dictionary of Biography*, Vol. 11 (Melbourne: Melbourne University Press, 1988), pp. 32–3.

12. *Race, Nation, and Religion in the Americas*, ed. H. Goldschmidt and E. McAlister (Oxford: Oxford University Press, 2004), p. 132.
13. C. Hall, *Civilising Subjects: Colony and Metropole in the English Imagination, 1830–1867* (Chicago, IL and London: University of Chicago Press, 2002), p. 16.
14. For Indigenous Christian leaders elsewhere in Australia at this time including David Unaipon, Eddie Atkinson, Bobbie Peters, Moses Tjalkabota, see J. Harris, *One Blood: 200 Years of Aboriginal Encounter with Christianity* (Sydney: Albatross Books, 1994).
15. H. Reynolds, *This Whispering in Our Hearts* (Sydney: Allen & Unwin, 1998), pp. 141–2.
16. H. Reynolds and D. May, 'Queensland', in *Contested Ground: Australian Aborigines under the British Crown*, ed. Ann McGrath (Sydney: Allen & Unwin, 1995).
17. H. Reynolds, *North of Capricorn* (Sydney: Allen & Unwin, 2003), pp. xv–xvi.
18. K. Ellinghaus, 'Absorbing the "Aboriginal Problem": Controlling Interracial Marriage in Australia in the Late 19th and Early 20th Centuries', *Aboriginal History*, XXVII (2003) 183–207.
19. D. B. Rose, *Nourishing Terrains: Australian Aboriginal Views of Landscape and Wilderness* (Canberra: Australian Heritage Commission, 1996); R. Ganter, *Mixed Relations: Asian-Aboriginal Contact in Northern Australia* (Perth: University of Western Australia Press, 2006).
20. *Griffith Review*, no. 9, p. 230.
21. R. M. Frappell, 'The Australian Bush Brotherhoods and their English Origins' *The Journal of Ecclesiastical History*, XLVII, no. 1 (1996) 82–98.
22. H. M. Carey, 'Attempts and Attempts': Responses to Failure in Pre- and Early Victorian Missions to the Australian Aborigines', in *Mapping The landscape. Essays in Australian and New Zealand Christianity*, S. and W. Emilsen (New York: Peter Lang, 2000); Harris, *One Blood*, p. 32.
23. R. McGregor, *Aboriginal Australians and the Doomed Race Theory, 1880–1939* (Melbourne: Melbourne University Press, 1997), p. 9.
24. A. O'Brien, *God's Willing Workers: Women and Religion in Australia* (Sydney, University of New South Wales Press, 2005), p. 120.
25. A. Haebich, *Broken Circles: Fragmenting Indigenous Families 1800–2000* (Perth: Fremantle Arts Centre Press, 2000), ch. 6.
26. Oral testimony given to the 1997 commission provides the basis of *Many Voices: Reflections on Experiences of Indigenous Child Removal*, ed. D. Mellor and A. Haebich (Canberra: National Library of Australia, 2002).
27. Memoirs of Gilbert White, Early recollections, p. 102 (typescript) ML MSS Australian Board of Mission. Gilbert White, papers, Box JJ, 36. M77.
28. R. Frappell, 'Imperial Fervour and Anglican Loyalty, 1900–1929' in *Anglicanism in Australia: A History*, ed. B. B. Kaye (Melbourne: Melbourne University Press, 2002), pp. 76–99.
29. J. Bayton, *Cross Over Carpentaria* (Brisbane: Smith and Paterson, 1965), p. 83; D. Wetherell, 'The Anglicans in New Guinea and the Torres Strait Islands', in *Vision and Reality in Pacific Religion*, ed. P. Herda et al. (Canberra: Pandanus Books, 2005), pp. 216–17.
30. Bayton, *Cross Over Carpentaria*, p. 77; Others include Matthew Hale (1811–1895), Bishop of Brisbane, George Frodsham (1863–1937) Bishop of North Queensland and the Jesuit missionary, Donald McKillop (1853–1925).

31. G. White, *Some Problems of Northern Australia* (Melbourne: Urquart & Nicholson, 1907), p. 11.
32. Cited in N. Loos, *White Christ Black Cross: The Emergence of a Black Church*, (Canberra: Aboriginal Studies Press, 2007), p. 64.
33. H. Goodall, *Invasion to Embassy* (Sydney: Allen & Unwin, 1996), p. 104.
34. G. White, *Answer Australia* (Sydney: Australian Board of Missions, 1927), pp. 17, 20.
35. 'Motion to be moved in General Synod by the Bishop of Carpentaria', with letter from Gilbert White to Alfred Deakin, Papers of Alfred Deakin, National Library of Australia, Series 15.5.1.1.1. Item 15/2095–6; Bayton, *Cross over Carpentaria*, p. 107.
36. White, *Thirty Years in Tropical Australia* (London: Society for Promoting Christian Knowledge, 1918), p. 158.
37. White, *Thirty Years*, pp. 72, 117, 162; White, *Answer Australia*, p. 26; White's views on race hierarchies were not restricted to the non-European 'other'. On one of his journeys he shared a coach with a 'fat German Jew' who, according to White, made himself 'as unpleasant as his tribe usually manage with the best intentions to do'. White, *Thirty Years*, p. 17.
38. White, *Some Problems*, pp. 9–13.
39. See Loos, *White Christ Black Cross*, pp. 67–8, for a discussion of White's report for the ABM's Aboriginal sub-committee of 1919 where he outlined these views which became ABM policy, pp. 67–8.
40. 'Motion to be moved in General Synod'.
41. B. Lawton, 'Australian Anglican Theology' in *Anglicanism in Australia: A History*, ed. B. Kaye (Melbourne, Melbourne University Press, 2002), pp. 180–1.
42. J. Shelton. S. Reed, 'Ritualism rampant in East London: Anglo-Catholicism and the Urban Poor', *Victorian Studies*, XXXI, no. 3 (Spring 1988) 375–403.
43. D. Hilliard, 'UnEnglish and Unmanly: Anglo-Catholicism and Homosexuality', *Victorian Studies*, XXV, no. 2 (1982) 181–200.
44. A. O'Brien, 'Missionary Masculinities, the Homoerotic Gaze and the Politics of Race: Gilbert White in Northern Australia', *Gender and History*, XX, no. 1 (April 2008) 68–85.
45. White, *Some Problems*, p. 1.
46. G. White, *Thirty Years*, p. 41.
47. White, *Answer Australia*, p. 9.
48. See in particular, 'Australia 1913' in *The Poems of Gilbert White* (London: Society for Promoting Christian Knowledge, 1919), pp. 46–7.
49. A. Johnston, *Missionary writing and Empire, 1800–1860* (Cambridge and New York: Cambridge University Press, 2003), pp. 187–99.
50. G. White, 'The Relation of Missionary Effort to Social Life from an Australian Standpoint' *The East and the West* VIII, no. 30 (April 1910) 142.
51. M. Lake, 'On Being a White Man, Australia, circa 1900' in *Cultural History in Australia*, ed. Hsu-Ming Teo and Richard White (Sydney: UNSW Press, 2003), pp. 98–112.
52. L. Davidoff and Catherine Hall, *Family Fortunes: Men and Women of the English Middle Class, 1780–1850* (Chicago, IL: University of Chicago Press, 1987), pp. 111–12.
53. O'Brien, 'Missionary Masculinities', pp. 71–2.

54. For a full discussion of White's relationships with the leprosy sufferers on Friday Island, see A. O'Brien, 'All Creatures of the Living God': Leprosy and Religion in Turn of the Century Queensland, *History Australia* 5.V, no. 2 (December 2008).
55. Draft letter (unsigned) to Bishop of Salisbury, 18 January 1934, White, papers, Box JJ, 36. M77.
56. White's account of this incident is in *Thirty Years*, pp. 109–13.
57. J. Richards, 'Moreton Telegraph Station 1902: The Native Police on Cape York Peninsular', in *Policing the Lucky Country* (eds), ed. M. Enders and B. Dupoint (Sydney: Hawkins Press, 2001) p. 104.
58. M. Finnane and J. Richards, 'You'll Get Nothing Out of It'? The inquest, Police and Aboriginal deaths in Colonial Queensland', *Australian Historical Studies*, no. 123 (April 2004) 84–105.
59. Sworn statement by the Bishop of Carpentaria, Queensland State Archives, A/58850.
60. White, *Some Problems*, p. 14.
61. Richards, 'Moreton Telegraph Station' pp. 99–100; John Whitehall, 'Dr W E Roth: Flawed Force of the Frontier', *Journal of Australian Studies*, no. 75 (2002) 59–69.
62. P. Grimshaw, 'The Fabrication of a Benign Colonisation? Keith Windschuttle on History', *Australian Historical Studies*, no 123 (2004) 122.
63. T. Griffiths, 'The Language of Conflict' in *Frontier Conflict: The Australian Experience* (Canberra: National Museum of Australia, 2003), p. 138.
64. Chapter 6 of Loos, *White Christ, Black Cross* is a meticulous and sensitive discussion of this massacre and its role in stimulating Australia's 'history wars'.
65. T. Roberts, *Frontier Justice: A History of the Gulf Country to 1900* (Brisbane: University of Queensland Press, 2005).
66. *ABM Review* (1 November 1911) 138.
67. *National Boomerang* (August 1986) 2.
68. G. Higgins, *James Noble of Yarrabah* (Lawson: Missions Publications of Australia, 1981), p. 4; Kociumbas, 'Noble, James'.
69. *Indigenous Peoples and Religious Change*, ed. P. Brock (Leiden and Boston, MA: Brill, 2005).
70. *Official Report of the Ninth Australian Church Congress* (Melbourne: Congress Committee, Diocesan Registry, 1925), p. 329.
71. *Official Report of the Ninth Australian Church Congress*, p. 329.
72. Letter from James Noble to Ernest Gribble, 31 March 1932. E R Gribble papers, ML MSS 4503, Add on 1822, Box 9, 11/8.
73. Higgins, *James Noble*, pp. 19–20.
74. C. M. Halse, 'The Reverend Ernest Gribble and Race Relations in Northern Australia' (PhD thesis, University of Queensland, 1992), p. 228.
75. N. Green, *The Forrest River Massacres* (Perth: Fremantle Arts Centre Press, 1995), p. 108.
76. *ABM Review*, 1 November 1911, 138.
77. *Official Report of the Ninth Australian Church Congress*, p. 329.
78. E. R. Gribble, *A Despised Race: The Vanishing Aboriginals of Australia* (Sydney: Australian Board of Missions, 1933), p. 133.

79. Letter from Ernest Gribble to Ernest C Mitchell, Inspector of Aboriginals, 3 May 1929, ML MSS 4503 Add on 1822, Box 10 G9 14/3.
80. Green, *The Forrest River Massacres*, p. 118.
81. C. Halse, *A Terribly Wild Man* (Sydney: Allen and Unwin, 2002), p. 189.
82. *Royal Commission of Inquiry into the Alleged Killing of Burning of Bodies of Aborigines in East Kimberley and into Police Methods when affecting arrests*. WA V&P, 1, 1927, no. 3, p. 40.
83. Halse, Rev. Ernest Gribble, p. 363.
84. Higgins, *James Noble*, p. 48.
85. Gribble, *A Despised Race*, p. 131.
86. *Daily News*, 30 January 1925.
87. See Goodall, *Invasion to Embassy*, chs. 12–18.
88. Green, *The Forrest River Massacres*, p. 108.
89. J. Cox, *Imperial Fault Lines: Christianity and Colonial Power in India, 1818–1940* (Stanford: Stanford University Press, 2002), pp. 7–19.

Further reading

Brock, P., ed. *Indigenous Peoples and Religious Change* (Leiden and Boston, MA: Brill, 2005).

Harris, J. *One Blood: 200 Years of Aboriginal Encounter with Christianity* (Sydney: Albatross Books, 1994).

Kaye, B., ed. *Anglicanism in Australia: A History* (Melbourne: Melbourne University Press, 2002).

Loos, N. *White Christ Black Cross: The Emergence of a Black Church* (Canberra: Aboriginal Studies Press, 2007).

McGrath, A., ed. *Contested Ground: Australian Aborigines under the British Crown* (Sydney: Allen & Unwin, 1995).

McGregor, R. *Aboriginal Australians and the Doomed Race Theory, 1880–1939* (Melbourne: Melbourne University Press, 1997).

O'Brien, A. *God's Willing Workers: Women and Religion in Australia* (Sydney: University of New South Wales Press, 2005).

Reynolds, H. *This Whispering in our Hearts* (Sydney: Allen & Unwin, 1998).

Thomson, J. *Reaching Back: Queensland Aboriginal People Recall Early Days at Yarrabah Mission* (Canberra: Aboriginal Studies Press, 1989).

Part II

Colonies and Mission Fields

Friends of the Native? Universalism and Its Limits

10

'The Sharer of My Joys and Sorrows': Alison Blyth, Missionary Labours and Female Perspectives on Slavery in Nineteenth-Century Jamaica

John McAleer

During the eighteenth century, Britain became the leading slave-trading country in the world, and her American and Caribbean colonies relied on slave labour for their economic prosperity. However, towards the end of the century, opposition to the slave trade grew, based on a combination of religious conviction, humanitarian concern, economic self-interest and strategic political manoeuvring. In 1807, Britain introduced an empire-wide ban on the slave trade. However, slavery continued in British colonies and only ended after the House of Commons passed an Emancipation Act in 1833, which nominally 'freed' some 750,000 slaves throughout the British Empire on 1 August 1834. Even then, full emancipation was only achieved in 1838, following the collapse of the 'apprenticeship' system of unpaid labour that had been instituted to bridge the transition from slave to free labour.[1]

Lowell Ragatz referred to one of the best-known journals of this period as giving 'an utterly inimitable and imperishable picture of planter society' in nineteenth-century Jamaica.[2] While his critical approval was reserved for the celebrated account by Lady Nugent, the journal of Mrs Alison Blyth, the subject of this chapter, provides a similarly unique insight into European life and politics in Jamaica in the early nineteenth century when Britain's relationship with colonial slavery was under intense scrutiny and debate. Both texts provide a female perspective on the social, economic and political milieux of the

Caribbean island in this period. If not exactly contemporary, they are certainly complementary in terms of the range of subjects on which the authors remark, the depth of their insights and the scope of their commentary.[3] But, while Lady Nugent wrote at a time when Jamaica and its economy was still one of the most valuable possessions in Britain's burgeoning empire, Mrs Blyth's personal diary entries focus on missionary activity among the enslaved, the agitation for reform and the social tensions that preceded the abolition of the institution of slavery. Mrs Blyth recorded her impressions of life and society in Jamaica in the form of lucid and articulate commentary kept in a personal journal. This chapter examines the contents of that diary, the place where Mrs Blyth detailed her views and insights on the island that she and her husband would make their home for 25 years. The diary provides evidence, in the form of personal testimony and transcribed dialogues, that simultaneously supports and problematizes the received view of Jamaica at this time. Furthermore, Mrs Blyth's journal provides telling insights into the fractious nature of missionary endeavour in the colonies, and how this interacted with anti-slavery rhetoric and campaigning at a crucial time in the British anti-slavery movement.

This chapter outlines how many of Mrs Blyth's comments sit within a conventional framework of European views of the island, its social systems and its people. However, it also explores important ways in which this document is a very personal response to the vexed question of slavery and to the people trapped in that system in the British Caribbean. The diary emerges as a valuable resource, not only in exposing the agency of a woman in a predominantly masculine society, but also in resurrecting the lost voices of those enslaved by, and ensnared within, a system of unrelenting cruelty and inhumanity.

When the Blyths landed in Jamaica in 1824, they arrived on an island that had undergone monumental social and economic changes under English and then British rule since it was first seized from Spain in 1655. The changes wrought over this period were due almost entirely to the slave-holding, plantation-based economy that allowed the island to supply sugar to sate the ever-sweetening taste of European consumers, while garnering immense profits in the process.[4] The third-largest island in the Caribbean, Jamaica had more arable, and therefore potentially sugar-producing land, than the other British-ruled islands combined.[5] The apparently insatiable European desire for sugar – by value, the most important import to Britain from the 1750s to the 1820s – fuelled the rapidly expanding economies of the Caribbean plantation islands, as well as increasing the demand for enslaved Africans to perform the hard labour of cultivation on the plantations themselves; nearly two-thirds of all enslaved people cut cane on sugar plantations.[6] They prepared the soil, planted and cut the cane, and carried it to the sugar works. Throughout the year, they worked from dawn until dusk in the sweltering Caribbean heat.[7] The importation of slaves was made illegal after 1807 when Britain abolished its slave trade, but the edifice of slavery itself, the bedrock of white Jamaican prosperity, lingered on until 1834.[8] In the meantime, the island was wracked by slave rebellions and uprisings, the most serious being the so-called 'Baptist War' of 1831–1832, which may have involved up to 60,000 slaves and whose epicentre was located close to the Blyths' home.[9] It was in the midst of this period of particular uncertainty and instability that the Blyths arrived on the island.

In this society, the place of the enslaved was constricted. But, so also was that of the few white women who lived on the island. From politics to plantations, from missionaries to merchants, the Jamaican social and economic scene was dominated by men. It is in this context, in spite of obvious differences in terms of content, the social status of their authors and their experiences, that the work of Mrs Blyth and Lady Nugent are distinct as examples of nineteenth-century Jamaican social commentary from a female point of view.[10] In particular, Mrs Blyth's detailing of missionary work in the parishes of Trelawny and St James in northern Jamaica is especially valuable because, as can be seen by the plethora of male-authored missionary narratives, this was generally recorded by men.[11] Even in the missionary recollections of her own husband, Rev. George Blyth, she is kept tantalizingly at the edges of the text, acknowledged as 'the sharer of my joys and sorrows' and praised for her work within the mission station, but invariably removed from the discussions of slavery and emancipation and never once mentioned by name.[12] The description of missionary wives as helpmates and supporters in the missionary process was common. The wife of Rev. George Johnston, a Wesleyan Methodist missionary who served in the Caribbean in the early nineteenth century, was remembered in similar terms. She died of 'swamp fever' in Dominica, and was lamented by her husband as one who had been 'a true help-meet [sic] for me in body and soul'.[13] This evidence bears out Dorothy Driver's assertion that 'the place of women in the colonies was carefully defined and circumscribed within what was an avowedly masculine enterprise'.[14]

The necessity and value of reconstructing female perspectives on, and experiences of, slavery is a subject that has been bedevilled, as Lucille Mair observed, by the male metropolitan orientation of much of the historiography.[15] Nevertheless, in recent years the literature has burgeoned.[16] Mrs Blyth's status as both a woman in the male-dominated arena of missionary activity, and as a European advocate of slave rights in a slave-holding society, positions her as an important witness to the events occurring around her. Her record of people, places and events defines her as a proactive subject, and in so doing, also restores agency to those denied it by the slave-holding system. As much as her words provide useful individual examples and charming details, they also present a more general theme of agency and subjecthood.

Mrs Blyth and the representation of Jamaican society

The manuscript diary in which Mrs Blyth recorded her comments is held at the British Empire & Commonwealth Museum in Bristol.[17] The original text, consisting of irregular daily entries for the years 1826–1831, was written in Alison Blyth's own hand in a small octavo-size book. The book was apparently given to her by her brother Henry in January 1826 as the flyleaf bears an inscription to that effect. There are approximately 54 pages written in ink that is somewhat faded in many places, as well as about 200 blank pages.[18]

Most of the events in the diary took place in the years 1826 and 1827, although the last few pages correspond to events in 1831. It was written by Alison Blyth, the wife of the above-mentioned Rev. George Blyth, a Scottish Presbyterian missionary working in northern Jamaica.[19] The inside front cover of the diary is recorded as being written at Goodwill, in Trelawny parish, in January 1826.[20] This was a missionary-established village in the north west of the island. In his memoirs, George Blyth refers to it as 'one of our lately established villages'.[21] The Blyths set sail for Jamaica under the patronage of the Scottish Missionary Society (SMS), the official Presbyterian missionary body, on board the *Lune* in January 1824 and remained on the island some 25 years.[22] This manuscript probably represents one of the earliest opportunities for Mrs Blyth to record her thoughts about life in Jamaica.

The journal refers to a number of missionary stations as well as plantations on which Rev. Blyth worked. The Blyths seemed to have spent much of their time at or near Hampden Estate. It is first mentioned on the eighth page of the journal. The Stirling family of Hampden and their Scottish compatriot, William Stehart, of the Dundee Estate, had invited the SMS to conduct missionary work in the area and provided part of Hampden Estate as a site for a church as well as offering to pay half the expense of erecting it.[23] Accordingly, a missionary station was established by Rev. Blyth on the plantation upon his arrival in Jamaica. The increased importance of non-conformist religious missions is reflected in the Blyths' activities and those of many other missionaries on the island. Before the Blyths arrived in Jamaica, the SMS had sent a minister and two catechists to Kingston in the 1800s but they did not survive the tropical climate. However, the Blyths proved a success. Assessing their pioneering achievement, another Presbyterian missionary in Jamaica, Rev. Hope Waddell remarked that 'by abundant labours, fidelity, and prudence, he [Blyth] formed a flourishing congregation'.[24]

Hampden Estate straddles the border between the parishes of Trelawny (sometimes spelled 'Trelawney') and St James, and it is situated six miles from Falmouth and 15 from Montego Bay.[25] Located in the north-western part of the island, these areas were known for their high concentration of sugar plantations and their economic productivity.[26] The main town of Trelawny was Falmouth, which had gradually supplanted Martha Brae to become the principal economic entrepôt in the parish. In subsequent years, it became known as 'the cradle of the abolition of slavery' for its place at the forefront of resistance, in no small part due to the actions, example and encouragement of dissenting missionaries in the town and its hinterland.[27] The slave trade had been illegal for almost 20 years but it would be another 10 years before slavery itself was abolished. In the meantime, the political machinations of the plantocracy (the white Europeans who owned Jamaica's plantations and, to a large degree, controlled the island), the anti-slavery agitation of her acquaintances and the resilience of the still-enslaved people that she encountered gave Mrs Blyth much material for her journal.

One of Mrs Blyth's apparently greatest disappointments, to which she alludes frequently in her diary, is the debauched state of Jamaican society and how the Europeans have 'degenerated'.[28] Among the most active Europeans to partake in, and benefit from, the plantation economy and the slave-holding system that sustained it were Scots.[29] Edward Long, a contemporary historian of Jamaica, remarked on the presence of Scots, and their effect on the island.[30] Despite their contribution to the economy of the island, her fellow countrymen had not, in Mrs Blyth's eyes at least, avoided the corrupting and degrading influence of slavery. She begins her journal by railing against her compatriots, lamenting:

> I always thought that wherever I went I would be proud of my country but here I feel almost ashamed to say I am a native of Scotland, when I see how her sons have degenerated. (2)

Mrs Blyth was not alone in finding Jamaica's white inhabitants less than virtuous. Her observations correspond very closely with many others, writing both before and after her. At the end of the seventeenth century, Francis Crow remarked that the island was full of 'sin'. About the same time, William May wrote to the Bishop of London concerning the incompetence and immorality of its leading inhabitants.[31] Rev. William Knibb, a Baptist minister working in Falmouth, also decried the behaviour of Europeans in Jamaica. Writing to his friend Samuel Nicholls in Bristol, Knibb declared: 'I have now reached the land of sin, disease and death, where Satan reigns with awful power, and carries multitudes at his will.'[32] Lady Nugent even referred to Jamaica as a 'sad immoral country'.[33] Thus, in her condemnation of European corruption, Mrs Blyth follows one of the most long-standing and serious charges laid against white planters in the Caribbean.[34] However, she goes further and immediately qualifies her observations by considering how this behaviour affected enslaved Africans:

> What for instance can a *married* negro [think], who is following on to know the Lord & endeavouring to walk in his commandments & ordinances blameless[ly]. What can such a man think of an attorney or master who is living with three or four women and laughing at every thing sacred. (17–18)

This indication of concern for the enslaved population of Jamaica and their interaction with Europeans, rather than simply commenting on the moral depravity of the latter, is a strong theme running through the manuscript. The diary, therefore, represents something of the spirit of the 'Empire of Reform' that commentators have suggested begins to characterize the British engagement with the wider world in the nineteenth century.[35]

The representatives of religion did not escape the opprobrium of commentators either. In fact, if anything, they provided an even more shocking example of how white Jamaican society in the eighteenth and nineteenth centuries was corrupted by debauchery and decadence. Again, Mrs Blyth follows and confirms patterns of criticism that were well rehearsed by the time she was writing. The representatives of the Established Church in Jamaica were notorious for their laxity, incompetence and immorality. David King remarked that 'Holy orders were readily given to men who were imperfectly educated and of indifferent moral character.'[36] In a similar vein, Mrs Blyth writes of what 'an awful blame must attach to those who could ordain such men as the present set of Jamaica parsons' (11). She closely parallels the thoughts of John Riland, who lamented that 'the clerical office in Jamaica was a sort of dernier [sic] resort to men who had not succeeded in other professions'.[37] The Church of England had only created a separate Episcopal See for Jamaica in 1825, the year before these remarks were written by Mrs Blyth.[38]

Anglican interest in the island was a reaction to the increased activity of other denominations in Jamaica, such as the SMS to which the Blyths were attached, and ran counter to the previous prevailing ecclesiastical practice and religious observance.[39] As early as 1671, Governor Thomas Lynch complained that 'the condition of the church is so low and the Number of the Ministers here so few that they are not worth taking notice of'. Charles Leslie, writing in 1740, gave a 'dismal account' of the 'Church Affairs of the Island', finding the clergy 'of a character so vile' as to be unmentionable and concluding that 'they are generally the most finished of our debauchers'. The pattern continued into the nineteenth century. In 1813, the *Jamaica Magazine* commented on the lamentable state of religion, and placed the blame squarely on the 'selfish and mercenary clergy' with their 'cold indifference' to the moral state of the island.[40] And, even had the clergy been inclined to contribute to the religious and spiritual welfare of Jamaicans, they would have been hard pressed to do so due to a chronic shortage of Anglican missionaries. Mrs Blyth records the comment of her husband who contemptuously dismissed the Church of England's efforts in Trelawny:

> Look at this parish, there are two clergymen for a population of 30,000! Though they could bear the glad tidings of salvation on angel's wings I would still say their number was insufficient. (9–10)

In the period from 1770 to 1820, there was on average only one Anglican clergyman for every 18,000 of the total population.[41] The perceived conflict between the Established Church and its dissenting co-religionists had major political repercussions as non-conformist missionaries were increasingly identified with anti-slavery sentiment and were targeted accordingly by the pro-slavery lobby.

The Blyths arrived in Jamaica at a time when increasing numbers of non-conformist religious denominations were moving to establish stations on the island.[42] The Moravians were the first to arrive, in the parish of St Elizabeth, in 1754. Methodism was introduced by Dr Thomas Coke in 1789, while Baptists from Britain arrived at Montego Bay in the north of the island in February 1814. As we have seen, the Presbyterian missions of the SMS were introduced at the beginning of the nineteenth century.[43] Appealing to the enslaved, and using a style of preaching that incorporated strong resonances of African religious practices, these preachers and their churches became very successful in converting the enslaved population. Indeed, many later commentators have adduced the rise of such Jamaican spiritual codes as Revivialism and Pocomania to the combination of forms of western European dissenting Protestantism and traditional African religions.[44]

The immediate effect of the influx of missionaries to Jamaica was a perceived improvement in the moral demeanour of the enslaved majority. Of course, this is in direct contradistinction to the effect of the planters' debauched activities on the enslaved population, as recounted by Mrs Blyth and others. Rev. James Phillippo, a prominent Baptist missionary, commented that in former times 'the sanctions of marriage were almost unknown' and 'every estate on the island – every Negro hut – was but a common brothel; every female a prostitute; and every man a libertine'.[45] The example of marriage (or lack of it) given by Phillippo was corroborated by Alfred Caldecott, writing a history of the Church in the West Indies in 1898, who remarked that 'the slaves were not allowed to marry'.[46] Mrs Blyth's journal is replete with examples of the good influence and edifying effect of Christian religion, and the ministrations of the Blyths in particular, on the enslaved population of Jamaica. For example, she recounts meeting a couple from Orange Valley Estate that had been married the previous Sunday.[47] She continues to describe the programme of her husband who encouraged 'all the O[range] Valley people to [go to] the Rector to be married' with the result that 'almost every week lately I have a piece of nice cake sent' (20). In this instance, the sympathies of Mrs Blyth are firmly with the enslaved majority. She fulminates against the attorneys and overseers who corrupt those slaves 'following on to know the Lord & endeavouring to walk in his commandments & ordinances blameless[ly]' (18). The diary entries repeat the rhetoric deployed by anti-slavery campaigners in Britain, where a sentimental view of the enslaved as child-like and innocent victims of European debauchery and corruption was promulgated.[48] For example, Charles Campbell commented:

> The slavish condition of the Negroes, and the total absence of moral and religious instruction, is a sufficient excuse for their ignorance. They know nothing about religion, and yet they have more piety than their masters. They are not so deficient in intellectual energy, as is sometimes asserted. A West Indian slave is every whit as rational

> a creature as a Scots peasant or mechanic, and tinged with less vulgarity.[49]

Clearly, Mrs Blyth conformed to the notion that the enslaved people of Jamaica were recoverable from the moral and spiritual morass into which their masters had sunk. Indeed, she records one conversation with an enslaved woman which seems to corroborate this:

> June 25. I had a long & interesting conversation today with an old negro woman. Among other things she said 'Mistress since dis blessed light come to we, we no like the same. Beforetime when we meet together for we talk all about badness & den Buckra say we talk about rising upon dem & so we talk it for true Mistress; but me no think negar talk so again for now him consider upon de commandments.[50] (42–3)

The diary illustrates not only how Mrs Blyth reproduced the received image of Jamaica and its people that was presented to nineteenth-century Britons, but also how this perception is problematized and disrupted, through the dialogues she transcribes and the contexts that she describes.

Mrs Blyth's remarks about the Jamaican Assembly provide another example of how her comments give a personal insight into the general political affairs of the time. In her recollections of the year 1826, she discusses the merits of the latest parliamentary provisions being put before the local House of Assembly: 'I hope the slave-evidence Bill will pass [at the] next meeting of Assembly and then a slave can ask [for] lawful redress for himself' (5–6).[51] During the period covered by the journal, a series of tortuous legal and legislative wrangling was in progress, in response to increasing pressure to modify the slave-holding system and legal structures of the island.[52] Through a combination of missionary activity, increased resistance by the enslaved throughout the Caribbean and mounting hostility to slavery in Britain, the 1820s was a decade of considerable political upheaval and instability in Jamaica.

Jamaica's House of Assembly was first convened in January 1664. Even then, the members asserted their right to speak and act on behalf of the colony, to control local taxation and to supervise expenditure within the colony. By 1728, most of the Assembly's claims to institutional integrity had been accepted by the British government at home.[53] The governor of the colony was charged with inducing this strong-willed body of men with specific interests to accept the measures recommended to them in accordance with his instructions from London. As many of these governors discovered, and as the many rejections of British law attest, this was not as straightforward as it appeared.[54] The Jamaican Assembly was equally as resistant to guidance from London on the issues of rights for the enslaved and codes to be implemented to protect them. Abolitionists in Britain argued that West Indian slavery denied basic human and civil rights to slaves because, to all intents and purposes, they had no civil presence in the eyes of the law.[55]

The bill to which Mrs Blyth alludes failed to be passed by the Assembly, meaning that it was still unlawful for a judge to accept the evidence or testimony of a slave in a legal suit against a European. The intransigence of the political classes is remarked upon by Mrs Blyth on 25 October, in a prescient assessment of the state of affairs:

> An awful crisis seems to have arrived when, either the proprietors of this island cannot yield a little to the wishes of the government at home & to the prospective mental improvement of the slaves or a dreadful scene of anarchy & confusion, perhaps bloodshed, must follow. (13)

Further discussion and debate in the legislature followed, however, and on 27 December Mrs Blyth tells us that the Assembly members agreed 'to revive the "Slave Code" & to make such alterations as they might think safe' (23). This revised 'Slave Code' incorporated some minor improvements for the enslaved. However, it also imposed more restrictions on dissenting preachers. For example, they were prohibited from holding meetings between sunset and sunrise (a particularly difficult restriction to circumvent as this was the only time when slaves were not labouring) and from receiving payment from slaves for religious instruction. It was perhaps for these reasons that the embryonic law was disallowed by the Colonial Secretary in May 1827.[56]

Subverting the stereotype? Missionary endeavour and enslaved voices

In these instances, Mrs Blyth conforms to views that were propounded by a plethora of missionary publications in Britain. Her condemnation of European planters and their morals and a sympathetic, if paternalistic, concern for the enslaved population were replicated by many commentators. However, her place as a Christian missionary on the island also provides opportunities to explore a more complex picture behind this conformity.

Fundamentally, the challenge presented by the non-conformist missionaries was their attitude to the political status quo on the island. While the Anglican clergy were largely content to preach submission to higher powers and to buttress the slave-holding system on the island, the dissenting missionaries presented a different outlook. Bringing views of a God who was a righteous judge and in whose eyes all men were created equal, the dissenting clergyman inherently challenged the institution of slavery at a time when the morality of this system was one of the foremost political questions of the day.[57] In this context, I will now turn to what Rev. Blyth and others say in their memoirs and then examine the evidence of their anti-slavery sentiments as recorded in the journal of his wife. By plotting the divergences between the private journal of a woman in the male-dominated worlds of colonial society and missionary endeavours and the publicly sanctioned memoirs of her husband and other missionary labourers, one can appreciate the veracity of Anna Johnston's assertion that published missionary texts are 'fundamentally and frankly propagandist in nature'.[58] Equally, the nature of missionary activity and its relationship with the rhetoric of empire can be further textured through the analysis of private journals such as this one.

Rev. George Blyth's *Reminiscences of Missionary Life, with Suggestions to Churches and Missionaries* (1851) portrays the officially endorsed view of Presbyterian missions in Jamaica to an audience in Britain familiar with the process of slave emancipation and concerned with effecting the same result in the United States. Thus, the rhetoric permeating Rev. Blyth's text is one that promotes missionaries as leading the charge for abolition in Britain's colonies: '[W]e were uniformly the friends of the slave. We assisted him to bear his chains, and when other missionaries were forced to retire to Britain, where they exposed the system throughout the length and breadth of the land, we were enabled to comfort their bereaved flocks.'[59] Rev. Blyth is unequivocal in his condemnation, declaring that 'the slave trade was one of the greatest evils and one of the most unjust systems that Satan ever invented', before going on to glorify the omniscience of God in bringing about its abolition. Blyth is not unusual in this response. Looking back in 1863, the Irish Presbyterian minister, Rev. Hope Waddell, denounced slavery as, 'the lowest stage of human society, and in its own nature debasing and essentially barbarous, detrimental, morally and physically, to both masters and slaves'.[60]

Furthermore, these texts do not spare the United States and its continuing slave-holding culture. Blyth in particular rails against the 'pollution of God's sanctuary' by religious ministers defending the institution. Like so many others published in this period, these texts represent the role of missionaries as being both pivotal and consistent in the struggle to abolish slavery in Britain's Caribbean colonies. But, these texts were published after the situation was fixed firmly in favour of abolition and emancipation. They entered the cultural bloodstream at a time when Christianity and missionary endeavour, as represented at its most militant by David Livingstone in Central Africa, was seen as the great British bulwark against the evils of slavery and foreign, tribal despotism.[61] Audiences had become acculturated to, and even expectant of, a liberal anti-slavery consensus being preached by their spiritual pastors.

Mrs Blyth's response complicates this assessment. Her diary presents a more complex situation, in which European missionaries tread a tenuous line between accepting and working within the strictures of the system, and challenging it. The striking difference between published accounts and everyday journal entries is not solely due to the time differential involved, although most of the memoirs were published about three decades after Mrs Blyth had written her journal. It is also dependent on her recording of incidents of personal import and the voices of the enslaved. Private diaries and impressions are necessarily more immediate than those published for public scrutiny many years after the event. Nevertheless, the Blyth journal provides a more rounded understanding of the way that slavery could be challenged, overturned and subverted from below, as well as the decidedly more ambiguous stance of the missionaries towards emancipation when the issue really was a matter of debate. She gives accounts and records sentiments that might seem to contradict the published protestations of non-conformist missionaries. Often, her version of events appears to fly in the face of the received missionary position of those clergymen working in Jamaica who, in many instances, created their own historiography through the publication of memoirs and recollections.

The first example in the diary occurs as Mrs Blyth describes her worries and fears of attending a public meeting held on Wednesday 27 September 1826 in Falmouth. As mentioned above, this town in northern Jamaica was known as the 'cradle of emancipation', partly because of the activism of the religious missionaries here. This meeting was one of the first campaigning for reform on the island and, as such, is an important instance of the role played by missionaries in anti-slavery agitation. However, her account gives us an indication of her husband's ambivalent attitude towards slavery and emancipation: 'Mr B at first declined the invitation to speak thinking public opinion not prepared for such a thing, but upon second thoughts if the attempt *was* to be made, he thought it his duty to assist. May the Lord give direction' (7). It would appear that George Blyth only went to the meeting under duress. And Mrs Blyth goes on to give her own feelings on the events. Rather than presenting herself as a radical anti-slavery protester, she records her fears at this dabbling in politics and an awareness of the dangerous territory onto which her husband has been forced to stray. She recounts that 'when we came to the chapel Mr B went on the platform & I was put in a seat with the other ministers' wives' (8). As the meeting got under way, she recalls that when her husband started speaking, 'I got a little afraid (knowing *spies* were present) when he got so near as this parish' (8). Her husband's words show that, far from making radical pronouncements on the evils of institutionalized slavery, George Blyth's denunciations are limited to the rather hackneyed clichés about established religion. While at first glance it might seem that attending such a meeting was a striking blow for emancipation, in fact it was limited to denominational disputes.

Contemporary evidence suggests that Mrs Blyth's views, rather than her husband's later protestations, were more in tune with the opinions and attitudes of dissenting missionaries at the time. They were looked on suspiciously by European plantation owners as fomenters of rebellion and intrigue among the enslaved majority. The Establishment, both political and clerical, disliked these missionaries' close personal contact with the enslaved, bringing not whips but books and preaching not of the value of submission but of the equality of all in the eyes of the Almighty. A less radical, more conciliatory approach towards the slave-holding system was almost a prerequisite for safety. Indeed, Presbyterian missionaries were expressly warned:

> Never converse with the natives on political subjects. Such conversation, you may be almost certain, will be misrepresented and turned as an engine against you.[62]

Although missionary supporters in Britain advocated emancipation, missionaries might find themselves in a delicate position, relying on the support of the planters to carry out their work. In this regard, Rev. Blyth was criticized for his friendship with Mr McQueen, a determined defender of slavery, and he appears to have also been criticized for carrying out missionary work with slaves, thereby giving credence to slavery itself.[63]

The reason for such compromise was clear. As we soon learn from the diary, Mrs Blyth's fears for her family's safety were well founded:

> A fortnight after the above meeting, a paper appeared in the *Cornwall Courier* against the Methodists in which the meeting was alluded to. Mr Drew was called a *mountebank* & the speakers his clowns, a most shameful string of abuse followed, in which they were charged with encouraging 'the slaves to rob their masters' to give them money etc etc. (11)

The editor of *The Cornwall Courier* referred to by Mrs Blyth was William Dyer, a local Justice of the Peace. He was noted for advising his readers on various occasions to tar and feather any missionary that they met.[64] Another inflammatory incident recorded by Mrs Blyth was 'the attack upon the Methodist chapel at St Ann's, at Christmas last [1826]' (38). Knibb records that this parish in the north of the island was notorious for its rough treatment of non-conformist missionaries. He recounts an occasion also detailed in Mrs Blyth's journal when shots were fired into the Methodist manse, as the minister and his family lay asleep: 'I have seen the bullet holes and handled the bullets and my blood thrilled through my veins at the sight. Will not God visit for these things?'[65] Therefore, Mrs Blyth's journal is a useful tool for shedding light on both the later testimony of missionaries, writing for a virulently anti-slavery constituency in Britain, and the political practicalities that those same missionaries had to navigate when faced with an Establishment diametrically opposed to their professed views.

Mrs Blyth's journal also provides a more nuanced appraisal of, and conciliatory position towards, the Sunday market phenomenon than one would expect from the average missionary. As a deeply committed Christian, Mrs Blyth would be expected to condemn this gross transgression of the Sabbath. In fact, one of the few issues on which pro-slavery planters and anti-slavery missionaries generally agreed was their united opposition to this institution:

> Sunday 11th [February, 1827]. I am left here to day alone, & I think if five minutes in this town on the Sabbath would not convince any one of the necessity of missionary exertions I don't know what would. (32–3)

Here, Mrs Blyth remarks on one of the characteristic features of British Caribbean slave-holding society, as enslaved people traded the produce that they had grown and harvested from their paltry patches of land. The Sunday market had become an integral part of the domestic Jamaican economy by the late eighteenth century. In Kingston in the 1780s, an estimated 10,000 people attended the market every week, which was held on a Sunday so as not to interfere with estate cultivation.[66] The produce sold at these events was grown on the provision grounds – plots of lands (on average about half an acre per slave), sometimes up to 10 miles from the estate house, which were granted by the estate owner to his slave labourers. They were initially intended to be land where slaves could grow crops such as potatoes, ackee, yams and plantains to supplement their meagre allowance.[67] However, by the 1720s, most of the enslaved had moved beyond mere subsistence agriculture on these plots and they were selling the surplus at markets to purchase items such as salt-beef, salt-fish and pork.

Despite her moral qualms about the market, Mrs Blyth must have been aware that these market activities were crucial for the survival of a 'protopeasant' economy in the West Indies that supplied most of the island's food and crafts.[68] According to Edward Long in 1774, 20 per cent of the island's circulating coin, and a large number of the small silver, was in the hands of slaves who operated at these markets.[69] Charles Rose Ellis, the owner of Montpelier estate told the Jamaican Assembly on 6 April 1797 that 'the markets of Jamaica are almost exclusively supplied with vegetables, fruit, and poultry by the negroes' and that 'all the industrious do by these means acquire property'.[70] A typical load of goods taken to market fetched between 10 and 12 shillings for the person who grew or raised it.[71] Indeed, Mrs Blyth's sympathies seem to be with those who were forced to work on a Sunday because their masters were eager for the plantation agricultural process to continue uninterrupted. In this, she diverges from the default position of the missionaries in published texts, which emphasized the general immorality of the phenomenon and those involved. Mrs Blyth's account paints a more nuanced picture, where the markets are recognized as an intrinsic necessity for those labouring under the yoke of slavery.

While most European commentators recognized the industriousness of those involved, the majority of the authors also condemned the practice for one reason or another. Pro-slavery authors tended to take issue with the hustle and noise that characterized such large gatherings, together with the ever-present fears about the possibility of violent insurrection. Missionaries, in contrast, were much more concerned with the transgression of the Sabbath. Mrs Blyth conforms to this criticism: 'It is really a disgrace that the vile Sunday markets are not entirely stopped' (33–4). However, regarding a possible change to the law, Mrs Blyth reflects, 'I heard many of the poor creatures as they passed saying this law would never do, unless Buckra gave them some time in the week' (41).[72] Therefore, despite her comments railing against this sacrilege of the Sabbath, she is inherently sympathetic to those whose perilous situation and legal anonymity made the Sunday market a lifeline. She understands the invidious position in which slaves found themselves, where transgressing Christian codes became the lesser of the many evils to which they were exposed. She identifies the ultimate cause of the problem as the exigencies of the plantation system rather than the general immorality of the population.

These examples serve to highlight the complex negotiations of political positions and moral values that missionaries and enslaved alike had to navigate in Jamaican society. Rather than presenting a clear-cut picture of the social circumstances of the island at the time, Mrs Blyth's diary is a testament to the human complexity of the situation. Frequently, the power of the Blyths' humanity comes across in the small snippets of conversation that Mrs Blyth transcribes in her diary. These may not be the grand, overarching anti-slavery pronouncements that some later published accounts would ascribe to the missionaries. However, in their humanization of the enslaved, from an amorphous abstract noun to a group of real people with common human concerns and for whom she has a lasting affection, they are telling.

In recounting the conversations that she had with many of her husband's flock, she provides unique and sympathetic insights into the lives of these people. The evidence of the journal illustrates how they engaged in culture-building forms of resistance to slavery, how they adapted Western Christianity to their own particular needs, how cultural preservation worked and how acutely aware they were of the limitations of the political process. So, for example, she had a 'long & interesting conversation' with 'an old negro woman' who tells Mrs Blyth how Christianity has changed her outlook: 'Mistress since dis blessed light come to we, we no like the same' (42). Later, she also records her interest in the continuity of African cultures that were forcibly transplanted to the Caribbean, during a conversation with another 'old Negro woman' [sic]:

> I was talking with her also about the different tribes of Africans, & there [sic – 'their'] conduct on arriving here. (43)

She writes about the prospect of slave emancipation. The dialogue noted by Mrs Blyth suggests a sophisticated understanding among the enslaved population regarding the prospects and exigencies of economic survival after the end of slavery and points to the important role that land would play in the development of post-emancipation Jamaican society:

> The word *Liberty* has not the charms to them as which it has to a British peasant. The common answer if you hint at freedom is 'if me free where me to get house & grounds'. (14)

Conclusion: Slavery, emancipation and missionaries in the Caribbean

The entries in her journal illustrate Mrs Blyth's confrontation with the realities and inherent ambiguities of a society founded on the enslavement of its majority community. Her record provides a unique insight into Jamaica at the cusp of full slave emancipation, highlighting the fragility of the institution of slavery itself, as well as the endurance of the social legacies to which it contributed.

If Lady Nugent's diary records the social occasions and activities of planter society, Mrs Blyth's details the missionary work undertaken by her husband and others in northern Jamaica at a time of great uncertainty and social upheaval. Lady Nugent's circle of acquaintances encompassed the pinnacle of white European planter society in Jamaica. In contrast, Mrs Blyth frequently describes the lives and experiences of the majority underclass of enslaved labourers on whose incessant toil the wealth of the island was founded.Despite their differences, these journals represent and symbolize Jamaican society in all its complexity and instability in the nineteenth century. Mrs Blyth and her brand of religious observance and activity was, in many ways, anathema to the easy-going lifestyle of the planter class over which Lady Nugent's husband sought to keep control. The rise of missionary activity as indicated by the very existence of Mrs Blyth's diary coincided with more virulent opposition, both in Jamaica by the enslaved majority and in Britain by European sympathizers, to the slavery which sustained the planters' prosperity. Mrs Blyth's narrative provides a rare insight into this aspect of Jamaican history and society. The trajectory of Jamaican history, apparently so inevitable to a modern readership, was not at all clear during the Blyths' time in the Caribbean, and the insecurity and instability of the political scene reverberates in Mrs Blyth's words, sitting side by side with comments and observations that illustrate the very human choices and situations confronting those who lived and moved in a pre-emancipation Jamaican society.

Recent scholarship has highlighted absences of archival evidence as inhibiting a thorough assessment of female experiences of missionary activity, thus separating women's contributions from the broader narrative of religious missions and their place in the history of empire.[73] The survival of Mrs Blyth's diary is a valuable resource in correcting this imbalance, and in expanding our understanding of missionary activity and, ultimately, the place of religion in the development of the British Empire.

Religious discourse and missionary enterprise sometimes provided channels through which imperial control followed, but at other times, they subverted, undermined and challenged this authority.[74] Mrs Blyth's observations illustrate the heterogeneous character of missionary activity in the British Empire and the complex interaction between religion, mission and empire. Her experiences, and the evidence of her diary, reflect this relationship at a crucial time in both British imperial and missionary history.

Notes

1. For further information, see C. L. Brown, *Moral Capital: Foundations of British Abolitionism* (Chapel Hill, NC: University of North Carolina Press, 2006); J. R. Oldfield, *Popular Politics and British Anti-slavery: The Mobilisation of Public Opinion against the Slave Trade, 1787–1807* (Manchester: Manchester University Press, 1995); D. Turley, *The Culture of English Antislavery, 1780–1860* (London: Routledge, 1991); R. Anstey, *The Atlantic Slave Trade and British Abolition, 1760–1810* (London: Macmillan, 1975); R. Blackburn, *The Overthrow of Colonial Slavery, 1776–1848* (London: Verso, 1988).
2. L. J. Ragatz, *The Fall of the Planter Class in the British Caribbean, 1763–1833* (New York: Century Company, 1928). Quoted in Maria, Lady Nugent, *Lady Nugent's Journal of Her Residence in Jamaica from 1801 to 1805*, ed. Philip Wright (Kingston, Jamaica: Institute of Jamaica Press, 1966), p. xi.
3. For an analysis of Lady Nugent's anti-slavery sentiments, see C. Hall, 'Gender and Empire', in *Empire, The Sea and Global History: Britain's Maritime World, c. 1760–c. 1840*, ed. D. Cannadine (London: Palgrave, 2007), pp. 134–52, 139–44.
4. J. R. Ward, 'The British West Indies in the Age of Abolition, 1748–1815', in *The Oxford History of the British Empire: The Eighteenth Century*, ed. P. J. Marshall (Oxford: Oxford University Press, 1999), pp. 415–39.
5. R. B. Sheridan, 'The Formation of Caribbean Plantation Society, 1689–1748', in *The Oxford History of the British Empire: The Eighteenth Century*, pp. 394–414, 395.
6. H. Thomas, *The Slave Trade, The History of the Atlantic Slave Trade, 1440–1870* (London: Picador, 1997), p. 275.

7. See D. Hamilton, 'Slave Life in the Caribbean', in *Representing Slavery: Art, Artefacts and Archives in the Collections of the National Maritime Museum*, ed. Robert J. Blyth and Douglas Hamilton (Aldershot: Lund Humphries, 2007), pp. 50–61.
8. In Jamaica, the population of slaves outnumbered that of Europeans by a ratio of ten to one. See G. Heuman, 'The British West Indies', in *The Oxford History of the British Empire: The Nineteenth Century*, ed. A. Porter (Oxford: Oxford University Press, 1999), pp. 470–94, 472.
9. See M. Craton, *Testing the Chains: Resistance to Slavery in the British West Indies* (Ithaca, NY: Cornell University Press, 1982), p. 291.
10. Across the Caribbean, there are a limited number of examples of female writers in this period. Mrs A. C. Carmichael's *Domestic Manners and Social Conditions of the White, Coloured and Negro Population of the West Indies*, 2 vols (London: Whittaker, Treacher & Co., 1833) and Janet Schaw's *Journal of a Lady of Quality; Being the Narrative of a Journey from Scotland to the West Indies, North Carolina, and Portugal, 1774 to 1776*, ed. E. W. Andrews and C. Andrews (New Haven, CT: Yale Historical Publications, 1921) are exceptions to this general trend.
11. As well as the memoirs of her husband, Rev. George Blyth, some examples of missionary writing directly pertinent to Jamaica include: Rev. Richard Bickell, *The West Indies as They are* (London: J. Hatchard & Son, 1825), Rev. James M. Phillippo, *Jamaica, Its Past and Present State* (London: John Snow, 1843) and Rev. Hope Masterson Waddell, *Twenty Nine Years in the West Indies and Central Africa: A Review of Missionary Work and Adventure, 1829–1858* [London: Nelson, 1863], 2nd edn with an introduction by G. I. Jones (London: Cass, 1970).
12. Rev. George Blyth, *Reminiscences of Missionary Life, with Suggestions to Churches and Missionaries* (Edinburgh: William Oliphant & Sons, 1851), p. 36.
13. Quoted in P. Grimshaw and P. Sherlock, 'Women and Cultural Exchanges', in *Missions and Empire*, ed. Norman Etherington (Oxford: Oxford University Press, 2005), pp. 173–93; 174.
14. Quoted in A. Johnston, *Missionary Writing and Empire, 1800–1860* (Cambridge: Cambridge University Press, 2003), p. 42. For further discussion of these issues, see P. Grimshaw, 'Faith, Missionary Life and the Family', in *Gender and Empire*, ed. Philippa Levine (Oxford: Oxford University Press, 2004), pp. 260–80.
15. See L. M. Mair, *The Rebel Woman in the British West Indies during Slavery* (Kingston, Jamaica: Institute of Jamaica, 1975).
16. A good introduction to the critical issues involved can be found in V. A. Shepherd, 'Gender and Representation in European Accounts of Pre-Emancipation Jamaica', in *Caribbean Slavery in the Atlantic World: A Student Reader*, ed. V. A. Shepherd and H. McD. Beckles (Oxford: James Currey, 2000), pp. 702–12.
17. Accession Number 2005/001/143. All quotes preserve the language, spelling and punctuation of the original manuscript. Likewise, all emphases are in the original text.
18. The manuscript was acquired by the British Empire and Commonwealth Museum (BECM) in 2005. The acquisition of such documents recording personal testimonies and insights, whether through donation or purchase, is a rare occurrence and makes this diary an important resource.

19. For further information on the Scottish Missionary Society, see Chapter 5 by Esther Breitenbach in this volume.
20. The journal obviously lay in disuse for some time because the opening diary entry on the first page is noted as being made at Endeavour, a missionary outpost, on 1 September 1826. Endeavour was another of the missionary stations in Trelawny to which Rev. Blyth ministered.
21. Blyth, *Reminiscences of Missionary Life*, p. 43.
22. Ibid., pp. 36–7.
23. J. Besson, *Martha Brae's Two Histories: European Expansion and Caribbean Culture-Building in Jamaica* (Chapel Hill, NC and London: University of North Carolina Press, 2002), pp. 64–5.
24. Rev. H. M. Waddell, *Twenty Nine Years in the West Indies and Central Africa: A Review of Missionary Work and Adventure, 1829–1858* [1863], 2nd edn with an introduction by G. I. Jones (London: Cass, 1970), p. 25.
25. P. Wright, *Knibb 'the Notorious': Slaves' Missionary, 1803–1845* (London: Sidgwick & Jackson, 1973), p. 69.
26. C. Hamshere, *The British in the Caribbean* (London: Weidenfeld and Nicolson, 1972), p. 140.
27. Besson, *Martha Brae's Two Histories*, p. 102.
28. *Journal of Mrs Alison Blyth*, p. 2. All subsequent references to the diary will give the page number in the original manuscript as held at BECM in parenthesis in the text following the quote.
29. See T. M. Devine, *Scotland's Empire, 1600–1815* (London: Penguin, 2004), pp. 235–8. For the most comprehensive study of this phenomenon, see D. J. Hamilton, *Scotland, the Caribbean and the Atlantic World, 1750–1820* (Manchester: Manchester University Press, 2005).
30. E. Long, *The History of Jamaica, or General Survey of the Ancient and Modern State of that Island*, 3 vols (London: T. Lowndes, 1774), ii, p. 286. See also A. L. Karras, *Sojourners in the Sun: Scottish Migrants in Jamaica and the Chesapeake, 1740–1800* (Ithaca, NY and London: Cornell University Press, 1992), p. 54.
31. See citations in O. Patterson, *The Sociology of Slavery: An Analysis of the Origins, Development and Structure of Negro Slave Society in Jamaica* (London: MacGibbon & Kee, 1967), p. 40.
32. Cited by Wright, *Knibb 'the Notorious'*, p. 24.
33. Lady Nugent, *Lady Nugent's Journal*, p. 172. Quoted in Hall, 'Gender and Empire', p. 143.
34. In addition to those instances cited in the text, the moral turpitude of Europeans in Jamaica gets particular airing in publications such as E. Long's *The History of Jamaica*. Further illustration of the points made here may be found in the life and journals of the notorious Thomas Thistlewood. See D. Hall, *In Miserable Slavery: Thomas Thistlewood in Jamaica 1750–1786* (Basingstoke: Macmillan, 1989) and T. Burnard, *Mastery, Tyranny & Desire: Thomas Thistlewood and his Slaves in the Anglo-Jamaican World* (Chapel Hill, NC and London: University of North Carolina Press, 2004).
35. See A. Porter, 'Trusteeship, Anti-Slavery and Humanitarianism', in *The Oxford History of the British Empire: The Nineteenth Century*, ed. A. Porter (Oxford: Oxford University Press, 1999), pp. 198–221.

36. D. King, *The State and Present Prospects of Jamaica* (London: Johnstone and Hunter, 1850), p. 79.
37. J. Riland, *Memoirs of a West India Planter* (London: Hamilton, Adams & Co., 1827), p. 106.
38. See Waddell, *Twenty Nine Years in the West Indies and Central Africa*, p. 23.
39. See Introduction chapter in this volume by Hilary Carey.
40. See Patterson, *The Sociology of Slavery*, pp. 40–1.
41. E. Brathwaite, *The Development of Creole Society in Jamaica, 1770–1820* (Oxford: Clarendon Press, 1971), p. 25.
42. See A. Porter, *Religion versus Empire? British Protestant missionaries and overseas expansion, 1700–1914* (Manchester: Manchester University Press, 2004).
43. Brathwaite, *The Development of Creole Society in Jamaica*, pp. 252–3.
44. Patterson, *The Sociology of Slavery*, p. 212, n. 2.
45. Phillippo, *Jamaica, Its Past and Present State*, p. 218.
46. Quoted in Shepherd, 'Gender and Representation in European Accounts of Pre-Emancipation Jamaica', p. 708. Marriage only received 'legislative encouragement' through the Consolidated Slave Law of 1816, and its various revisions. See *Journals of the Assembly of Jamaica*, 14 vols (St Jago de la Vega, Jamaica: John Lunan, 1829), xiv, p. 758.
47. *Journal of Mrs Alison Blyth*, p. 19.
48. The most pervasive example of this phenomenon is probably the image of the kneeling African supplicating for freedom. It was famously used on the medal produced by Josiah Wedgwood for the Society for Effecting the Abolition of the Slave Trade and was subsequently reproduced on a variety of objects in many cultural contexts. For a discussion of this, see J. N. Pieterse, *White on Black: Images of Africa and Blacks in Western Popular Culture* (New Haven, CT: Yale University Press, 1992), pp. 52–63.
49. C. Campbell, *Memoirs of Charles Campbell, At Present Prisoner in the Jail of Glasgow, Including His Adventures as a Seaman and as an Overseer* (Glasgow: J. Duncan & Co., 1828), p. 19.
50. 'Buckra' is a Jamaican English term for a white man. It is also spelled 'backra'. See *Dictionary of Jamaican English*, ed. F. G. Cassidy and R. B. Le Page, 2nd edn (Cambridge: Cambridge University Press, 1980), p. 18.
51. A major debate on this issue took place on 12 October 1826 in the Jamaican Assembly. See *Journals of the Assembly of Jamaica*, xiv, pp. 595–601.
52. Porter, 'Trusteeship, Anti-Slavery, and Humanitarianism', p. 203.
53. See Brathwaite, *The Development of Creole Society in Jamaica*, p. 8.
54. See Lady Nugent, *Lady Nugent's Journal*, p. xxiii.
55. See Patterson, *The Sociology of Slavery*, p. 85.
56. See Wright, *Knibb 'the Notorious'*, p. 39.
57. Porter, 'Trusteeship, Anti-Slavery, and Humanitarianism', pp. 201–2.
58. Johnston, *Missionary Writing and Empire*, p. 4.
59. Blyth, *Reminiscences of Missionary Life*, p. 68.
60. Waddell, *Twenty Nine Years in the West Indies and Central Africa*, pp. iv–v.
61. See A. C. Ross, *David Livingstone: Mission and Empire* (London: Hambledon and London, 2002).
62. Quoted in M. Turner, *Slave and Missionaries: The Disintegration of Jamaican Slave Society, 1787–1834* (London: University of Illinois Press, 1982), pp. 9–10.

Similar strictures were imposed on missionaries of other denominations. See Porter, *Religion versus Empire?*, pp. 86–7.

63. See Chapter 5 by Breitenbach in this volume.
64. Wright, *Knibb 'the Notorious'*, p. 103.
65. Quoted in Wright, *Knibb 'the Notorious'*, p. 65. So inflammatory was this incident that the Governor, the Duke of Manchester (1808–1827), had to answer a Parliamentary question on the debacle.
66. Patterson, *The Sociology of Slavery*, p. 226.
67. The ackee is the fruit of a tree, originally from West Africa but introduced to the Caribbean, which is eaten as a vegetable. A plantain is a type of banana, which is high in starch.
68. The market phenomenon is, according to Sidney Mintz, evidence of a society transforming itself from one type of labouring population to another (i.e., from a slave-holding system to a peasant economy). See S. W. Mintz, *Caribbean Transformations* (Baltimore, MD and London: John Hopkins University Press, 1974), p. 158.
69. Besson, *Martha Brae's Two Histories*, p. 87.
70. Quoted in B. W. Higman, *Montpelier, Jamaica: A Plantation Community in Slavery and Freedom, 1739–1912* (Kingston: The Press-University of the West Indies, 1998), p. 191.
71. Ibid., p. 243.
72. The introduction of a law preventing Sunday trading after ten o'clock in the morning was being debated in the Jamaican Assembly at this time. See *Journals of the Assembly of Jamaica*, xiv, p. 598.
73. Grimshaw and Sherlock, 'Women and Cultural Exchanges', p. 175.
74. See A. Porter, 'Religion, Missionary Enthusiasm, and Empire', in *The Oxford History of the British Empire: The Nineteenth Century*, ed. A. Porter (Oxford: Oxford University Press, 1999), pp. 222–46; 222–3.

Further reading

Anstey, R. *The Atlantic Slave Trade and British Abolition, 1760–1810* (London: Macmillan, 1975).

Besson, J. *Martha Brae's Two Histories: European Expansion and Caribbean Culture-Building in Jamaica* (Chapel Hill, NC and London: University of North Carolina Press, 2002).

Brathwaite, E. *The Development of Creole Society in Jamaica, 1770–1820* (Oxford: Clarendon Press, 1971).

Craton, M. *Testing the Chains: Resistance to Slavery in the British West Indies* (Ithaca, NY: Cornell University Press, 1982).

Hamilton, D. J. *Scotland, the Caribbean and the Atlantic World, 1750–1820* (Manchester: Manchester University Press, 2005).

Karras, A. L. *Sojourners in the Sun: Scottish Migrants in Jamaica and the Chesapeake, 1740–1800* (Ithaca, NY and London: Cornell University Press, 1992).

Ragatz, L. J. *The Fall of the Planter Class in the British Caribbean, 1763–1833* (New York: Century Company, 1928).

Shepherd, V. A. and H. McD. Beckles, eds. *Caribbean Slavery in the Atlantic World: A Student Reader* (Oxford: James Currey, 2000).

Temperley, H. *British Antislavery, 1833–1870* (London: Longman, 1972).
Turner, M. *Slaves and Missionaries: The Disintegration of Jamaican Slave Society, 1787–1834* (London: University of Illinois Press, 1982).
Whyte, I. *Scotland and the Abolition of Black Slavery, 1756–1838* (Edinburgh: Edinburgh University Press, 2006).
Wright, P. *Knibb 'The Notorious': Slaves' Missionary, 1803–1845* (London: Sidgwick & Jackson, 1973).

11

Richard Taylor and the Children of Noah: Race, Science and Religion in the South Seas

Peter Clayworth

Race, religion, science and empire[1]

The nineteenth century was a time of expansion for the British Empire and the Christian missions. The period from the 1830s through to the 1870s was also a time in which a change in racial paradigms occurred in British intellectual life. In the 1830s, thinking on race was dominated by Christian paternalistic humanitarianism, as reflected by the abolitionist and missionary movements. By the 1870s the views of many thinkers on race had hardened, with contemporary science being used to support a vision of defined racial hierarchies resulting from natural evolutionary processes, accompanied by the inevitable decline and extinction of 'weaker races' in competition with Europeans. The weakening of the Christian humanitarian view on race has been seen as reflecting the mid-nineteenth-century triumph of a scientific world-view over one based on religion. This chapter seeks to question this supposed polarization of science and religion. It looks at how one man, Richard Taylor (1805–1873), a missionary and a naturalist, sought to reconcile the issues raised by contemporary scientific discoveries with his own religious beliefs. It examines Taylor's attempts to use both science and religion to argue against racial inequality and for the idea that a truly Christian British Empire would mitigate the damaging effects of imperialism by spreading civilization and the word of God.

Richard Taylor was an Anglican clergyman who from 1839 worked as a missionary to the Maori of New Zealand; he was also a prominent naturalist with a strong involvement in the fledgling scientific community of the New Zealand colony. Taylor was therefore part of two strongly interconnected British global networks: the protestant missionary movement and the scientific community of ethnology and natural history. Missionaries, along with traders, explorers, military and naval personnel, colonial administrators, surveyors, prospectors and settlers, collected and supplied information on natural history and ethnology to scholars in the Imperial centres, who in turn used this information to develop and refine scientific theories. Taylor's writings demonstrate that, just as missionaries were by no means passively obedient to orders from their metropolitan superiors, colonial naturalists were not merely passive collectors and providers of information. He tried to influence British thinking on the Empire to see imperialism not as the action of a superior race, but as a godly exercise through which all races could and would acquire a British standard of civilization and Christianity. Taylor made a genuine attempt in his writings to reconcile new scientific discoveries with the belief that the opening chapters of the *Book of Genesis* could be taken seriously as an account of early human history.

Taylor worked through a period when the relationship between science and religion was undergoing a revolution. In the 1830s, the Anglican Church still dominated much of British scientific thought through its control of the Universities of Oxford and Cambridge, and the strong involvement of 'parson-naturalists' in natural history research. Nature was still largely understood through the prism of Natural Theology, but changes were underway. By the early nineteenth century, even clergymen-scientists like the geologist Adam Sedgewick were coming to accept that contemporary geology proved that the Earth was millions of years old and had been inhabited by creatures now extinct. The publication of Darwin's *Origin of Species* in 1859 led to a further revolution in thinking, propelling the debate on evolution and its mechanisms. Darwin's theory was used as a weapon by a group of young scientists, led by Thomas Henry Huxley, botanist Joseph Hooker and physicist John Tyndall, determined to break the control of the church over science, and place their own new breed of young professionals in command.[2]

The work of ethnologists and archaeologists like John Lubbock, along with philologists such as Max Müller, indicated to the Victorian British that the lives of 'uncivilized' peoples around the world reflected the stages their own ancestors had gone through on the path to civilization. New theories of physical and social evolution added strength to a long-established view that 'native' fauna, flora and peoples were inevitably displaced by more robust European imports. The expansion of the British Empire, combined with the apparent decline of indigenous peoples, was seen by many as proof of British (or sometimes specifically Anglo-Saxon) racial superiority. Support for ideas of white supremacy and racial struggle was enhanced by reaction to events such as the Indian Mutiny of 1857, the Jamaican Rebellion of 1865 and the New Zealand Wars of 1860–1872; seen by some Britons as the actions of incorrigibly violent and ungrateful dark races attacking their white benefactors. Settler communities throughout the Empire pressed the idea on the imperial centre that priority must be given to the establishment of 'civilization' in the settler colonies, by force if necessary.[3]

The period Taylor spent in New Zealand, from 1839 to 1873, was one of profound changes. When Taylor arrived in New Zealand, the islands were still independent with a mainly Maori population. The small Pakeha resident population were a mix of traders and whalers, scattered around the country and often married into Maori tribes, along with a small number of missionary families concentrated in Northland.[4] The year 1840 saw the signing of the Treaty of Waitangi, which meant that at least nominally New Zealand became part of the British Empire. At the same time, systematic Pakeha colonization began. A large number of mostly English settlers arrived in New Zealand, part of a private venture under the auspices of Edward Gibbon Wakefield's New Zealand Company. At first Maori generally welcomed Pakeha colonization, while missionaries had very mixed feelings towards the process. By the 1860s, the Maori population had declined, while the Pakeha immigrant population, now including substantial numbers of Scots and Irish, had begun to outstrip them in numbers. Maori attempts to control the loss of their land and maintain a degree of political independence led to a series of wars with British and colonial forces, from 1860 to 1872. Missionaries came under attack from both sides; accused by Maori of tricking them into signing the Treaty of Waitangi, while Pakeha settlers often regarded missionaries as dangerous apologists for Maori 'savages'.[5]

Humanitarian visions

Richard Taylor was a Yorkshireman from a comfortable farming background, but was orphaned by the time he was 13 years old. After graduating from Queens' College, Cambridge, with a BA, Taylor was ordained as an Anglican minister in 1829. The Church Missionary Society appointed him as a missionary to New Zealand in 1835. Taylor and his wife Caroline left England with their family in February 1836. They arrived in New Zealand in September 1839 after a sojourn in New South Wales.[6] Two significant events bracketing Taylor's departure from

Britain and his arrival in New Zealand provide a telling background to the ideas and attitudes Taylor was to advocate in his writings. In 1835, the Select Committee on Aborigines began hearing evidence in London. The Select Committee, chaired by anti-slavery campaigner Thomas Fowell Buxton and dominated by Christian humanitarians, was set up to investigate the impact of colonization by British settlers on indigenous peoples. Although most of the evidence heard was from the South African and Canadian colonies the committee formed a general conclusion that throughout the empire innocent 'native' peoples were suffering through the actions of unscrupulous and ungodly Europeans. The Select Committee's findings were released in two volumes over 1836 and 1837. They argued that settlers and natives alike needed to be reformed into new moral beings through the teachings of the gospel, with the ultimate aim of an assimilated community of virtuous Christians of all races. A first step was for the British Crown to give greater protection to colonized peoples. It is not certain that Taylor ever read the report of the Select Committee, but its conclusions closely matched his own views on the disastrous results of colonization unmitigated by Christian morality. He agreed that the British Empire was selected by God to bring the benefit of civilization to the world but that the empire's moral standing was now threatened by the actions of ruthless settlers. These were common opinions among the missionaries and humanitarians associated with Exeter Hall. From 1836 to 1847, their political influence was at its height with James Stephen, an Exeter Hall activist, serving as Permanent Under-Secretary for the Colonies.[7]

The morality of James Stephen and the Exeter Hall evangelicals was one of the forces behind the Colonial Office sending Captain William Hobson to New Zealand in 1839, with instructions to draw up a treaty with the Maori chiefs. On 6 February 1840, less than six months after his own arrival in New Zealand, Taylor was a witness and occasional scribe at the signing of the Treaty of Waitangi. Those Maori leaders who signed conceded that the British Crown could establish a Governor over New Zealand, in exchange for British guarantees of protection for Maori lands and treasured possessions, plus the granting to Maori of the rights and privileges of British subjects. Many contemporary observers and later commentators saw the treaty as a product of missionary influence on the Colonial Office, although others have argued that it was simply a device to gain a foothold in New Zealand for British governance while avoiding the expensive process of deploying military forces.[8] The Treaty of Waitangi was controversial from the start, with Wakefield settlers referring to it as 'a device to amuse savages'. The missionaries had originally advised Maori to sign the Treaty. Maori now looked to them to ensure that the Crown kept its side of the bargain, but remained sceptical as to how much effort the churchmen would make on their behalf.[9] Throughout his career, Taylor upheld the values expressed in both the report of the Select Committee on Aborigines and in the Treaty of Waitangi believing that Maori and other indigenous peoples could and should become civilized Christians with the rights and privileges of British subjects. These remained his views long after the influences of both the settler lobby and ethnological theorists had eroded much of the humanitarian influence of the missionaries throughout the British Empire.[10]

Missionary naturalist

From 1843, Taylor and his family moved permanently to the Whanganui district on the west coast of New Zealand's North Island. Pakeha settlers soon arrived in the area and came into conflict with Maori. Taylor, as a clergyman who spoke Maori, often tried to act as a mediator between the groups. He had to steer a course between Pakeha settlers who saw him as too soft on the natives and Maori who saw him as an agent of the settlers. While Taylor attempted to keep peace between Maori and settler, he did not seek conciliation on all fronts. A staunch evangelical, Taylor did not have good relations with his own Bishop, Augustus Selwyn, regarding him as a dangerously pro-Roman High Churchman. When it came to other denominations in the Whanganui area, Taylor's relationship with the Wesleyan missionaries might be described as uneasy, while he was openly antagonistic to French Roman Catholic missionaries.[11]

Taylor travelled extensively in the Whanganui district and beyond, combining his pastoral duties with the study of ethnography, geology, fauna and flora. He was actively involved in the birth of a scientific community in New Zealand being elected in 1852 as a member of the New Zealand Society, the short-lived forerunner of the New Zealand Institute. He was also involved with the Wanganui Acclimatisation Society. The observations he made during his travels, combined with his wide reading, formed the basis for his voluminous writings.[12]

Throughout his time in New Zealand Taylor corresponded with some of the most prominent British naturalists, sending information and specimens, including fossils to Adam Sedgewick, plants to Joseph Dalton Hooker at Kew and moa bones to Richard Owen.[13] Taylor also kept abreast of the latest scholarly writings coming out of Europe and North

America. His wide reading is indicated in the 1867 pamphlet *Our Race and its Origin*. In this publication, he quoted from Richard Owen on anatomy, from Max Müller on philology, from Darwin's *The Origin of Species* on evolution, from J. C. Prichard's *Natural History of Man* on ethnology and from John Crawfurd of the Ethnological Society on the subject of polygeny. He used observations from the travel writings of Du Chaillu and Bishop Heber, and from the medical writings of the medical researcher Dr Devay of Lyons, to illustrate his own points.[14] In an 1866 pamphlet, *The Age of New Zealand*, Taylor quoted from Hooker; explorer Thomas Brunner, the geologists Ferdinand von Hochstetter, Hugh Miller, Charles Lyell and Phillip Sclater, a geologist and zoologist who wrote on Madagascar.[15] During his return visits to Britain from 1855 to 1856 and from 1867 to 1871, Taylor frequented a number of museums; meeting and discussing science with Owen, Hooker and the brothers Gray, naturalists based at the British Museum. Taylor was a full participant in the nineteenth-century global community of scholars, whose networks mirrored those of contemporary commercial and political empires.[16]

The Semitic Maori

The expansion of Western European power and trade led to continual encounters between Europeans and the 'others'; the peoples of Africa, Asia, the Americas and the South Seas. Two biblical stories proved particularly useful for Western Christians trying to explain how such a bewildering array of peoples became scattered across the Earth. The story of the migrations of the sons of Noah explained how a variety of people could be descended from a common origin. The story of the ten lost tribes of Israel helped place the behaviours of even the most exotic of peoples within a Judaeo-Christian world-view, through perceived similarities with customs from the scriptures.[17] The idea that exotic peoples outside of the Mediterranean and northern European world might in fact be descended from the ten lost tribes became popular among European Christians in the middle Ages. From the fifteenth century onwards, Europeans and their American descendants were to speculate that a wide variety of peoples, from Aztecs through to Zulus, were in fact displaced Hebrews. Vague similarities between languages and customs of these peoples and those of the biblical Hebrews were seized upon as proof. Tudor Parfitt argues that Europeans used their image of a familiar other, the Jew, to help explain and domesticate the otherwise bewildering range of human behaviours found throughout the globe.[18]

In New Zealand historiography, the term 'Semitic Maori' has been applied to the idea that the Maori and other Polynesians were descended from Middle Eastern Semitic people linked to the Hebrews of the Old Testament. The Polynesians were supposedly once at a higher level of civilization and had degenerated through a process of migration and isolation. Proponents of aspects of the Semitic Maori idea included such early writers on New Zealand as traveller J. L. Nicholas, missionary Samuel Marsden, Governor Robert FitzRoy, trader Joel Polack and naturalist Ernst Dieffenbach. Taylor was to suggest a version of the Semitic Maori theory in his first major publication.[19]

In 1855, Taylor returned to Britain with the permission and financial support of the Church Missionary Society. One purpose of his trip was to oversee the publication of *Te Ika a Maui or New Zealand and its Inhabitants*, the book in which he set out his conclusions on the natural history, ethnography and history of New Zealand.[20] In this first edition of his work, Taylor set out his current thinking on Maori history and migration. His ideas were drawn from the Bible but reflected the careful study of many eminent contemporary scientific thinkers. He thought the 'New Zealanders' were a Semitic people who had made a slow, roundabout migration to New Zealand. Taylor believed the Polynesians had undergone a cultural decline during their migrations. Taylor explained degeneration as the loss of knowledge resulting from emigration away from the original sources of civilization and religious knowledge.[21] He compared the Maori to the prodigal son of the parable:

> May not this beautiful parable have its literal fulfilment in the history of the New Zealand race; in it may we not behold one of the lost tribes of Israel, which, with its fellows, having abandoned the service of the true God, and cast aside his Word, fell step by step in the scale of civilization; deprived of a fixed home, became nomade [sic] wanderers over the steppes of Asia, a bye-word and a reproach among nations, and gradually retreated until... they finally reached New Zealand, and there fallen to their lowest state of degradation[22]

Taylor devoted a chapter of *Te Ika a Maui* to Maori 'Origin as traced by language'. He acknowledged some linguistic and cultural links between the Maori language and Sanskrit, suggesting Indian connections. He also believed that there was a Japanese or Chinese element in Maori ancestry.[23] Taylor finished his speculations on the origins of the Polynesians by advocating their descent from the ten lost Hebrew tribes of the Bible.[24] He proposed that descendants of the Hebrew tribes had ventured out through Central Asia to America, the Sandwich Islands, Easter Island and eventually New Zealand.[25]

Taylor appears to have believed all humans had equal potential, but that some had been severely degraded by circumstance. He went to considerable lengths to show that both the Maori and the Australian Aborigine had shown great intelligence and innovation in many features of their ways of life. He argued that Europeans under similar circumstances might have also ended up developing 'savage' or 'barbaric' practices. Taylor's views on race and environment reflected contemporary evangelical beliefs in the common origins of humanity and the potential of all people to achieve salvation and civilization. As such, they also encapsulated the tension between the idea of the innate equality of humanity and the need to convert and civilize the fallen heathen, as discussed elsewhere in the current volume by O'Brien, Breitenbach and MacKenzie.[26]

The children of Noah and the birth of civilization

The biblical foundations of Taylor's ideas on racial migration were clearly illustrated in his 1867 pamphlet *Our Race and its Origin*.[27] Taylor described the various human 'races' as the descendants of the three sons of Noah: Shem, Ham and Japhet.[28] Taylor's use of the term race shows how ambiguous this term was in nineteenth-century scholarship. In this paper he referred to the 'New Zealand race', the 'brown race', the 'Negro race', the 'Celtic, Saxon and Patagonian races', the 'Hametic race' and the Human race, as well as referring to the gypsy and Jewish 'races'.[29] In the same pamphlet, Taylor attacked the views of the polygenists, as represented by the ideas John Crawfurd had advocated in a paper to the Ethnological Society. Polygenist thinkers rejected both the biblical account of creation and evolutionary ideas of the descent of all humans from common ancestors; they instead viewed different races as being in essence separate species evolved from distinct ancestors.[30] Taylor disputed Crawfurd's contention that variations in skin colour and physical appearance indicated the various human races were not descended from common forebears. In Taylor's view, all people were descended from Adam and Eve, who were probably of red or brown skin colour. Subsequent changes in skin colour and appearance were due to the effects of climate as humans migrated around the globe.[31]

Taylor also rejected the evolutionary theories put forward by the anonymous author of *Vestiges of the Natural History of Creation*, published in 1844, and by Darwin in *The Origin of Species*, published in 1859. Taylor referred to both books as 'ingenious', but remained unconvinced by their central arguments. While he accepted that there could be a wide range of variation within a species, Taylor argued neither geology nor observations of living creatures revealed any compelling evidence of the transition from one species to another. He believed geology showed that humans were created in their current form and in relatively recent times. Taylor re-emphasized the special creation of humans, but acknowledged that they shared a great many physical characteristics with their fellow creatures. He set out biblical arguments against evolution, but only after he had presented scientific arguments derived from contemporary comparative anatomy and geology.[32]

For all his interests in science, Taylor's essential view of human history was based on the Bible. He maintained that all the human races had a common point of origin, with Noah's family, the sole survivors of the Great Flood. Taylor argued that comparative philology confirmed the original unity of the human race. Here he acknowledged his debt to the work of Max Müller, in particular Müller's *History of Ancient Sanskrit Literature*.[33] Taylor saw Indo-European language connections as proving that 'Hindu, Greek and Teuton', including the English, shared common ancestors before the 'first separation of the Northern and Southern Aryans.'[34] The evidence of the words and character of their languages also showed that the Polynesians were related to the original speakers of Sanskrit.[35] As Ballantyne has pointed out Taylor pioneered the idea of India as one of the original homes of the Polynesians. An examination of *Our Race and Its Origin* does not, however, support Ballantyne's further argument, that Taylor largely abandoned the Semitic Maori idea. Taylor clearly held that the Polynesians had settled in India, but he still traced their main line of descent back through the lost tribes to Noah's son Shem.[36] Even when describing the assumed Indian connections of certain Maori customs Taylor traced their origins back to the Semites in the Holy Land: 'Thus the sacred groves of the Druid may be followed from Canaan to India, and thence through America and Polynesia, to the *Wahi Tapu* of New Zealand, so encircling the entire globe.'[37]

The long Western tradition of explaining human identities through the migrations of Noah's family was continued by British thinkers, such as Taylor, keen to explain the diversity of peoples encountered in the processes of empire building. In the latter part of the eighteenth century Orientalist scholars such as Jacob Bryant and William Jones had sought to combine the Genesis account with contemporary linguistic studies, to trace the migrations, ancestry and interconnections of peoples, particularly the inhabitants of India.[38] Taylor followed the common belief that Ham was the ancestor of the darkest races, Shem was the ancestor of the Semitic peoples and Japhet was the ancestor of the 'white' Europeans. Taylor did not assign any particular colours to Shem, Ham and Japhet themselves, holding that that their descendant's skin colours were products of environmental factors.[39]

According to Genesis, Noah had cursed Ham's son Canaan to be the slave of the descendants of Shem and Japhet, because of Ham's disrespect to Noah while the patriarch lay drunk. Despite the fact that this account makes no mention of skin colours of Ham and his brothers, by the nineteenth century many Western Christians accepted that Noah's curse had condemned the dark-skinned descendants of Ham to eternal slavery. Scholars have noted that Noah's curse did not originally have racial overtones, the modern European concepts of race, nation and geography being non-existent in biblical times. The curse may have been used to justify Hebrew conquest of and domination over the Canaanites. The earliest representations of Ham as black appear to have been in rabbinic literature of the fourth to sixth centuries. Braude points out that up until at least the sixteenth century Christian European representations of Ham and his descendants were highly variable, with Hamitic peoples being portrayed as white and Mongolian, as well as black. Ham's descendants were sometimes portrayed as living in Asia rather than Africa. While Jewish and Moslem representations of Ham influenced the way he came to be portrayed by Christians, it would appear that the development of the transatlantic slave trade was the decisive factor in the fixing of a black and African identity on the 'children of Ham'. By the time Taylor was writing many Christians believed that these ideas had scriptural authority. Braude notes that as the idea of Ham as father of the black races became solidified in Christian mythology, so did the idea of Japhet as father of the white Europeans.[40]

Taylor believed that the biblical account of the migrations of Noah's children was an accurate account of early human history. He argued that the Hamites had founded the oldest civilizations of Egypt, Nineveh and Babylon. He believed Genesis indicated that Ham built the city of Babel, while his descendent Nimrod had founded the first Mesopotamian kingdom. In developing these kingdoms the Hamites had created the first codes of laws, the earliest regular systems of government and many of the sciences, including astronomy. They had built the pyramids and Mesopotamian monuments, and the ancient 'cyclopean' works in Europe. They were the inventors of agriculture, milling, flour, bread, beer, wine, paper, printing, hieroglyphics and writing. They had pioneered painting, sculpture, dentistry and funeral customs and tamed the horse, camel, elephant, dog and cat. Taylor believed the Hamites had introduced civilization to Greece, India, Canaan and North Africa.[41]

Taylor's vision of the fate of Hamitic Civilizations reflected his degenerationist view of human history. He believed that civilizations produced by human intellect without knowledge of the Judeao-Christian God could only reach certain heights before collapse. The great Hamitic civilizations in Mesopotamia, Egypt and India had all collapsed eventually; their knowledge lost to such an extent that 'Ham's degenerate descendants' could now only build 'clay hovels'.[42] Taylor also believed Ham's descendants suffered under Noah's curse. In the Middle East, India and Africa they had been conquered and enslaved by the Semites and the Japhetic Europeans. Taylor believed that a similar historical process had occurred in New Zealand. He believed the Chatham Island Moriori was descended from the original inhabitants of New Zealand, a dark Hamitic people, degenerated through isolation and migration. With the arrival of the Semitic Maori, the original inhabitants had been conquered and enslaved, a process repeated with the Maori invasion of the Chatham Islands in 1835.[43]

Taylor saw the revelation to the Israelites of the knowledge of the one true God as the only real intellectual achievement of the Semites, the children of Shem. The Hamitic Canaanites and Phoenicians had possessed a more culturally advanced civilization, but they were spiritually inferior to the invading Semitic Israelites. Taylor believed that the Semites had produced all the great prophets, including Zoroaster and 'Mahomet, the false prophet'.[44] Taylor pictured the Semitic kingdoms as short-lived compared to those of Ham's children. Meanwhile the descendants of Japhet had settled in Europe and in parts of America living as hunters and gatherers. They later rose in the levels of civilization until their empires conquered those of the Semites. This process began with Alexander conquering the Semitic Persians and continued with the Roman Empire. The modern Christian empires of Europe were now completing the process, God having now made the sons of Japhet 'keepers of his word'. Taylor viewed the British Empire as a tool of God's will, spreading civilization, knowledge and humane behaviour through the world. Taylor held that British culture was superior to that of the rest of the world, but only due to the dispensation of God, rather than any innate British racial or moral qualities. His frequent criticisms of British morality, of discriminatory racist behaviour in the British Empire, and of the decimation of indigenous people in the American and Australian colonies, show that he did not regard the British as by any means perfect. His views on the decline and fall of the empires of the Hamites and Semites could also be seen as a warning to the British that the British Empire would only last if it became and remained a truly Christian empire.[45]

Taylor saw the migrations of the Children of Noah as part of God's plan for humanity, with New Zealand as a microcosm of this plan. The final stage was the expansion of the children of Japhet, carrying the knowledge of European Christianity and its moral code, along with modern science and technology. Taylor believed this process was carrying out God's will by bringing 'freedom, spiritual as well as temporal, to the children of Ham', through the missions and the anti-slavery movement.[46] While Taylor believed that the enslavement of Hamitic peoples fulfilled Noah's prophecy, this was not a justification for slavery. He was a staunch abolitionist who looked forward to a future Christian world of racial equality. God's historical plan for New Zealand was fulfilled with the arrival of the British, bringing with them civilization and true religion to benefit all races. Such an idea is in contrast with many late nineteenth-century thinkers who saw migration and conquest as part of social evolutionary processes leading inevitably to the extinction of weaker groups. The idea of inevitable progress was implicit in Taylor's thinking, but as a fulfilment of God's plans for the world. Taylor was therefore an advocate of British immigration to New Zealand. European, and especially British, imperial expansion brought both scientific and religious knowledge to the rest of the world, creating a situation where 'the three sons of Noah shall again be united as the members of one family'[47] Thus despite his beliefs in equality, his racial theories advocated the divine justification of British Imperialism and Pakeha colonialism.[48]

The vision revised

Taylor's second edition of *Te Ika a Maui*, released in 1870, was a major rewrite of the original. In this new edition, Taylor clarified his ideas on the identity of Maori and of the supposed pre-Maori population of New Zealand. The third chapter of the new edition was entitled 'Our Race and Its Origin'. This was a revised version of the 1867 pamphlet with the same conclusions regarding human origins and migrations.[49] In the new edition of *Te Ika a Maui*, Taylor further developed his theories on the darker, Hamitic people who he believed were the first colonists of the South Pacific. In a completely new second chapter entitled 'The Two Races Which Peopled Polynesia', he clearly identified this earlier people as Melanesians. Realizing that his ideas ran counter to the general contemporary Pakeha view that Maori were New Zealand's first colonists, Taylor presented a wide range of evidence to prove that Melanesians were the original inhabitants of New Zealand.[50] Taylor had not previously speculated on the origins of the Chatham Island Moriori, but in the 1870 *Te Ika a Maui* he portrayed the Moriori as a Hamitic Melanesian people, originally resident in New Zealand. He described the Moriori as another 'degraded race' who went naked, with primitive housing and technology.[51] Taylor's awareness of the Melanesians had been increased by the activities of the Auckland based Melanesian Mission under Bishops Selwyn and Patteson. The fact that Taylor was informed by Patteson's contacts with Melanesia is borne out by the citing of Patteson's journals in the footnotes of chapter two of *Te Ika a Maui*. Such information had not been available when Taylor was writing in the 1850s.[52]

In the 1870 edition of *Te Ika a Maui* Taylor removed most of his references to the Polynesians being one of the ten lost tribes of Israel.[53] It is unclear whether he had ceased to believe that the Polynesians were descended from the Jews or whether he simply had enough doubts about the matter to place little emphasis on it. Taylor repeated his theory that the ancestors of the Polynesians had travelled from the Middle East, through Central Asia: one section branching off to India and the other going through America and eventually reaching Polynesia from the East. In the process of migration, the Polynesian culture degenerated as they moved further from the centres of civilization. Ballantyne shows that Taylor did come to see a strong connection between the Polynesians and India, and placed less emphasis on the possibility that the Polynesians were literally one of the ten lost tribes of Israel. It does appear, however, that Taylor continued to believe that the Maori were a people of Semitic origin, who had reached New Zealand after a long period of wandering.[54]

Conclusion

In Richard Taylor we see an example of a committed Christian clergyman who was also a serious student of natural history, a classic clergyman naturalist. Taylor's Christian beliefs provided the foundation of his world view; the Bible was not simply a source of moral teachings but also a work of history, describing actual events from the early part of the human story. Significantly, Taylor never relied on the scriptures alone in any argument concerning the natural world, as he always looked for the latest scholarly evidence to complement his biblical arguments. He tended to look for the scientific flaws in contemporary theories of social development, evolution and polygeny, rather than attack them solely on religious grounds. In his own attempts to reconcile religion and science, Taylor does not appear to have undergone any crisis of faith in either discipline. His approach evokes that of Paley's natural theology, which saw nature as God's second book. Taylor was obviously aware that many saw the new ideas on geology, evolution and social development as creating a tension between science and religion, but believed that all supposed differences were reconcilable. Taylor took a similar view to that presented in Brietenbach's account in this current book of the Scottish missionary discourse; that scientific thought was in fact reflective of a superior European Christian mode of thinking, aiding rather than contradicting Christianity.[55]

Lester has pointed out that in the 1830s and 1840s humanitarians and settler advocates were involved in a 'discursive struggle' over the meanings of being 'British'. Humanitarians emphasized that the British were in effect older siblings to 'natives'. Colonized people could rise to achieve a level of civilization equivalent to that of the British, if given benevolent rule, fair dealing, respect and Christian teaching. The settler discourse on the other hand emphasized the differences between the British and colonized peoples. The colonized were intrinsically racially inferior and could only be ruled with an appropriate degree of force. Lester argues that by the 1860s the settler discourse had come to dominate both the metropolitan and colonial view of the 'British' and 'native' identities, with the settler discourse complemented by the writings of polygenists and of many evolutionary scientists and ethnologists.[56]

Whether or not the settler discourse had become dominant, Taylor did not abandon his faith in the Christian values of the earlier part of the century. In addition to trying to reconcile science and religion, Taylor tried to explain Maori and other indigenous peoples to the metropolitan British and to Pakeha settlers. Taylor's vision was strongly contrary to Spencerian ideas of the survival of the fittest that were becoming fashionable from the middle of the nineteenth century. He sought to promulgate his views that all humans were descendents of common ancestors, tracing all living human races back to the family of Noah. Colonizer and colonized were members of the same family who through the Christian faith could live together peacefully under British rule.

While strongly holding to the evangelical's belief in the innate equality of all humanity, Taylor also argued that some branches of the human family had degenerated. It is clear that he believed such degradation was a result of circumstance and that Europeans too could fall to such depths under similar conditions. Taylor acknowledged the current superiority of the European children of Japhet, but believed that they had been savages when the Hamites dominated the civilized world and made the innovations on which civilization was based. Both the Hamites and the Semites had eventually gone into decline, due in part to their abandonment of God. Taylor believed the height of Japhetic civilization was represented by the British, who would not fall as long as they were doing God's work, bringing all the children of Noah together in civilization and Christianity. In contrast to the settler idea of the incorrigible savage, Taylor saw all peoples as redeemable in this world and the next. Taylor was well aware that the British had committed terrible crimes in the colonization of America and Australia and believed they must atone for such behaviour by following God's laws. Britain's current dominance was due to the dispensation of God, rather than the innate superiority of the British themselves, and that dispensation could be removed. Taylor's accounts of the rise and fall of civilizations were a warning to the reader that if the British failed to follow God's directions the British Empire too could fall as had the empires of the Hamites and Semites before them.

It is open to question how much influence Taylor had. His ideas may have been too strongly based on the Bible and Christian mythology to be taken seriously by the new generation of scientific scholars emerging in the middle years of the nineteenth century. Taylor had promoted the idea that an earlier population lived in New Zealand before the Maori. This view was widely adopted in the early twentieth century, but its main proponents, Stephenson Percy Smith and Elsdon Best of the Polynesian Society, used Maori sources to support the idea. They rejected Taylor's material, considering his ethnology unreliable due to what they saw as his poor command of the Maori language.[57] The idea of a Semitic origin for the Polynesians was largely rejected by Pakeha scholars from the mid-nineteenth century, but was taken up enthusiastically by many Maori. This Maori view of Semitic origins was due to their own interpretations of the scriptures, rather than any influence from Taylor.[58] Taylor's idea of an Aryan origin for the Polynesians was adopted by many Pakeha scholars and might in fact have been his strongest contribution to the later debates on Polynesian origins.[59]

To Taylor the New Zealand situation was a microcosm of world history. The earliest settlers were dark-skinned Melanesian children of Ham, degraded through many generations of wandering far from the original centres of their civilization. They were conquered by the lighter-skinned

Semitic Maori, a people who Taylor came to believe had migrated through India and America into the South Seas. The final stage of what Taylor saw as God's historical plan for New Zealand was the arrival of the Japhetic British, the lightest skinned and most advanced group of all bringing with them civilization and true religion. For all his beliefs in racial equality Taylor remained an apologist for the Pakeha colonization of New Zealand, seeing the British Empire as part of God's plan for the good of all the world's peoples.

Notes

1. This chapter came about through work on my thesis, which concerned the development of ideas on pre-Maori populations in New Zealand. I must thank my supervisor Michael P. J. Reilly, who pointed out to me the considerable differences between the 1855 and 1870 versions of Richard Taylor's book *Te Ika a Maui*, an observation from which this chapter eventually developed.
2. There is a wide range of literature on science, religion and race in the nineteenth century. Relevant works include N. Stepan, *The Idea of Race in Science* (London: Macmillan, 1982); G. W. Stocking, *Victorian Anthropology* (New York: Free Press, 1987); *Science and Religion in the Nineteenth Century*, ed. T. Cosslett (Cambridge: Cambridge University Press, 1984); J. H. Brooke, *Science and Religion: Some Historical Perspectives* (Cambridge: Cambridge University Press, 1991); R. M. Young, *Darwin's Metaphor: Nature's Place in Victorian Culture* (Cambridge: Cambridge University Press, 1991); J. H. Brooke and G. N. Cantor, *Reconstructing Nature: The Engagement of Science and Religion* (Edinburgh: T & T Clark, 1998); *Disseminating Darwinism: The Role of Place, Race, Religion and Gender*, ed. R. L. Numbers and J. Stenhouse (Cambridge: Cambridge University Press, 1999); A. N. Wilson, *God's Funeral* (London: John Murray, 1999); P. Brantlinger, *Dark Vanishings: Discourse on the Extinction of Primitive Races, 1800–1930* (Ithaca, NY: Cornell University Press, 2003). Still of real value and covering exactly the period of the current article is W. E. Houghton, *The Victorian Frame of Mind, 1830–1870* (New Haven, CT: Yale University Press, 1957). On the clash between the theologians and the young scientists over the professionalisation of science see in particular J. R. Lucas, 'Wilberforce and Huxley: A Legendary Encounter', *The Historical Journal*, XXII, no. 2 (June 1979) 313–30; F. M. Turner 'The Victorian Conflict between Science and Religion: A Professional Dimension', *Isis*, LXIX (1978) 356–76; S. Gilley, 'The Huxley-Wilberforce Debate: A reconsideration', in *Religion and Humanism: Studies in Church History, Vol. 17*, ed. K. Robbins (Oxford: Basil Blackwell, 1981), pp. 325–40; and A. Desmond, *Huxley: The Devil's Disciple* (London: Michael Joseph, 1994).
3. For discussions of the hardening of racial attitudes, including theories of polygeny or distinct racial origins, see G. W. Stocking, 'What's in a Name? The Origins of the Anthropological Institute, 1837–1971, *Man*, VI (1971) 370–90; C. Bolt, *Victorian Attitudes to Race* (London: Routledge, 1971); D. A. Lorimer, *Colour, Class, and the Victorians* (Leicester: Leicester University

Press, 1978); N. Stepan, *The Idea of Race in Science*, 1800–1960; R. J. C. Young, *Colonial Desire: Hybridity in Theory, Culture, and Race* (London: Routledge, 1995); Brantlinger, *Dark Vanishings*. On the influence of the settler lobby see A. Lester, 'British Settler Discourse and the Circuits of Empire', *History Workshop Journal*, LIV (2002) 24–48.

4. Throughout this chapter I have used the Maori word 'Pakeha' for the European settlers in New Zealand. This name was used by Maori from at least the 1830s and has since been adopted by many New Zealanders of European descent as a self-descriptive term. H. W. Williams, *Dictionary of the Maori Language* (Wellington: Legislation Direct, 1971), p. 252.
5. For recent general accounts of New Zealand history over the period in question see J. Belich, *Making Peoples: A History of the New Zealanders: From Early Polynesian Settlement to the End of the Nineteenth Century* (Auckland: Penguin Books, 1996): M. King, *The Penguin History of New Zealand* (Auckland: Penguin Books, 2003). In 1840 it is estimated that the Maori population of New Zealand was around 70,000 to 90,000, while the Pakeha population was less than 2,000. By 1860 the Maori population had declined to around 60,000, while the Pakeha population increased rapidly through immigration. See I. Pool, *Te Iwi Maori: A New Zealand Population Past, Present and Projected* (Auckland: Auckland University press, 1991).
6. J. M. R. Owens, 'Richard Taylor', in *The Dictionary of New Zealand Biography, Vol. 1, 1769–1869* (Wellington: Allen and Unwin/Department of Internal Affairs, 1990), pp. 437–8. For a more in-depth account of Taylor's life, see J. M. R. Owens, *The Mediator: A Life of Richard Taylor, 1805–1873* (Wellington: Victoria University Press, 2004).
7. E. Elbourne, 'The Sin of the Settler: The 1835–1836 Select Committee on Aborigines and the Debates over Virtue and Conquest in the Nineteenth Century British White Settler Empire', *Journal of Colonialism and Colonial History*, IV, no. 3 (2003), http://muse.jhu.edu/journals/journal_of_colonialism_and_colonial_history/ (accessed 2 August 2006); Lester, 'British Settler Discourse', pp. 25–30.
8. I. Wards, *The Shadow of the Land: A Study of British Policy and Racial Conflict in New Zealand, 1832–1852* (Wellington: Historical Publications Branch, Department of Internal Affairs, 1968), chs. 1–2.
9. C. Orange, *The Treaty of Waitangi* (Wellington: Allen and Unwin/Port Nicholson Press, 1987), p. 126. See also ibid., chs. 4–6; R. Burrows, *Extracts from a Diary Kept by Rev R Burrows during Heke's War in the North, in 1845* (Auckland: Upton and Co., 1886), pp. 6, 9.
10. Lester, 'British Settler Discourse', pp. 40–4.
11. The first Church of England mission to New Zealand was established in Northland by the Reverend Samuel Marsden, chaplain to the New South Wales penal colony, in 1814. The mission operated under the auspices of the Church Missionary Society. The first Maori converts were not made until over ten years after the first mission was established. The first Wesleyan mission was established in New Zealand in 1824, by the Wesleyan Missionary Society. The CMS and WMS missionaries generally operated in an uneasy alliance. Roman Catholic missionaries did not arrive in New Zealand until 1838 when French Marists under Bishop Jean Baptiste Pompallier established a base in New Zealand. Both Anglican and Wesleyan missionaries were

generally horrified at the arrival of French 'papists'. See Belich, *Making Peoples*, pp. 134–7, 164–9; King, *History of New Zealand*, pp. 131–50. On Taylor's Church Missionary Society in New Zealand see *Mission and Moko: Aspects of the Work of the Church Missionary Society in New Zealand*, ed. R Glen (Christchurch: Latimer Fellowship of New Zealand, 1992). For a broader context of the Christian missions to New Zealand see A. K. Davidson, *Christianity in Aotearoa: A History of Church and Society in New Zealand,* 3rd edn (Wellington: Education for Mission, 2004) and I. Breward, *A History of the Churches in Australasia: Oxford History of the Christian Church* (Oxford: Oxford University Press, 2001). *Christianity, Modernity and Culture: New perspectives on New Zealand History*, ed. J. Stenhouse (Adelaide: ATF Press, 2005) is also valuable and includes discussion on Natural Theology in New Zealand.

12. Owens, 'Richard Taylor', 437–8. It should be noted that the Maori spelling of the river and surrounding district is Whanganui, whereas the settler town, district and associated institutions are spelt Wanganui. For an account of Taylor's travels see A. D. Mead, *Richard Taylor: Missionary Tramper* (Wellington: A. H. & A. W. Reed, 1966).
13. The British naturalists' appreciation of Taylor's work was shown at a very early stage. In 1840, Adam Sedgewick, a personal friend, successfully nominated Taylor as a Fellow of the London Geological Society. Taylor's co-nominators with Sedgewick were the Rev. William Branwhite Clarke, a pioneer of Australian geology, and Charles Darwin. See Owens, *The Mediator,* pp. 67–8.
14. Paul Du Chaillu (1835–1903) was a French-American explorer and ethnologist who travelled extensively in West Africa in the 1840s–1860s and in Lapland in the 1870s. Bishop Reginald Heber (1783–1826) was an English missionary and hymn writer who travelled extensively in India. Dr Francis Devay was a French medical doctor who in the 1850s was attached to the Hôtel Dieu and in the 1860s was Professor of Clinical Medicine at Lyons.
15. Thomas Brunner (1821?–1874) was an English surveyor who won fame for his explorations of the rugged West Coast of New Zealand's South Island in the 1840s. Ferdinand von Hochstetter (1829–1884); he was a German geologist who sailed with the Austrian *Novara* expedition in 1857.
16. A. Anderson, *Prodigious Birds: Moas and Moa Hunting in New Zealand* (Cambridge: Cambridge University Press, 1989), 11–13; R. Taylor, *Our Race and Its Origin* (Auckland: George T Chapman, 1867), pp. 4–6, 8, 16–18, 28; R. Taylor, *The Age of New Zealand* (Auckland: George T Chapman, 1866), pp. 15, 18–22; Owens, *The Mediator*, pp. 67–8, 187, 201–2, 251, 271.
17. On the use of ideas from scripture to explain human differences see W. M. Evans, 'From the Land of Canaan to the Land of Guinea: The Strange Odyssey of the Sons of Ham', *American Historical Review*, LXXXV (1980) 15–43.; A. David, 'Sir William Jones, Biblical Orientalism and Indian Scholarship', *Modern Asian Studies*, XXX, no. 1 (1996) 173–784; B. Braude, 'The Sons of Noah and the Construction of Ethnic and Geographical Identities in the Medieval and Early Modern Periods', *William and Mary Quarterly*, XLIV (1997) 103–42.
18. T. Parfitt, *The Lost Tribes of Israel: The History of a Myth* (London: Weidenfeld and Nicolson, 2002).

19. M. P. K. Sorrenson, *Maori Origins and Migrations: The Genesis of Some Pakeha Myths and Legends* (Auckland: Auckland University Press, 1979), pp. 14–16. On the debates over the settlement of the Pacific see also K. R. Howe, *The Quest for Origins: Who First Discovered and Settled New Zealand and the Pacific islands?* (Auckland: Penguin Books, 2003). The idea of a Semitic origin for Maori was largely rejected by Pakeha scholars by the late nineteenth century, but gained a widespread acceptance among nineteenth century Maori, an acceptance that has continued in some circles to this day. This was particularly the case among prophetic movements such as Pai Marire and Ringatu.
20. Owens, *The Mediator,* p. 202.
21. R. Taylor, *Te Ika a Maui or New Zealand and its Inhabitants* (London: Wertheim and Macintosh, 1855), pp. 6–8.
22. Ibid., p. 8.
23. Ibid., pp. 184–90.
24. Ibid., p. 190.
25. Ibid., pp. 8, 189–91. Ballantyne argues that Taylor was not convinced of the Semitic origin of the Polynesians, citing that Taylor's use of terms such as 'venturing', 'seem' and 'hint' in proposing a Semitic origin to the Maori indicate Taylor's lack of faith in the theory. T. Ballantyne, *Orientalism and Race: Aryanism in the British Empire* (Basingstoke, Houndmills: Palgrave, 2002), pp. 64–5.
26. Taylor, *Te Ika a Maui* (1855), pp. 2–9.
27. The paper appears to have been developed from an earlier lecture on 'races of men', which Taylor gave in 1864. Owens, *The Mediator,* p. 253.
28. Japhet is referred to in the King James Bible as Japheth, but Taylor in his published works always uses the name Japhet.
29. Taylor, *Our Race,* pp. 6, 9–10, 14, 16–17, 19, 22, 25, 33–4.
30. In Britain polygenist ideas were particularly championed by members of the Anthropological Society. The more moderate Ethnological Society was dominated by thinkers with a monogenist view of human development, but included a few polygenists such as the Society's President, Crawfurd. See Stocking, 'What's in a Name?', pp. 370–90; Stepan, *The Idea of Race in Science,* pp. 41–6; Lorimer, *Colour, Class, and the Victorians,* p. 138; Young, *Colonial Desire,* pp. 14–16.
31. Taylor, *Our Race,* pp. 8–17, 25. It should be noted that there was little support for polygenist views within the small New Zealand scholarly community.
32. Ibid., pp. 4–87, 8, 22–3, 34; Taylor, *Te Ika a Maui or New Zealand and Its Inhabitants,* 2nd edn (London: William Macintosh, 1870), p. 65. The fact that *Vestiges of the Natural History of Creation* was written by the Scottish publisher Robert Chambers (1802–1871) was not revealed until 1884, when the twelfth edition of *Vestiges* was released. For more detail on the reception of Darwinian theories in New Zealand see J. Stenhouse, 'The Wretched Gorilla Damnification of Humanity', *New Zealand Journal of History,* XVIII, no. 218 (1984) 143–62 and J. Stenhouse, 'Darwinism in New Zealand, 1859–1900', in ed. Numbers and Stenhouse, *Disseminating Darwinism,* pp. 61–90.
33. Taylor, *Our Race,* pp. 9, 25–8, 28n.
34. Ibid., p. 28.
35. Ibid., pp. 23–4.
36. Ballantyne, *Orientalism and Race,* p. 66; Taylor, *Our Race,* p. 26.

37. Ibid., p. 27.
38. Ballantyne, *Orientalism and Race,* pp. 28–9; David, 'William Jones', pp. 173–84.
39. Taylor, *Our Race,* pp. 24–6.
40. Braude, 'Sons of Noah', pp. 103–42, see in particular pp. 134, 138; Evans , 'Land of Canaan', pp. 33–4, 39–43. For an example of ante-bellum Southern acceptance of the scriptural basis of the 'curse of Ham' on black Africans see S. R. Haynes, 'Original Dishonor: Noah's Curse and the Southern Defense of Slavery', *Journal of Southern Religion,* III (2000) http://jsr.as.wvu.edu/honor.htm (accessed 16 June 2004).
41. Taylor, *Our Race,* pp. 25–7, 31–4.
42. Ibid., p. 34.
43. Ibid., pp. 16, 34–6. Taylor, *Te Ika a Maui* (1855), pp. 7–8. The Moriori of the Chatham Islands or Rekohu, located 870 kilometres to the east of New Zealand, are in fact a Polynesian people closely related to the Maori. In 1835 the islands were invaded by Ngati Tama and Ngati Mutunga Maori, originally from Taranaki, who enslaved and decimated the indigenous population. See Michael King, *Moriori: A People Rediscovered* (Auckland: Viking, 1989).
44. Taylor, *Our Race,* p. 35.
45. Ibid., pp. 29–31, 36. Taylor, *Te Ika a Maui* (1855), pp. 2, 9.
46. Taylor, *Our Race,* pp. 34–6.
47. Ibid., p. 36.
48. Ibid., pp. 26–7, 36; Taylor, *Te Ika a Maui* (1855), pp. 268–9; Taylor, *Te Ika a Maui* (1870), pp. 89–90.
49. Ibid., ch. 3.
50. Ibid., pp. 13, 16–17.
51. Ibid., pp. 7, 12–13, 16–19.
52. Taylor, *Our Race* , pp. 16, 34; Taylor, *Te Ika a Maui* (1870), pp. 15–21, 16n; K. R. Howe, *Where the Waves Fall* (Sydney: Allen and Unwin, 1984), pp. 303–7.
53. Taylor, *Te Ika a Maui* (1855), pp. 8, 190–2. The corresponding passages, without the lost tribes references, in *Te Ika a Maui* (1870), are pp. 7, 392–3. There is one reference in ibid., p. 82, to the ancestors of the Malays and Polynesians as being of 'the ten tribes'. Given that other references to the lost tribes have clearly been removed, this surviving reference is a puzzle.
54. Ibid., pp. 33–59. Ballantyne, *Orientalism and Race,* pp. 64–6.
55. See E. Breitenbach's chapter, (chapter 5) in this volume.
56. Lester, 'British Settler Discourse', pp. 42–4.
57. On the discussions over a pre-Maori population in New Zealand see P. Clayworth, '"An Indolent and Chilly Folk": The Development of the Idea of the Moriori Myth' (PhD Thesis, University of Otago, 2001).
58. Elsmore, *Mana from Heaven*; Mikaere, *Te Maiharoa*; Binney, *Redemption Songs.*
59. Ballantyne, *Orientalism and Race,* pp. 62–6.

Further reading

Ballantyne, T. *Orientalism and Race: Aryanism in the British Empire* (Basingstoke, Hampshire: Houndmills, 2002).

Belich, J. *Making Peoples: A History of the New Zealanders: From Early Polynesian Settlement to the End of the Nineteenth Century* (Auckland: Penguin Books, 1996).

Brooke, J. H. *Science and Religion: Some Historical Perspectives* (Cambridge: Cambridge University Press, 1991).

Cosslett, T. M., ed. *Science and Religion in the Nineteenth Century* (Cambridge: Cambridge University Press, 1984).

Davidson, A. K. *Christianity in Aotearoa: A History of Church and Society in New Zealand,* 3rd edn (Wellington: Education for Mission, 2004).

Glen, R., ed. *Mission and Moko: Aspects of the work of the Church Missionary Society in New Zealand* (Christchurch: Latimer Fellowship of New Zealand, 1992).

Howe, K. *The Quest for Origins: Who First Discovered and Settled New Zealand and the Pacific Islands?* (Auckland: Penguin Books, 2003).

Numbers, R. L. and J. Stenhouse, eds *Disseminating Darwinism: The Role of Place, Race, Religion and Gender* (Cambridge: Cambridge University Press, 2001).

Owens, J. M. R. *The Mediator: A Life of Richard Taylor, 1805–1873* (Wellington: Victoria University Press, 2004).

Parfitt, T. *The Lost Tribes of Israel: The History of a Myth* (London: Weidenfeld and Nicolson, 2002).

Taylor, R., *Te Ika a Maui or New Zealand and its Inhabitants* (London: Wertheim and Macintosh, 1855).

12

From African Missions to Global Sisterhood: The Mothers' Union and Colonial Christianity, 1900–1930

Elizabeth E. Prevost

Recent studies of feminist networks in the British Empire have highlighted the difficulties of reconciling universal claims of sisterhood with competing hierarchical categories of race, class and nation. Although the discourse of 'imperial feminism' which emerged in Britain in the late nineteenth and early twentieth centuries projected womanhood as a common basis of oppression and source of empowerment, many scholars have argued that British women's reform movements legitimized the 'civilizing mission' through the oppositional image of non-white and working-class woman's degradation, by claiming that only fellow women could rescue their sisters from the Indian *zenana* or the East End slum.[1] Since feminists, philanthropists and missionaries could only establish their own authority outside the home by casting British womanhood against the uncivilized female 'other', white, middle-class women won their own emancipation at the expense of their 'heathen' sisters, preventing any real possibility of partnership between them. Scholars of gender and empire have thus underscored how these missions of sisterhood established a divisive legacy of British women 'speaking for', rather than with, non-Western women.

However, this chapter considers the case of one international women's organization to suggest that historicizing the discourse of imperial sisterhood from the *periphery* rather than the metropole yields a more complicated story – one of solidarity and empowerment as much as fragmentation and subjugation. Furthermore, it takes religion as a starting point rather than a backdrop for approaching women's transnational humanitarian interventions, to understand the specific Christian framework in which women construed their social and professional 'emancipation.'[2] Using gender and spirituality as central categories of analysis reveals that empire constituted a pervasive but not exclusive or static idiom in which British women promoted the globalization of Christianity and the emancipation of women. In fact, Anglican women eventually used a feminized version of Christianity to interrogate and critique the moral justifiability of empire. Thus, although missionary and humanitarian projects often worked to promote the empire at home, religious institutions also shaped imperial discourse by highlighting the economic and racialized inequalities of colonialism.[3]

The Mothers' Union (MU) offers an illustrative way of approaching the gendered relationship of Christianity and empire, for as the largest Anglican women's organization in Britain and worldwide, it was in a unique position to promote and mediate women's interactions between centre and periphery. Founded in 1876 by Mary Sumner as an Anglican women's voluntary association, the MU placed a high priority on marriage and motherhood as the foundation of church, nation and empire.[4] The MU's objectives and its limitation of full membership to married women earned it a widespread reputation for anchoring the moral and spiritual foundation of the body politic; further, the overseas expansion of the MU from 1900–1930 vested family life with explicitly imperial responsibility.[5] After the end of the South African War in 1902, the MU began exporting the organization as a way of reaching out to isolated colonial wives and consolidating the British diaspora; membership grew quickly thereafter in the Dominions (South Africa, Canada, Australia and New Zealand). By the time of the first Worldwide Conference in 1930, the MU listed its membership at over half a million worldwide, encompassing 62 dioceses and 10,000 branches in Britain, and 108 dioceses and 2,200 branches overseas.[6] These latter figures also included large numbers of non-white women. Around 1900, individual missionaries working through other Anglican societies had begun to launch indigenous MU branches in Antigua, Jamaica, Nigeria, South Africa, India, Burma, Japan and Hong Kong, among others.

One of the earliest and most successful indigenous branches was initiated by the Madagascar mission, and the Madagascar MU subsequently took a leading role in globalizing the Union. A focus on Madagascar offers both an instructive and unique case of how British women construed female authority and solidarity through an explicitly Christian rather than nationalist lexicon. The island did not fall within the boundaries of the British Empire, which perhaps made it easier for

Anglican missionaries to critique the social consequences of European conquest.[7] The French colonization of Madagascar in 1895 prompted British women to institute the MU as a way of combating the alleged secularism and moral laxity of colonial rule (manifested particularly in prostitution and divorce). Yet the MU's membership and leadership quickly grew independent of British women's agenda: the number of Anglican women missionaries working in Madagascar in the early twentieth century amounted to no more than a dozen at any given time, but the MU garnered a diocesan membership of at least 1,000 Malagasy women in its first few years, and Malagasy women took responsibility for running most of the local branches. Therefore, while the MU had its roots in Western norms of femininity and social relations, it also facilitated indigenous women's religious authority.[8]

The evolving relationships between Malagasy and British women reveal how the tension between colonial constructions of difference and Christian constructions of universalism, also examined elsewhere in this volume,[9] had deeply gendered implications. British and indigenous women in Madagascar used the MU to claim a shared spiritual connection that was capable of forging personal and corporate relationships across space and culture; on the other hand, uncertainty about whether Christian womanhood constituted an innate or learned condition posed a perpetual challenge to this unanimity. The possibilities and contradictions which grew out of women's work in the Madagascar mission field ultimately informed the collective MU's response to the upheavals of the early twentieth century, at a time when other missionary agencies were seeking similar ways of reconciling international conflict by reinventing the global meaning of Christian community.[10]

This chapter discusses the global expansion of the MU to show more broadly how the Christian periphery functioned as a site of both the making and the unmaking of British imperial identity. First, the MU shifted the locus of female evangelism from an imperial civilizing mission to an international association of Christian women. The impetus for this ideological shift came from both missionary and African agency: the colonial MU formulated an idiom of universal sisterhood and a network of spiritual solidarity which were subsequently appropriated by the metropolitan MU. Second, the spiritual focus of the MU dynamically shaped British and indigenous women's encounters in both mission field and metropole. Although they did not take place on equal footing, these encounters were mutually transformative, creating a hybridized version of Anglican Christianity which legitimized new possibilities for women's religious authority and leadership in an otherwise patriarchal structure. The case of the MU thus illustrates that Western women's religious interventions produced neither universal sisterhood nor insurmountable hierarchy, but rather paved a complex and shifting middle ground between these two extremes.

The contested 'Bonds of Motherhood' in Madagascar

The colonial church was not confined to the areas of British red on the map, and missions were often caught in the crossfire of the European partition of Africa. The MU entered the evangelistic arena in Madagascar in 1902, a time when the French displacement of the Protestant Merina state had discredited pre-existing models of female evangelism and prompted a shift in the means and the objectives of Anglican women's mission work. The French instatement of secular and non-vernacular language policy in schools undermined the model of Anglican girls' education on which missionaries had previously relied to convert women and facilitate female ministry. Instead, missionaries used the MU to redirect the tenor of evangelism onto marriage and motherhood, enabling both British and Malagasy women to redefine the scope of the Anglican religious community and to confront the challenges of colonial rule and culture. Gertrude King was the most visible force behind this effort. The sister of the bishop of Madagascar, she worked as a missionary in Madagascar from 1900 until 1920, when she returned to England and worked as the overseas secretary for the MU in London.

The first two decades of the twentieth century in Madagascar were marked by the French colonial government's discrimination against Protestant mission work. Already perceived as a threat because of its support of the Merina state, the Anglican mission also became linked with British nationalism and imperial competition.[11] The Malagasy themselves bore the brunt of the persecution; a special prayer published with a list of other intercessions in 1911 entreated 'that it may please Thee to overrule for good the present restrictions upon our religious work, and in due time give our natives that liberty to meet for worship in their homes which is now denied them.'[12] The year 1911 was also a difficult year for missionaries: Anglican schools were closed and church building projects halted, while other missionary organizations – notably the London Missionary Society (LMS) – were forced to curb evangelism altogether.[13] The hostility of the colonial government to Protestant evangelization thus created a counter-cultural, sometimes underground movement to sustain Anglican practices.

For missionaries, the challenges rendered by formal colonial policies were manifested in the deteriorating moral fabric of colonial society. The onset of French control transferred the object of missionary anxiety from 'heathenism' to multicultural secularism. The social and economic dislocation rendered by war and famine in the mid-and late 1890s, the creolization of coastal and urban centers and the growth of colonial garrisons were all construed by missionaries as a moral threat. In particular, the increase in prostitution and colonial concubinage instigated a female missionary campaign to preserve the racial and cultural purity of the Malagasy through the vehicles of Christian marriage and domesticity. In Africa as in Britain, 'mothercraft' provided a means of regulating social, moral and racial health.[14]

Out of this dual sense of religious and moral crisis emerged a new narrative of female evangelism, posing the MU as the vanguard of a spiritual movement which maintained continuity between the pre-colonial and colonial phases of mission work. A report from Deaconess Blanche Porter in 1916 deployed the militant language of nineteenth-century 'muscular' Christianity in the service of twentieth-century women's work: armed against the secular policy of the French colonial state by their education in mission schools during the previous generation, MU members were fighting to reinstate a Christian community. The report posed the MU as a powerful anti-colonial agent which afforded women the means to critique and transform contemporary society.

> With the coming of the French the children had been taught atheism in the Government schools... The French had brought with them a great wave of worldliness, and only those who took part in it were counted as anything at all. The Mothers' Union was trying to stand for belief in Christ and purity of life.[15]

This characterization encapsulated the shifting terms and objectives of women's mission work in Madagascar in the early twentieth century. Unlike earlier rationales that had situated Madagascar's evangelization in schools and education, the MU's most powerful leverage now came through prayer and intercession, tools that required no literacy and which assumed women's innate spirituality. Furthermore, women's evangelism took responsibility for counteracting the secularism of imperial policy and the 'worldliness' of colonial society, empowering the Malagasy re-appropriation of Christianity through the female bond created by the MU. But women's encounters also forced a re-evaluation about what constituted that bond, and what Christian womanhood looked like.[16]

The MU began on the initiative of missionaries already working in conjunction with the Women's Mission Association of the Society for the Propagation of the Gospel (SPG) in Madagascar.[17] Gertrude King started the first branch in August 1901 in the capital, Antananarivo (Tananarive during the colonial period). Within the first few months 80 'earnest Christian mothers' were attending probationers' classes.[18] After translating the members' cards into Malagasy, King appealed to Mary Sumner to have them printed in England, claiming 'it would be the very greatest link between our colored women and their white sisters'.[19] By 1906 there were 250 members in four branches 'who meet monthly for prayer and instruction' in and around the capital; and by 1912 there were 1,000 members in 17 branches in Imerina and on the eastern coast.[20] This rapid spread was not connected to education, unlike in the previous generation of female evangelism when there had been a strong corollary between the Merina state's education policies and missionary services. And unlike the evangelical Church Missionary Society, the SPG did not tie its mission work so closely to the written word or require converts to read as a prerequisite of baptism.

The MU missionaries therefore relied on the spoken word, prayer and meditation to forge a religious connection with Malagasy women and to encourage indigenous female leadership.[21] In fact, according to King, the earliest MU work did not even require a common language. Unlike educational and medical dimensions of mission work which required professional skills and linguistic proficiency, King argued that the MU's brand of evangelism relied only on a gendered emotional connection:

> Nine faithful Malagasy mothers... sat in a circle on the rush-mats and listened with eyes as well as ears to the Mothers' Union message. West and East met in the bond of Motherhood, and what the faltering lips of the Western messenger could not frame in a foreign tongue, the Eastern mothers learnt by intuition.[22]

King herself was not married, but clearly she felt herself to be an equal member of a global 'tribe' of women who were united in a 'bond of motherhood'. In the course of their devotional encounters with African women in church services and prayer meetings, missionaries discovered a shared sense of womanhood that mediated language barriers and encompassed divergent cultural systems. King experienced a particular commonality with African women at the communion rail, through their shared participation in the sacrament: 'I cannot express to you the overpowering joy of the fellowship of womanhood that has never failed to come when we are kneeling side by side at the Sunday Eucharist.'[23] Therefore, if the MU began in Madagascar as a didactic instrument for cultivating European forms of femininity and domesticity, it also projected a universal ideal of female spirituality that had unexpected outcomes for *British* women's conception of Christian womanhood.

The MU's growing commitment to female unity and leadership opened the possibility of circumventing cultural difference and moulding a mutual space for Christian women's religious authority. Yet this premise also raised the problem of how to connect *through* womanhood in the context of different ideas about *what* womanhood constituted, and in practice, missionaries found it difficult to reconcile their claims of spiritual equality with the MU's culturally specific understanding of morality and femininity. Nowhere was this tension more charged than in the issue of divorce. The MU in Madagascar determined membership candidacy according to an individual's communicant status, in the belief it would be 'in itself a guarantee of moral fitness'.[24] 'Communion' amounted to good standing with the church, and it required confirmation as well as baptism. The diocese of Madagascar also adhered to all rules laid down by the Lambeth Conference of Anglican bishops in determining communicant standing, which after 1908 included a policy concerning marriage and divorce.[25] The church and the MU operated uniformly in disqualifying divorcees from either communion or membership, which, according to the bishop, earned the Anglican Church the reputation of maintaining the strictest moral code of any church operating in Madagascar.[26] The Madagascar MU was the first diocesan affiliate outside Britain to articulate this contingency of membership on communicant status, a policy which the central MU eventually adopted in codifying a worldwide standard of moral discipline in 1913 (and which proved controversial among missionary branches of the MU in other parts of the world, as the next part will discuss).[27] Yet divorce posed an ideological inconsistency in the Madagascar mission in that it did not *irrevocably* bar women from communion the way it did from the MU: women who had been through the divorce courts could potentially be readmitted to communion but not to the MU, even in cases where the wife had been the unwitting victim of desertion.[28] The unbending position the MU took in excluding divorcees thus tended to undermine its vision of corporate Christian womanhood.

This discrepancy between church and MU policy did not escape the notice of Malagasy women, who objected to the MU's doctrinaire claims of moral authority. In 1916, the first conference of Malagasy branch leaders in the Tananarive district, representing about 900 members, challenged that 'fallen members' could return to communion but not to MU membership.[29] George King responded to Malagasy perplexity on the divorce issue by qualifying that 'Communion is necessary to salvation, and belonging to the MU is not. Also the MU exists for a specific purpose – the upholding of the Sanctity of Marriage – and therefore cannot re-admit to membership those who have acted in such a way as to lower it.'[30] The stakes of salvation were higher than membership in a women's organization and were therefore more difficult to impose. This explanation de-emphasized the MU's ecclesiastical function, separating it from the church's sacramental authority. Yet Bishop King also verified that church standing was indeed more tolerant than the MU's requirements for marital sanctity.[31] Thus, in forcing Anglican leadership to acknowledge the faultlines of moral and doctrinal authority, the divorce debate also exposed the contested hierarchies of race and gender.

The MU's rigid stance on divorce sat unevenly with its attempts to reconcile divergent understandings of marriage and family life through an overarching spiritual affinity. Early discussions surrounding the adoption and adaptation of the membership 'rule' and prayer conveyed missionaries' genuine hope that the MU would ultimately embrace cultural diversity, rather than trying to fit Malagasy Christianity into a metropolitan framework. The first nine Malagasy women to join the MU in 1901 met with King 'to consider the objects of the Union and a rule suited to their own lives which could affiliate them with English mothers over the seas'.[32] Membership was therefore a matter of negotiation from the start, and it was understood that the Malagasy MU was intended to be an 'affiliate' rather than a replication of the British MU, tied together by prayer and devotion rather than cultural prescriptions. King contrasted the MU Prayer, which translated 'most beautifully into Malagasy', with the MU Rule, which laid down a series of culturally specific regulations (such as 'public-house dangers' and 'European sleeping accommodations') which 'have no place in their daily lives!'[33] King felt that while it was unrealistic to follow the official English Rule, 'our objects are the same and the simple rule is one which every communicant could and should keep of daily prayer, ... regular communion, purity in body and mind, and great care of their children'.[34] In reinterpreting the regulatory tenets of the MU, King cast prayer, devotion, morality and motherhood as gendered rather than acculturated expressions of Christianity.

King and others worked to define different forms of authority to reflect these principles and to empower the MU's expansion independent of European control. A number of prominent Malagasy women (including wives of clergy and teachers) assumed secretarial positions; however, because many of the earliest MU members could not read or write, secretariats did not always present a ready outlet for leadership, particularly in rural branches.[35] This limitation necessitated creating other leadership positions appropriate to local situations. In 1916, King organized the first conference for 22 Malagasy leaders of country branches, deliberately designating them 'leaders' instead of secretaries because the authority they represented was not contingent on literacy.[36] In 1918 she convened another meeting for 20 *mpitarika* or 'roll-keepers' from country villages, along with ten MU 'Visitors' who were responsible for home evangelism.[37] Mission reports credited the spread of membership and leadership to the MU's cultural flexibility: '[T]he real usefulness of the Mothers' Union is found in its adaptability to the social customs of the Malagasy, and this makes it, as it were, quite "at home" wherever it is introduced.'[38]

The expanding scope of Malagasy women's participation in the mission was therefore critical in developing an explicitly religious female community, and, moreover, in forging a relationship between this community and the MU worldwide. King always took pride in what she considered the leading example of Madagascar, as 'the first "Black Branch" of the MU', in promoting both indigenous evangelism and a larger sense of Anglican sisterhood.[39] King wrote a report for *Mothers in Council* describing a typical meeting that would end with prayers for 'fellow-members throughout the world, that we may all follow Christ with pure and obedient hearts and minds. Then, in the prayer of silence which closes the meeting, our hearts fly across the seas, and we feel very near to our English fellow-members'.[40] In 1910 the Madagascar MU began formalizing this connection with the metropole through an official 'Link' with a British branch.[41] Soon thereafter, the central MU in London adopted Madagascar's scheme as a way of bringing its increasingly global membership into closer contact with the metropole. Under this system, either a British or overseas branch could initiate an epistolary relationship with the other to exchange prayers, learn about one another's culture and create a sense of Christian community across vast distances.

Although the missionary and metropolitan MU launched the initiative, African women took a leading role in fostering and consolidating these links. Malagasy members asked British correspondents for prayers on behalf of their families and communities, sent gifts of handiwork for devotional use in Britain and expressed condolences when they received news of bereavement. African members established a level of spiritual intimacy with British women through the mutual experience of prayer and the ultimate promise of salvation, thus overriding the temporal distance of the present life with the unity of the next. One letter from a Malagasy branch to the MU founder Mary Sumner in 1912 entreated 'that we may all have one meeting place in the Rest of Paradise when we shall see each other every day'.[42] This strategy of collapsing distance and difference through the promise of salvation was typical of Malagasy women's correspondence with MU members in Britain. Conversely, MU missionaries sometimes invoked the distinction between temporal and spiritual equality to circumvent the ideological problem of maintaining racialized hierarchies. On one occasion, King felt compelled to respond to the metropolitan concern that the expansion and consolidation of the MU threatened the dominance of white womanhood in Christianity. 'Our Malagasy Mothers are very humble, loving, faithful members of the Union, and they value their link with the English Mothers without trying to put themselves on the same level.'[43]

Thus, in the years before the Great War, the Madagascar MU fostered connections first between white and African women in the mission field, and then between the Anglican periphery and centre. The colonial MU laid a foundation for women's Christian solidarity based on the individual and corporate capacity of women to connect through prayer, and in the process, fostered unprecedented opportunities for indigenous women's leadership in the Anglican Church. Yet this was neither a coherent nor a stable process. The MU established a feminized alternative to church and mission authority and articulated an expansive vision of corporate community, but it placed limits on who merited inclusion in that community. The Great War and its aftermath provided an unexpected opportunity to interrogate on a wider scale these contradictions between universal claims of womanhood and rigid moral censure, between global unity and colonial hierarchies.

From temporal to spiritual empire

In Britain as in Madagascar, the climate of anxiety which legitimized a racialized discourse of national and imperial health also led Anglican women to seek cultural transformation and regeneration through the locus of female Christianity. The First World War gave new urgency to this objective and called into question the relationship between Britain's temporal and spiritual empire. By asserting women's conciliatory influence in contrast to the fractured conditions which men had rendered through empire-building and warfare, the MU articulated a vision of female religiosity as a vital intermediary between temporal displacement and divine redemption. The colonial MU took a leading role in articulating this discourse, which in turn led to an unprecedented effort to build an international coalition of Anglican women. At the same time, this effort exposed some of the more problematic tensions in the MU's universalist vision of Christian sisterhood.

Before the war, the MU cast its spiritual networking in an imperial framework of Protestant responsibility and British paternalism. Maud Montgomery, wife of Bishop Montgomery, the Secretary of the SPG, argued in 1912 that

> the Mothers' Union would not be true to her Imperial ideals if she were not making some effort to keep in touch with the scattered branches in the larger Great Britain over the seas ... We glory in the fact that the Mothers' Union consists of mothers of every race and color, and it has been found to be the most effective means of building up the Christian mothers in the faith and teaching them the Christian ideals of motherhood and the sanctity of family life.[44]

The MU in Britain almost apotheosized the sacred role of home, family and motherhood in formulating women's importance for national and imperial stability, reflecting the uneasy coexistence of national confidence and social anxiety in Britain during the pre-war years.[45] MU discourse scripted this connection on two related levels: the home itself represented a kingdom or empire ruled by mothers, and women's responsibility for rearing future generations constituted a moral index and instrument of national character.[46] In urging the increase of active MU service in the colonies, MU literature and sermons of the early twentieth century also invoked the imperial figureheads of Queen Victoria and Queen Mary as models for how the 'sovereignty' of motherhood should advance global Christianity through faith and exemplary home life.[47]

The MU's polemics charged all Christian women with a divinely sanctioned mission of defending family life, whether or not they participated formally in mission work at home or abroad, and regardless of their class or professional status. Thus, MU members at home and abroad idealized the empire not as an object of state or economic expansion but as a conduit of piety and right living. The gendered terms in which MU members construed Britain's temporal and spiritual empire profoundly shaped the organization's response to the Great War. The leading role that Anglican women should take in post-war England was constantly discussed in MU circles, even before the war had ended. Mrs Wyndham Knight-Bruce, suffragist and wife of the former bishop of Mashonaland, asserted in 1916 that

> there lies before us, as we hope, after the war the regeneration of English life. How is that regeneration to be accomplished? ... Surely at last woman is to fulfill, by her help in making the new England, the great purpose for which God originally sent her into the world.[48]

Although Knight-Bruce's avowed feminism was not typical of the MU, most members agreed that women should take an active role in transforming religion and society. Women's growing participation in formal politics, which had begun locally even before women's enfranchisement on a national level, suggested to some that the MU should more actively inform political decision-making, and set an example in responsible citizenship.[49] Others felt that the way forward lay in rejuvenating the Church of England's devotional life, confronting the secular 'apathy' which pervaded post-war ethos.[50] Most of all, Anglican women looked to the MU to institute a spiritual, intercessory network that would mend not only post-war Britain but also the rest of the world through the regenerative power of womanhood.

Women in foreign mission communities shared the desire for an intercessory community of women to ameliorate the trauma of war, and they built on the system of Linked Branches in constructing such a community. In expressing her appreciation for Malagasy women's condolence letters to those who had lost husbands, brothers and sons, missionary Barbara Blair suggested that the MU made it possible for Christian womanhood to cut across cultural and geographical boundaries: 'the Malagasy are very ready to "weep with them that weep" as well as to "rejoice with those that do rejoice" ... the Mothers' Union supplies help and makes them and us realize our oneness in the great family in heaven and earth'.[51] King wrote a condolence letter on behalf of the Madagascar MU to Lady Chichester, whose son had died in the war, suggesting that the MU's worldwide bereavement had also forged a stronger fellowship among its women: 'You must be so proud in spite of all the sadness, and it brings you the untold joy of feeling with so many, so very many members of the MU all over the world.'[52]

This widespread wartime experience of both loss and solidarity motivated the central MU to host a Conference for Overseas Workers on King's recommendation. The Conference which materialized in London in 1920, however, was a far cry from King's hope for united vision or action. Instead, the discussions revealed that the MU's goals of moral purity and universality did not easily coexist. Yet even as the conference revealed tensions among fellow missionaries on the regulatory role of marriage in colonial branches, it also provided a chance for MU workers to assess how marriage would inform women's political, social and professional status in the wake of the war, in metropole and colony alike. King, who herself 'expressed diffidence as an unmarried woman', nonetheless invoked Mary Sumner's counsel that 'every true woman has a mother's heart, and if you are a true woman you have to mother the world'.[53] King went on to explain that her own unmarried status enabled her to endow married members of the MU with the responsibility of material and spiritual improvement, hence drawing upon the evangelistic strategy that she had formulated in the course of her work with Malagasy women. This rationale sat unevenly with the stringency of the membership criteria she required of non-white women, but it also enabled her to legitimize her own missionary authority through a naturalized discourse of femininity.[54] In the mission field, King had used this rationale to empower Malagasy Christian women through the MU; in London, she used it to bolster her own authority.

Other conference participants, however, did not find the ambiguity of married and unmarried womanhood to be an asset to the spread of Christianity or the consolidation of feminine spirituality in public life. In fact, some MU members used the conference as an opportunity to reclaim for married women the professional authority that they felt had become the exclusive privilege of single women. The third set of conference discussions, 'The Deepening of the Spiritual Life', wrestled with the evangelistic and civic roles of wives and spinsters in the post-war world. Mrs Knight-Bruce, one of the few MU leaders who had openly supported women's suffrage before 1919, situated herself in an older, 'cultural' feminist discourse in defining the MU's post-war, global persona: 'We have got a tremendous task before us; we have to insist upon the womanhood of women.'[55] The MU thus played a crucial role in avowing the 'womanhood of women' for civic virtue, over and against the sexless modern woman. Moreover, according to Knight-Bruce, if the woman's movement had thus far been dominated by single women, the time had come for married women to define modern womanhood and to use their maternal facilities to effect change at home and overseas. 'It is the wives and mothers who will help to build a good democracy.'[56] Knight-Bruce also invested wives and mothers with the same kind of Christian 'vocation' that missionaries in Africa had long construed for non-white marriage and motherhood as its own missionary vocation, and in some cases, as an explicitly professional form of evangelism. Hence, Anglican womanhood became the key not only to evangelizing the rest of the world, but also to re-Christianizing the West.

Knight-Bruce forcefully argued that women's newly enfranchised political status and their distinctive religious authority would be mutually constitutive agents of the world's spiritual renewal. Her version of solidarity was not reflective of all viewpoints; the MU struggled with the tension of uniformity and flexibility in women's coalition-building, and with metropolitan leadership and peripheral autonomy. The fact that the MU did not invite any non-white overseas workers to participate in the 1920 conference intensely underscored these contradictions. Yet this post-war moment also afforded the possibility for dialogue, if not consensus, about the shape of Christian women's leadership, a dialogue which sparked new trends in the MU's international expansion over the next decade and which culminated in an expanded conception of non-white women's authority.

Towards the globalization of Anglican sisterhood

The MU's post-war commitment to social and spiritual solidarity was most evident in the international women's prayer movement which grew out of the 1920 conference, called the 'Wave of Prayer'. Beginning in 1921 as a result of conference discussions that had centred on the 'deepening of spiritual life', the MU press published a monthly intercessory calendar for use in worship, at branch meetings, or in personal prayer.[57] The Wave gave all MU branches and members access to a comprehensive means of exchanging news and prayers across global networks, rather than just with a specific branch. This movement anchored a sense of corporate membership among the MU worldwide, fostering a common spiritual purpose among numerous branches in Britain, Europe, Africa, Asia, Oceana and North America.

The first MU Worldwide Conference in 1930 was a testimonial to this new cross-cultural solidarity which put female spirituality, rather than the metropolitan institutions of the church or MU, at the centre of Christianization. Unlike the 1920 Conference 'for Overseas Workers' which was limited to British women working abroad, the 1930 conference invited non-white MU members from around the world to participate in two weeks of discussions, devotional retreats, worship services and 'pilgrimages' to Anglican spiritual centres like Canterbury and Winchester Cathedrals. The discussions also covered a more expansive terrain than they had in 1920, with topics including local and worldwide governance, theological study, religion in the home and in schools and member recruitment, in addition to the subjects which had structured the 1920 conference, such as 'lapsed membership' and 'the deepening of the spiritual life'.[58] An anonymous conference participant who later gave her impressions of the proceedings attributed the powerful fellowship and sense of common mission forged by the conference to 'the volume of intercessory prayer that has preceded and surrounded its deliberations'. This referred not only to larger intercessory trends of the Wave of Prayer in the 1920s but also to the worldwide day of prayer and the seven retreats in different British diocesan centres that immediately preceded the conference meetings. Throughout the conference, the proceedings maintained a devotional and occasionally even revivalist character, and the chapel at Mary Sumner House (MU headquarters) was described as 'a veritable "power station" of prayer'.[59]

The presence of many provincial and non-white MU members in the metropole rendered the conference a moment of unprecedented cultural encounter for many English women. Such an encounter did not necessarily unfold on equal terms, as the official delegates and conference leaders were white MU members. At the same time, the anonymous participant noted a powerful sense of fellowship, dialogue and common mission forged by the conference. Non-white visitors enjoyed the full privileges and status of MU membership, and membership in turn assumed a commitment to upholding the sacredness of marriage and motherhood. The conference therefore enshrined these functions, long central to discourses of Christian womanhood, in a newly democratic context and along new lines of consensus. The observer felt that by putting a multiracial ideal of female spirituality at the centre of the conference objectives, the devotional tone of the proceedings projected a corporate sense of unity which worked to stabilize an otherwise tenuous moral and social climate in Britain and the world. The Conference therefore represented a turning-point in the trajectory of women's Christian leadership: 'As each delegate returned to her own sphere of influence, closely bound to us all by the ties formed in these wonderful days together, the real work of the Conference would begin.'[60]

That the MU assumed a prominent role in public discourse indicates that its priorities intersected with larger discussions of how to stabilize an otherwise unstable moral, religious and social climate. Such characterizations of the conciliatory influence of Christian womanhood and domesticity on the national and international order, drawing at once upon conservative and progressive strains of gender ideology, were not an uncommon response to the massive displacement experienced by Europeans in the wake of the Great War. Yet historicizing this discourse of regeneration in the specific context of mission work underscores the role of the MU in assigning a reconstructive and redemptive role to mission Christianity – not as a tool of cultural imperialism, but as a force for consensus between white and non-white women, and between metropole and periphery. The female solidarity that individual missionaries had long worked to establish in the mission field around mutual priorities of marriage and motherhood had come to carry stakes beyond individual conversion and emancipation: the stability and salvation of the international order now depended on the power of corporate womanhood in enacting change. By making home and family consonant with public service, and feminine piety a necessary condition of international healing, the conference sought to pave the way for a new era in which women's evangelism, broadly defined, would Christianize *both* the Western and non-Western world.

The MU Worldwide Conference of 1930 signalled a new phase in the nature of women's international humanitarian interventions, as the discourse of national decline which had caused many Anglican women to link imperialism with evangelism before the war gave way to a universalist language of regeneration. This shift also reflected and anticipated larger challenges to the enterprises of imperial and Christian institution-building, for the war exposed cracks in the edifice of empire and church alike, and demanded new modes of envisioning global unity. Moreover, the enfranchisement of British women after the war gave renewed purpose to imagining political space as both feminine and Christian. Anglican women thus approached the social and political upheaval that emerged out of the war years as a moment of opportunity for building a women's coalition of global citizenship, marshalling the feminized forms of Christianity that had developed in the mission field towards a new vision of international unity – one which did not find its only meaning in the ecclesiastical or imperial sovereignty of Britain.

This dynamic historical moment for the MU thus reveals more broadly how feminized discourses of spirituality worked to collapse the gap between women's 'public' and 'private' spheres of influence, and between the colonial centre and peripheries of Anglican Christianity. Of course, the subsequent history of the MU has not always born out the unity that participants in the Worldwide Conference idealized in 1930. Since then, branches in Britain and overseas have struggled to establish a common ground for divergent interpretations of the MU's commitment to 'upholding the sanctity of marriage'. Competing approaches to women's social issues have revealed the multiplicity of cultural and political frameworks that have shaped the MU's national and international following. Yet this contested history is reflective of the complex roots of the MU as a worldwide women's movement; and the continued growth of the global MU suggests that these early moments of solidarity like the 'Wave of Prayer' and the 1930 Worldwide Conference opened new possibilities for women's dialogue and social action across cultural and geographic borders.

Notes

1. Suffragists and missionaries maintained that the institutional seclusion of Hindu and Muslim women prevented any but white women from liberating them from their heathen conditions. See A. Burton, *Burdens of History: British Feminists, Indian Women, and Imperial Culture, 1865–1915* (Chapel Hill, NC: University of North Carolina Press, 1994); S. Thorne, 'Missionary-Imperial Feminism', in *Gendered Missions: Women and Men in Missionary Discourse and Practice*, ed. M. T. Huber and N. C. Lutkehaus (Ann Arbor, MI: University of Michigan Press, 1999), pp. 39–65; J. Haggis, 'White Women and Colonialism: Towards a Non-Recuperative History', in *Gender and Imperialism*, ed. C. Midgley (Manchester: Manchester University Press, 1998), pp. 45–75; J. I. Nair, 'Uncovering the *Zenana*: Visions of Indian Womanhood in Englishwomen's Writings, 1813–1940', *Journal of Women's History*, II, no. 1 (1990) 8–34.
2. In this respect I draw on a growing literature which focuses on the mutually constitutive dynamics of Christianity, feminism and empire. See, for example, R. A. Semple, *Missionary Women: Gender, Professionalism, and the Victorian Idea of Christian Mission* (Woodbridge: Boydell and Brewer, 2003); G. Francis-Dehquani, *Religious Feminism in an Age of Empire: CMS Women Missionaries in Iran, 1869–1934* (Bristol: Center for Comparative Studies in Religion and Gender, 2000); J. Haggis, '"A Heart That Has Felt the Love of God and Longs for Others to Know It": Conventions of Gender, Tensions of Self and Constructions of Difference in Offering to be a Lady Missionary', *Women's History Review*, VII (1998) 171–92.
3. Other authors in this volume who examine the impact of missions on imperial ideology and metropolitan discourse include Catherine Hall, Esther

Breitenbach, Shurlee Swain, and Peter Clayworth. For more extended analyses of how missions shaped imperial culture in Britain, see C. Hall, *Civilizing Subjects: Colony and Metropole in the English Imagination, 1830–1867* (Chicago, IL: University of Chicago Press, 2002), and S. Thorne, *Congregational Missions and the Making of an Imperial Culture in Nineteenth-Century England* (Stanford: Stanford University Press, 1999).

4. The MU's three objects were:
   1) To uphold the sanctity of marriage, 2) To awaken in mothers of all classes a sense of their great responsibility as mothers in the training of their boys and girls, the future fathers and mothers of England [changed to 'the Empire' in 1902, then dropped in the mid-1920s], 3) To organize every place a band of mothers who will unite in prayer and seek by their own example to lead their families in purity and holiness of life.

   For an overview of the Mothers' Union, see O. Parker, *For the Family's Sake: A History of the Mothers' Union, 1876–1976* (London: Mowbray's, 1976); F. Hill, *Mission Unlimited: The Story of the Mothers' Union* (London: Mothers' Union, 1988).
5. In 1904, for example, the Bishop of Stepney preached a sermon at a special service for the MU at St Paul's Cathedral, in which he corroborated the MU's premise that 'the vigor of civil and religious life depends upon the family.' *Mothers in Council* (*MIC*), July 1904, 130–3. The designation of married 'members' v. unmarried 'associates' was essentially rhetorical, since both enjoyed the same privileges and status within the organization.
6. *Mothers Union Journal (MUJ)*, March 1930, 15. See also D. Marshall, *Around the World in 100 Years: A History of the Work of the Mothers' Union Overseas* (London: Mothers' Union, 1977).
7. The imperial rivalry which eventually pulled Madagascar into the partition of Africa also continued a longer pattern of French diplomatic and military activity. C. Newbury, 'Great Britain and the Partition of Africa, 1870–1914', in *The Oxford History of the British Empire, Vol. 3. The Nineteenth Century*, ed. A. Porter (Oxford: Oxford University Press, 1999), p. 640; S. Feierman, 'A Century of Ironies in East Africa', in *African History: From Earliest Times to Independence*, ed. P. D. Curtin et al. (New York: Longman, 1995), p. 374; H. Brunschwig, 'Anglophobia and French African Policy', in *France and Britain in Africa: Imperial Rivalry and Colonial Rule*, ed. P. Gifford and W. R. Lewis (New Haven, CT: Yale University Press, 1971), pp. 4–15. On the French colonization of Madagascar, see S. Ellis, *The Rising of the Red Shawls: A Revolt in Madagascar, 1895–1899* (Cambridge: Cambridge University Press, 1985). On Christianity and the precolonial state, see P. M. Larson, 'Capacities and Modes of Thinking: Intellectual Engagements and Subaltern Hegemony in the Early History of Malagasy Christianity', *American Historical Review*, CII (1997) 969–1002.
8. John MacKenzie has discussed a similar shift in Scottish women's mission work in South Africa, from an emphasis on 'a western concept of the separation of gendered spheres' to a paradigm of 'black female professionalism'. See MacKenzie (Chapter 6), this volume.
9. See Anne O'Brien (Chapter 9), this volume.
10. See Ruth Compton Brouwer (Chapter 14), this volume.

11. For example, in 1911, when MU members in Britain tried to send commemorative cards of George V's coronation to missionary friends in Madagascar, they were told regretfully that the diocese could not endorse or circulate any printed material bearing the British flag. King commented that 'the very critical situation' arising from the 'constant blows that fall on work here' was a product of 'the anti-religious atmosphere of French politics'. King to Miss Thomas, 14 August 1911, Mothers' Union Archives, accessed by kind permission of Cordelia Moyse, Mary Sumner House, London, MU/OS/005/13/08.
12. *Madagascar Church Mission Association Quarterly Paper*, October 1911, MU/OS/005/13/08.
13. Byam to Maude (MU central secretary), 18 November 1911, MU/OS/005/13/08.
14. See A. Davin, 'Imperialism and Motherhood', *History Workshop Journal*, V (1978) 9–66; C. Summers, 'Intimate Colonialism: The Imperial Production of Reproduction in Uganda, 1907–1925', *Signs*, XVI (1991) 787–807.
15. *Mothers' Union Workers' Paper* (*MUWP*), September 1916, 174.
16. For a discussion of these contested norms of womanhood in Norwegian missions in Madagascar, see L. N. Predelli, 'Sexual Control and the Remaking of Gender: The Attempt of Nineteenth-Century Protestant Norwegian Women to Export Western Domesticity to Madagascar', *Journal of Women's History*, XII (2000) 81–103. On African and Western women's encounters around gender and domesticity, see K. T. Hansen, ed., *African Encounters with Domesticity* (New Brunswick, NJ: Rutgers University Press, 1992); J. Allman, S. Geiger and N. Musisi, eds.,*Women in African Colonial Histories* (Bloomington, IN: Indiana University Press, 2002); D. L. Hodgson and S. A. McCurdy, eds., *'Wicked' Women and the Reconfiguration of Gender in Africa* (Portsmouth: Heinemann, 2001); *Women and Missions: Past and Present, Anthropological and Historical Perspectives*, ed. F. Bowie, D. Kirkwood, and S. Ardener (Providence and Oxford: Berg, 1993).
17. The WMA was called the Ladies' Association until 1894; it became the Committee on Women's Work after 1904 when it was incorporated formally into the SPG.
18. *MIC* (April 1902) 127.
19. King to Mary Sumner, 12 July 1901, MU/OS/005/13/08. Sumner, founder and president of the MU, served as the point of correspondence before the offices of Central Secretary and Overseas Secretary were established.
20. *MIC*, October 1906, 253; typescript report, 9 January 1912, MU/OS/005/13/08.
21. The MU's focus on prayer meetings may well have resonated with older traditions of Malagasy Christianity; Pier Larson has shown how peasants' transformation of the evangelical ideology of 'belief' to an indigenous idiom of 'prayer' was central to the subaltern subversion of both missionary and elite Malagasy authority in the early nineteenth century. Larson, 'Capacities and Modes of Thinking', p. 982.
22. King, 'How the MU began in Madagascar', *Mothers Overseas*, 9–10.
23. 21 March 1902, United Society for the Propagation of the Gospel Archives, Rhodes House, Oxford, CWW 98.
24. King to Maude, 8 October 1910, MU/OS/005/13/08.

25. Although the 1908 Conference maintained the 1888 Lambeth resolution that no divorcees should be remarried in church, resolutions 39 and 40 also stipulated that the church should not deny 'innocent' parties in divorce cases communicant status if they chose to remarry under civil law. Lambeth Conference Resolutions Archive, 'Resolutions—1908' (Anglican Communion Office, 2005), http://www.lambethconference.org/ resolutions /1908/ (accessed 28 November, 2006).
26. This was a point of pride for the bishop, who stated that 'as to divorce ... our rules are stringent as to Communion in the Church–far more so than any other Christian body in the island, a fact which keeps down our numbers of Church people, and a woman not eligible on moral grounds for Communion is also not eligible for the Mothers' Union.' George King to Maude, 9 January 1913, MU/OS/005/13/08. In reality, the fact that Anglicans were a minority of Malagasy Christians was probably due to the French colonial government's continued discrimination against Protestant mission work, and to the relatively small proportion of Anglicans among the larger mission community in Madagascar, which included the London Missionary Society, Norwegian Lutherans, Quakers, Catholic sisterhoods and Jesuits.
27. Maude to Gertrude King, 13 November 1912, MU/OS/005/13/08. In 1913, the Central Council in London passed the resolution that (1) 'as a result of experience ... Native Branches should be run on a communicant basis', and that (2) divorce would disqualify any potential or standing member from MU membership, 'even if the disciplinary laws of the Church admitted her to Communion at the end of an interval of years.' In conveying this development to Bishop King, the central secretary added that 'the passing of this minute was really due to the example set by Madagascar'. Maude to George King, 13 March 1914, MU/OS/005/13/08; see also Violet B. Lancaster, *A Short History of the Mothers' Union* (London: Mothers' Union, 1958), p. 119.
28. This stipulation was confirmed by Gertrude King at the 1920 MU Conference for Overseas Workers in London (*MUWP*, October 1920, 195). However, it is important to clarify that the marriage and divorce standards set by either the church or the MU only took effect after baptism.
29. King to Maude, 31 July 1916, MU/OS/005/13/08.
30. George King to Maude, 31 July 1917, MU/OS/005/13/08.
31. The church supported the MU's stringency and relied on the women's organization to provide the moral guard that it (and the civil marriage registrar) could not. George King to Maude, 31 July 1917, MU/OS/005/13/08.
32. King's reflection on the first meeting of the MU, reprinted in the *Madagascar Church Mission Association Quarterly Paper*, October 1911, MU/OS/005/13/08.
33. King to Mary Sumner, 12 July 1901, MU/OS/005/13/08.
34. King, 16 July 1906, MU/OS/005/13/08.
35. Ibid.
36. King to Maude, 31 July 1916, MU/OS/005/13/08.
37. King to Maude, 15 August 1918, MU/OS/005/13/08.
38. Barbara Druitt Blair, *MUWP* (April 1919) 58.
39. King, 16 July 1906, MU/OS/005/13/08; King to Maude, 18 July (n.d. – probably 1910), MU/OS/005/13/08.
40. *MIC*, October 1912, 248.

41. King to Sumner, 18 July 1910, MU/OS/005/13/08. Soon thereafter the branch became linked with Holy Trinity, Twickenham, *The Church Abroad* (January 1912) p. 4.
42. Andevorante branch to Sumner (King's translation), 11 November 1912, MU/OS/005/13/08.
43. *MIC*, October 1912, 248.
44. *MIC*, January 1912, 33.
45. See Davin, 'Imperialism and Motherhood.'
46. 'Queens shall be Nursing Mothers of England', sermon preached by Archbishop of York, Sheffield Parish Church, 13 November 1911 (reprinted in *MIC*, April 1912, 68–71).
47. Ibid.; Mrs. William de Winton, 'The Influence of Englishmen and Englishwomen in the Colonies and in Foreign Lands', *MIC* (April 1907) 113–18.
48. 'Ideal Womanhood', *MIC*, October 1916, 208. Health prevented Knight-Bruce from accompanying her husband while he served in Africa but she worked as a fundraiser and publicist for the Mashonaland mission from Britain. Deborah Kirkwood, 'Wives of Missionaries Working with the Society', in *Three Centuries of Mission: The United Society for the Propagation of the Gospel, 1701–2000*, ed. Daniel O'Connor (London: Continuum, 2000).
49. Ibid., pp. 214–15.
50. The *MUWP*'s paraphrase of an address by the chaplain to the 1920 Conference for Overseas Workers illustrated this concern: '[T]he majority of the people who went to Church did not feel the worship in which they were taking part to be real.' *MUWP* (December 1920) 225.
51. Blair, *MUWP*, April 1919, 59. Although these letters highlight Malagasy empathy, many indigenous MU members may have experienced such loss directly. The French raised 45,000 troops from Madagascar; although Kitchener originally requested that they be used to help Britain defend its East African colonies, the Malagasy troops mainly fought in Europe. David Killingray, 'The War in Africa', in *The Oxford Illustrated History of the First World War*, ed. Hew Strachan (Oxford: Oxford University Press, 1998), p. 96; Anne Samson, *Britain, South Africa, and the East Africa Campaign, 1914–1918: The Union Comes of Age* (London: Tauris, 2006), pp. 54, 122.
52. King to Lady Chichester, 20 November 1914, MU/OS/005/13/08.
53. *MUWP*, October 1920, 195. Mary Sumner was the founder and longtime president of the MU until her death in 1921.
54. *MUWP*,(October 1920) 195.
55. Mrs Knight-Bruce, paper on 'The Value of the Mothers' Union', reprinted in *MUWP* (December 1920) 235–7.
56. Ibid.
57. Missionaries in Uganda encouraged the emulation of this noon prayer service among African branches. Letter from Mrs Denne to Mrs Maude about her travels in Uganda, July 1926, MU/OS/005/13/33.
58. *MUWP* (March 1930) 42–4.
59. *MUWP* (August 1930) 160–1.
60. Ibid., p. 163.

Further reading

Allman, J., S. Geiger, and N. Musisi, eds. *Women in African Colonial Histories.* (Bloomington, IN: Indiana University Press, 2002).

Bowie, F., D. Kirkwood, and S. Ardener, eds. *Women and Missions: Past and Present, Anthropological and Historical Perspectives* (Oxford: Berg, 1993).

Burton, A. M. *Burdens of History: British Feminists, Indian Women, and Imperial Culture, 1865–1915* (Chapel Hill, NC: University of North Carolina Press, 1994).

Francis-Dehqani, G. *Religious Feminism in an Age of Empire: CMS Women Missionaries in Iran, 1869–1934* (Bristol: Center for Comparative Studies in Religion and Gender, 2000).

Huber, M. T. and N. C. Lutkehaus, eds. *Gendered Missions: Women and Men in Missionary Discourse and Practice* (Ann Arbor, MI: University of Michigan Press, 1999).

Levine, P., ed. *Gender and Empire* (Oxford: Oxford University Press, 2004).

Midgley, C., ed. *Gender and Imperialism* (Manchester: Manchester University Press, 1998).

Morgan, S., ed. *Women, Religion, and Feminism in Britain, 1750–1900* (New York: Palgrave Macmillan, 2002).

Semple, R. A. *Missionary Women: Gender, Professionalism, and the Victorian Idea of Christian Mission* (Woodbridge: Boydell and Brewer, 2003).

Strobel, M. and N. Chaudhuri, eds. *Western Women and Imperialism: Complicity and Resistance* (Bloomington, IN: Indiana University Press, 1992).

Part III

Post-Colonial Transformations

13

Ireland's Spiritual Empire: Territory and Landscape in Irish Catholic Missionary Discourse

Fiona Bateman

Interest in the Irish diaspora and the influence of the Irish worldwide has often focused on the stories of those who left Ireland as victims of famine in the nineteenth century and the economic refugees of the twentieth century, but has also drawn attention to the Irish who were settlers in the colonies of the British Empire and those in the colonial service or otherwise allied with British expansion.[1] One readily identifiable group, neither unwilling economic migrants nor colonial officials, is that of the missionaries who travelled from Ireland with the goal of converting others to Christianity, primarily Catholicism. Historically, Ireland was known as the 'Isle of Saints and Scholars' due to the activities of Irish monks as missionaries throughout Europe, a missionary project that began soon after the dawn of Christianity.[2] But during the ninth century, what Neill describes as Ireland's 'great and beautiful Christian civilization' was destroyed in Viking attacks,[3] and, despite the international reputation of early Irish Christian missionaries such as Brendan and Columbanus, whose activities take on a legendary quality in popular history, the Irish had no distinctive profile of missionary activity from the close of the ninth century until the 1820s.[4] After achieving Emancipation in 1829, the Irish Church grew in confidence and, while the first priority to be addressed was the religious needs of the diaspora, the presence of some French missionary organizations in Ireland was bringing an awareness of the issue of the conversion of pagans to public attention, a cause which was also being promoted by the papacy. Irish men and women joined the missionary orders of other countries but it was not until the twentieth century that indigenous Irish orders were founded and the 'second spring' of Irish missionary activity was celebrated.[5] The growing Nationalist movement, with which the Church was strongly allied, brought additional questions of identity. With the foundation of specifically Irish missionary organizations, Irish priests and nuns had the opportunity to represent their own country in the mission field, and an Irish missionary project could be identified. In the 1920s, that resurgence of missionary activity in Ireland and the huge public participation in, and enthusiasm for the foreign missions, which occurred subsequently, was to become not only a religious movement but also a social and cultural phenomenon lasting for more than three decades. These twentieth century 'pagan missions' are the focus of this chapter.

Until this period, the primary usage of the term 'Spiritual Empire' had been about the many Irish emigrants to all parts of the world who brought their Catholic faith with them and to whom the Irish hierarchy were quick to supply missionary priests. By the 1920s, those emigrant communities had established their own identities and were providing their own priests; they no longer needed assistance (or interference) from the Irish 'mother' church. Therefore, Ireland had a surplus of missionaries and alternative mission fields had to be considered. Although, in the main, the spiritual empire referred to the Irish diaspora rather than any putative empire where Irish missionaries exercised control over an indigenous population, the meaning changed subtly as the overseas missions expanded to include missions to 'pagans' in countries such as Africa, China and India. It was noted by one commentator in the 1930s that 'Ireland has not a square inch of territory abroad, yet her spiritual empire is limited only by the bounds of the earth.'[6]

For the young Irish state, endeavouring to gain recognition as an independent country, with an identity separate from Britain, it was desirable to contribute to world affairs through some specifically Irish activity. The economy was not of a scale to attract international attention and Ireland was never going to have an empire in the material sense, but the high profile of a relatively small number of missionaries, allied with a history of missionary activity, gave Ireland an opportunity to take a place on the world stage. Historian Joe Lee describes the missionary achievement abroad as 'one of the more remarkable conquests of the age of imperialism'.[7] Indeed when one considers the traditional conservatism of Irish Catholicism, involvement in such an expansionist project seems paradoxical; but in its years of ministering to Irish emigrants, the Irish Church had constructed itself as 'mother country' to the English-speaking Catholic world and therefore it was not a huge leap to extend Ireland's 'spiritual empire' to include as much of the 'pagan' world as might be managed. Reference to 'those spiritual colonies of Ireland – the greater Ireland of the Catholic faith'[8] implies the expansion or growth of Ireland territorially by spiritual means.

In addition to the vocations that were being produced in Ireland at this time, other aspects of the social and political situation supported the new missionary project: after the political upheaval of 1916 and the subsequent civil war, there was a need for national heroes who would not cause division and missionary priests had a historic precedent to exploit that role. It was also essential to establish an Irish identity, separate and distinct from Great Britain and an activity in which Irish Catholicism, one of the main 'differences' could be to the fore, was ideal. As it happened, the conversion of the pagan was also a priority in the Vatican.

In this chapter, I will illustrate the fact that Irish missionary organizations utilized several key tropes and strategies of nineteenth-century British imperial discourse to describe and realize their project. Ironically, this discourse can be seen to replicate certain aspects of the discourse of Protestant missionaries and proselytizers in Ireland in the mid-nineteenth century, whom the Irish Catholic hierarchy viewed as their foes and competitors. It should not be surprising that elements of the discourse of British imperialism were well established and integrated into Irish cultural life in the late nineteenth and early twentieth century, and in fact some recent Irish cultural historians have argued that Irish nationalism itself often mirrored the values, vocabularies and strategies of its ostensible antagonist: the British empire.[9] Thematic elements of imperial discourse were replicated by missionary discourse including: ideals of masculinity; belief in the superiority of Western culture; focus on the bringing of civilization (light to darkness); issues of hygiene; a presumption of *terra nullius* (which gives rise to 'naming' and claiming territory); rivalry among competing groups; the infantilization and 'animalization' of indigenous peoples; and, inevitably, speaking for the 'Other'.[10] Irish newspapers provided up-to-date accounts of life in the colonies and the battles to win and maintain them. Ireland had been a part of the British Empire, although there was disagreement whether 'she' should be considered a subject colony or an element of the ruling 'Mother country'. Significantly, in British imperial discourse, the Irish were often depicted as an inferior race, scarcely considered 'white'. Irish primitivism was represented in caricature, and by the 1860s in England the 'representative Irishman' was depicted with 'simian or porcine features, and festooned with weapons, a vivid reminder that decent English citizens stood in mortal peril'.[11] Author

Charles Kingsley lamented the 'white chimpanzees' he encountered on his Irish travels; his writing highlighted the difficulty in dealing with a subject race whose skin colour was the same as that of its rulers, at a time when the savage nature of black and coloured races was largely unquestioned.[12] The image of the 'ape-like' Irishman rankled, and 'Fr. Kavanagh' appealed to the sense of humiliation it aroused, in an effort to stir up Irish indignation and support for the Boers, during the Anglo-Boer War. In the *United Irishman*, he referred to one particular contemporary insult:

> NO WONDER SHE CALLS US HOTTENTOTS. No wonder she dresses up apes in her museums in the costume of stage Irishman and exhibits them thus to her grinning yokels. There are two of these animals on exhibition in the London Zoo upon whom, by way of delicate compliment to our nation, they have given the name of Pat and Biddy, the names of our venerated patron saints.[13]

Hence, from an Irish perspective, a certain amount of re-framing of the discourse was essential, and the opportunity to adapt and rewrite a discourse in which the Irish might shoulder the 'White-man's burden', must have appealed.[14] The re-appropriation of a powerful discourse, to counter the stereotypes it generates is a strategy identified in post-colonial theory and typically, the generation of a new set of stereotypes is the outcome when that discourse is predicated on the Manichean dichotomy of 'self' and 'other'.

This chapter will focus on two aspects of the Irish Catholic missions' public discourse regarding Africa that draw strongly on that different imperial tradition. These are first, the representation of issues of territory and description of landscape – the ownership and control of geographical space. And, second, the adventure of travel in these territories, often associated with narratives of exploration and the mastery of savage and pagan lands.

Territory

Religious and civilizing motives have long been used as a justification for imperial adventures. The expansion of both Spain and Portugal during the Age of Discovery was presented as a project to spread the word of God, among other more materialistic motivations, and the Vatican supported such efforts to win territories and the souls who lived within these new worlds. As a part of the Protestant British Empire, Ireland was not in a position to support a Catholic missionary project, and it was only after Catholic Emancipation in 1829, that Irish missionary activity began to re-emerge. Unlike other missionary projects, Irish missionaries could hardly be accused of having imperial expansion as their motivation, although their activities may have enabled the process, and certainly some would have supported the imperial project. The same argument would apply to those new Irish missionary orders which emerged in the twentieth century, but despite the fact that the missionary project was concerned with 'harvesting' souls rather than claiming territory, discussion and description of the physical environment pervades missionary discourse. While conversion was the priority, the territory in which the missionaries could operate was regulated, and hence issues of territorial control were important as a means of access to the population. Authority over the land, both symbolically and physically, became an essential facet of the missionary effort. Debates about land, the rivalries which developed and how landscape was represented within missionary discourse may be compared to elements of colonial discourse regarding the discovery, mastery and control of terrain. Edward Said has commented that imperialism 'is an act of geographical violence through which virtually every space in the world is explored, charted, and finally brought under control'.[15] The discourse produced by Irish missionaries could certainly be described as participating in or mirroring that act.

For Irish missionary orders in the twentieth century, access to territories was the first step in converting Africa to Christianity. Permission to operate a mission was granted by the relevant colonial authority and so, in hugely competitive negotiations, involving missionaries, government representatives and Propaganda Fide, the papal organization responsible for the direction of Catholic missionary work, different missionary organizations divided the world. A 'sphere of influence' was a term used to designate an area where a particular religious group had 'rights' whether they were actively engaged in missionary work at the time, or not. Access was ensured, regardless of whether missionaries were available for a region. Once a sphere of influence had been established, other groups could not attempt to establish a mission there, even in the absence of any intervention by the group who had control over that territory. Promising regions were thus claimed and left in paganism, until the missionary organization had the resources to devote to the conversion of the population. There was competition for territories both between orders of the same religion and between Catholics and Protestants. In his account of the history of St Patrick's Missionary

Society, Thomas Kiggins describes the transfer of territory in the Vicariate of Zanzibar from the 'Holy Ghost Fathers' to the 'Consolate Fathers', in the early years of the twentieth century, as 'a move that in modern business circles might be termed a *hostile takeover*'.[16]

Maps of missionary organizations' 'spiritual empires' were displayed in many school classrooms in Ireland to illustrate the extent of their success. Similar to maps of the British Empire, coloured to illustrate control and authority, the maps provided a graphic image of the success and growth of missionary activity. Despite the protestations of missionaries that their project was not expansionist as the imperial project was, the control of territory was undoubtedly a fundamental part of their mission, and these maps related more directly to territories than to souls. And, while the control the missionaries exerted was largely religious, it included control over education, family life, marriages and other civil matters in a territory. As a visual expression of missionary success, the maps often exaggerated the scale of missionary influence. An organization might show an entire country coloured in, if there was a missionary presence somewhere in its vast area, in effect the 'presence' might be just two priests. Another use of the map is that on the children's pages of the *African Missionary* magazine, where maps of Ireland and Africa provide illustrations on either side of a verse, but the map of Ireland is larger than that of Africa,[17] perhaps reflecting the discourse which structures Ireland as an adult and Africa a child. In the tradition of travel writing, books which describe missionary journeys and adventures often include a map of the area, which may be referred to by the reader. Such illustrations enforce the factual nature of the account provided, although drawings of exotic animals and native masks which adorn some, enhance the popular rather than the scientific appeal.

Many bitter complaints regarding the success of rival missionary groups permeate the texts; competition was equally as fierce on this level of religious rivalry as among the nations who carved up Africa as colonial possessions. In 1935, a magazine reports a 'formidable scramble among the sects to seize the souls of the unconverted Africans'.[18] It was, essentially, a second scramble for Africa. The missionaries often make the point that they are not imperialists claiming land but a much more spiritually motivated movement interested in winning souls not territories: 'palpitating trophies' to be piled up to Our Lady of Africa.[19] Whatever the original motivation, lines drawn on a map separated different missionary efforts and created borders which might be breached or challenged. The language of 'ownership' regarding mission lands was naturalized and it was not remarkable for a missionary magazine to contain a statement such as the following: 'everybody seemed Christian and happy in this picturesque nook of our huge African territories'.[20]

Missionary organizations were most keen to get access to 'virgin' or 'untouched' land (meaning that there had been no missionary activity there previously). There was a conviction among Catholic missionaries that the conversion of Protestants, that is, Africans who had already been converted to Christianity, was more difficult than converting pagans. Effectively, whichever branch of Christianity first succeeded in introducing their religion in an area, tended to remain. In fundraising appeals, it is frequently emphasized that the situation is urgent, that it is necessary to act now before others move in. Irish Catholic antipathy to Protestantism had cultural and historical roots in the long experience of Catholic oppression and the proselytization that had occurred in the mid-nineteenth century; even so, the venom with which Protestantism is described in the missionary literature is quite extraordinary. The rival Christian faith was represented as equally pernicious as indigenous juju and superstition, recalling the Protestant attitude towards Catholicism evident in the proselytizing literature of the 1850s.

In a statement evocative of military reports of trench warfare, McGrath claims that, in Anua, 'non-Catholic missions have been crowded out of the immediate vicinity, and year by year their lines are being pushed further back'. He goes on to describe proudly how a non-Catholic hospital, ten miles away, which had been successful, is now closing down due to the success of 'our hospital'.[21] Another report triumphantly refers to the General Assembly (Protestant), which deplored 'the havoc the Catholics were working among their ranks'.[22] It is apparent that inter-religious warfare rather than spiritual or humanitarian interests takes precedence for some. The 'struggle' is not only against paganism; 'Mohammedanism' and Protestantism are both a threat. In this discourse, saving the Africans from the darkness and degradation of paganism is interchangeable with saving them from the scourge of Protestantism or Islam and, apparently, deserving of equal effort; some missionaries seem more engaged with the struggle with rival missionaries than the conversion of pagans.

Landscape

The description of landscape is another element of this broad topic as the missionaries attempted to portray the environment in visual terms for their supporters at home. In 1926, Stephen Brown described three means, all comparatively new, of arousing interest in the missions which had been found 'particularly serviceable': the press, exhibitions and congresses.[23] While magazines and other publications were popular, the written word had its limitations and the pages were often illustrated with photographs and sketches. British explorers and empire-builders had resorted to the magic lantern to describe the unknown beauties and dangers of Empire to the British public, with the aim of engendering support and a spirit of ownership in an otherwise apathetic population. Irish missionaries too were quick to appreciate the value of the image, whether in the form of slide, photograph, drawing or map and eventually film. As in the imperial project, photography and film were extremely valuable tools, making it possible to display the landscape as well as the people as a spectacle. From an early stage, until replaced by the film projector, the magic lantern was a familiar prop at public lectures. Numerous missionaries recount in their memoirs their recollections of such displays, which captured their imagination and prompted them to consider a missionary vocation. Fr James Mellett recounts that, 'at that time [1925] they were the first pictures of Africa seen in most places and they made a powerful impression'.[24]

While these images were a novelty and caught the attention of the public, words often provided more detailed and more nuanced descriptions than the visual images could. Academic Mary Louise Pratt explains that in travel writing, the landscape is typically described in terms of natural history; it is 'uninhabited, unpossessed, unhistoricized, unoccupied'.[25] Missionaries also employ, paradoxically, the idea of virgin territory, although they obviously have a basic requirement that the landscape be occupied. However, the space is still unhistoricized, uncivilized and lacking culture; the people do not possess or populate the land in any recognizable way, and if the land is 'untouched' by missionaries, it is considered 'virgin' territory.

Irish missionary accounts describe the African landscape (prior to the conversion of its population) as inhabited, but inefficiently inhabited: badly planned and shabby. Poor quality buildings, huddled in groups, are falling down; villages are untidy and dirty. Arriving in the town of Moshi, at the foot of Kilimanjaro, missionary priest Thomas Gavan Duffy remarks that it is 'a wilderness of shacks; tentative, impermanent, untidy'.[26] In the following description by Fr James, a Cork-based priest who spent three months on 'holiday' in Africa, it is implied that human neglect or ignorance has left its mark on the environment: 'the country was depressing ... villages [were] situated right on the verge of swamps and were filthy', huts were circular and 'badly made'.[27] A very similar description of a 'bedraggled landscape', with 'uncared for orchards', and towns which are 'dirty and smelly' is given by Fr John Lupton, a frequent contributor to the *African Missionary*.[28] Rather than look for a rationale, everything is measured against European norms and criticized if it does not meet expectations. Pratt notes that travel writing on America, particularly Spanish America, often portrays indigenous society as neglectful or lazy. The impulse to exploit resources is presupposed, 'making a mystery of subsistence and non-accumulative lifeways'.[29] The ideal landscape, one that has been 'touched' by missionaries is depicted by missionary Edward Leen, who describes the mission station at Emekuku as a veritable oasis, in which he finds shrubs, a hedge and a gravel path. He considers 'It was like being set down in a corner of Ireland.'[30] In their nineteenth-century texts, Protestant missionaries employed very similar descriptions of Irish landscape to those of the Irish missionaries in their accounts of Africa. Describing the physical environment of Connemara the emphasis is on a comparison between the 'wild' and the 'settled' or 'cultivated'. There is an assumption shared with the Irish missionaries that control over the environment is emblematic of civilization and Christianity. They are disturbed by an inhabited landscape that is not domesticated or tamed, as if the landscape itself is 'pagan' and it too must be converted. In 1855, Howard writes approvingly of 'the sides of a once barren mountain [which] are now adorned with cultivated fields or gardens'.[31] And on entering Achill, a 'wild island, covered with desolate looking heaths', he is 'cheered by the sight of an English settler's house, around which symptoms of cultivation were manifesting themselves, fields and plantations causing the wilderness to rejoice'.[32] 'Neat and pleasing' is the ultimate accolade, the opposite combination of adjectives is 'wild and uninteresting'.[33] Neither group of missionaries seems to consider that local factors such as climate, cultural and social norms or poverty may have influenced the landscape they so easily criticize; and, control over the environment is presented as emblematic of civilization and Christianity.

The African landscape is frequently described as eerie or weird, as well as threatening. In a heathen place, not only are the people 'other', but also the landscape, buildings and even nature are threatening, more primitive and more out of control than in civilized society; hence, the compulsion to seize control and impose order. As the terrain is brought under control and cultivated, so are the people tamed, civilized and Christianized. In the Bible, the idea that the ground is cursed and needs redemption is introduced: 'The earth also was corrupt before God, and the earth was filled with violence.'[34] A Catholic Truth Society of Ireland pamphlet refers to 'soil once fetid with devil-worship', indicating this bond between people and place.[35] A 1928 article regarding native ordination justifies the long 'trial of twenty years' before ordination, with a mention of 'the superstition and filth with which their land is impregnated'.[36] There is a suggestion that the land itself possesses some supernatural power: a mysterious attraction commonly known as the 'great lure of Africa'. Villages, which were converted and then neglected, 'seemed to straggle out of sight to hug their darkness'.[37] The discourse suggests that the landscape conspires with the dark forces to reverse the process of civilization and conversion, that some power draws the villages back to paganism.

Usually 'darkness' is the aspect of the landscape which is most abhorred. It is a description that is also applied to the people and their culture (if the existence of culture is acknowledged); the term is used both metaphorically and as a physical description. One tribe, the Ekoi, described as jungle folk, live in 'tiny huts 'neath the towering leafy giants that shut out even the African sun. They were primitive of the primitive, these spirit-ridden denizens of the forest'.[38] Their choice of habitat, the shade they prefer to live in, is presented as evidence of the attraction of 'darkness' to them, they are 'benighted'. Shadow implies secrets, dishonesty, the darkness of sin and a desire to hide. Civilization is bright and filled with light; all is illuminated and unambiguous. Africa is contrasted with Ireland, in terms of brightness and weather. In Ireland, there are the 'clean fresh breezes of a vigorous living faith'.[39] And 'the dark mysterious and sin-laden African jungle' is compared with 'the wide open fields of heaven'.[40] The weather is implied to be obstructing the missionaries' progress, because even it is pagan and opposed to Christianity. The following description suggests the weather is an army under Satan's control: 'wettings and soakings for hours at a time in tropical storms to the accompaniment of round after round of a black and angry sky's artillery. A Pagan sky impeding the advance of Christian forces by shelling your line!'[41] In a similar manner in *Heart of Darkness*, Joseph Conrad describes the land's resistance to colonization, 'as if nature herself had tried to ward off intruders'.[42]

The graphics used in missionary magazines, either on the covers or as illustration for articles, often make use of the same ideas and themes which recur in the printed texts. For example, a crest on the front cover of *The African Missionary* includes representations of Ireland and Africa which refer to the metaphorical use of darkness and light. This small drawing is divided in half, with Ireland and Africa represented on either side, joined at the top by a cross which appears to radiate light, and at the base with a planet Earth and the words: 'Fianna Críost Sinne' ('We are warriors of Christ'). Ireland is lit by an enormous Celtic revival-style sun, whose rays bathe the landscape in light; the figure of St Patrick and a small church appear in an open and serene landscape of water and hills with shamrock in evidence on the slopes. Africa, in contrast, is dark. Small, indistinct, dark figures by a round straw hut are dwarfed by the shadow-casting vegetation of huge trees, which seem to be moving in the breeze.[43] The dichotomy is clear, Ireland is light-filled, Africa is dark and shadowy.

The linking of the light of salvation with fire introduces another common metaphorical usage:

> Long ago ... our country stood out as a beacon-light in the pagan gloom of Europe. At her sacred fire the living torches of faith were kindled and carried abroad to many lands. Today ... farther afield and in deeper gloom the pagan races await its first penetrating rays.[44]

Another reference to fire where the continuity of Irish missionary activity, from the golden era of Colmcille to the present, is being rather tenuously proposed, notes that there were moments 'when, without ever being wholly quenched, the fire of missionary zeal burnt low'.[45] In this manner, missionary activity is represented as 'fire', an element that 'cleanses' as well as brings light.

> That fire has been kindled in Ireland. And if history speaks truly it is the mission of the Irish to fan it to a bright flame on the hearths of the homeland and to carry its embers abroad to light up the darkness of paganism.[46]

It is a challenge to the missionaries to describe the strangeness of the landscape, but rather than expressions of wonder at the novelty of the scene, the missionary's reaction is often unease and apprehension. The landscape is too threatening, too physically dangerous to be aesthetically savoured under the category of 'the sublime', and that fear is evident in various images. Fr James describes a sunset, the ultimate romantic cliché, in violent and intimidating language: 'For hours that sun will hold its place until the veldt goes suddenly to flames and a fire, like a devouring demon, races wildly along the hill-sides.'[47] The reference to fire in this 'pagan' context renders it a dangerous and frightening element, not something enlightening or positive.

The adverbs and adjectives used are negative or imply irrationality: 'the sun...madly finds its zenith', the beauty of a moonlit night is 'scarcely earthly'.[48] Sounds are eerie. Inscrutable is a word used often, expressing a fear of the unknown or of what one is not privileged to understand. Another emotion that might be identified in the accumulation of descriptive prose is that of fascination, and maybe a reluctance or fear to engage with the sensuousness of the landscape, which is overwhelming. The intensity of colour, the richness of the visual experience was a shock to the senses of these Irish priests. The impression from the texts is that the land quite simply defeats their conceptual apparatus and faced with this carnival of the senses their response is often apprehension rather than celebration. Occasionally, a description captures the beauty of the unfamiliar landscape without any implications of threat or danger. A multicoloured sunset is described thus by Holy Ghost missionary, Henry Aloysius Gogarty: 'The blue, velvet vault of sky was draped with lemon-tinted veils of vapour, which hung down over a pale green band of sky bordering a furnace-red horizon.'[49]

Travel narratives

Accounts of missionary travels by Irish missionaries, which might be termed 'travel narratives', describe Africa in terms of landscape, modes of travel, the existing mission stations and the improvements which have been made. Travel is frequently considered a triumph in its own right, involving as it does the mastery of the hostile landscape; and it is invariably traumatic, uncomfortable and dangerous.[50] Roads or tracks are characterized by a series of dangerous bends or perched on the edge of a cliff; they are often rutted, flooded, muddy or generally impassable:

> we were on the worst stretch of road that I had ever seen. It consisted of two (sometimes four) ruts which seemed almost a foot in depth, and the bottom of the ruts was both lumpy and slippery, being sunk in black cotton soil and often full of water.[51]

On the other hand, less dramatically: 'what went by the name of a road was certainly no road in any European sense'.[52] Rivers filled with crocodiles, hippopotami and rapids, may be crossed by ferry, raft or canoe (infrequent, late or cancelled at short notice, and invariably less than seaworthy) or a 'bridge'. The roads and bridges suggest a framework of civilization, an effort to master the landscape, but one which is clearly inadequate and underdeveloped. Added to which, the 'route' is surrounded by dense jungle containing a multitude of wild and dangerous organisms, ranging in ferocity from lions to mosquitoes and tsetse flies. The climate varies but is usually uncomfortably hot, cold, wet or dry.

An Irish missionary based in Pondicherry in India, Fr Thomas Gavan Duffy, wrote of his travels in various mission territories; *Let's Go* (from which I have quoted earlier in the chapter) is an account of his journey overland by car from Mombasa on the East Coast of Africa to Lagos and Dakar on the Atlantic-seaboard from November 1927 to March 1928. The book includes many photographs, some of which describe the actual process of travel and show the car stuck in mud or being ferried over rivers. This more journalistic type of writing is an obvious effort on the part of the missionary organizations to reach a wider world of readers than missionaries alone.

Graham Dawson remarks on the opportunities in abundance for British would-be heroes offered by Empire in 'a multitude of real landscapes transfigured by imaginative geography into adventure terrain'.[53] It would seem that missionary discourse closely mirrored this strategy. Like the accounts provided to the 'would-be heroes' of empire, these accounts are directed at potential missionaries, participants in the spiritual empire rather than tourists. Due to the rather exotic location, elements of adventure writing creep in, such as encounters with wild animals – typical imperial boyhood adventure material. For example, Fr James Mellett's memoir includes chapters entitled, 'I Sleep in a Leper's Bed', 'Danger from a Hippo' and 'I Am Asked to Pay Homage to the Devil'.[54] The adaptation of the conventions of the adventure genre is a logical strategy; however, having highlighted the danger of missionary life in an attempt to attract the attention of religious young men, looking for a life of excitement, the missionary organizations rather ingenuously complain that the missionary enterprise is too often 'regarded as a foolhardy attempt of reckless and quixotic daring'.[55] Drawing attention to the distinction between the spiritual and secular empires, Fr James O'Mahony defended the title of his account of three months in Africa in 1935, *African Adventure*, acknowledging that there was another book of that title: 'But then there is adventure and adventure. And if my rival was interested in the adventures of wild animal-hunting in Africa, I was thinking of quite a different form of adventure.'[56] He proceeds to describe how he slept uneasily in an isolated hut from which, just a short time previously, a snake had been removed. He had almost walked on five or six snakes in a three-mile stroll, and then transported through the bush in a shaky old train that on its last trip had been attacked by a lion. Then he states that 'there was nothing heroic or spectacular about my particular journeyings'.[57] He seems to be assuming that as long as he does not portray himself as the hero, the account is not an 'adventure' story but something else. His final statement is, if anything, an assertion that danger is an everyday presence.

In common with other women who participated in Empire, generally missionary nuns do not adopt the travel or adventure narrative; their focus is less on the journey and more on the practicalities of their work, with their main concern being the condition and treatment of women. They describe their journeys in a different style; while obviously sharing many of the same experiences their narratives are less 'heroic' in the telling. They provide an alternative perception of landscape and people, reacting maternally to the sight of 'adorable black babies'. They employ metaphors that are more domestic: a river is like 'an ocean of pea soup, steaming'.[58] Their unfamiliarity with their surroundings is exploited for comic effect rather than to create tension or demonstrate bravery. On one occasion, having travelled into the bush, a group of nuns is awakened by the drumming of 'tom-toms'; the writer relates that after a while she 'discovered it was for the Angelus – not for an attack in force'.[59] A 'hazardous drive' provides an opportunity to muse on the temperament of missionaries rather than a detailed account of the various hazards that have been encountered.

Travel was indeed hazardous, and missionaries often lost their lives in accidents (though more usually in accidents involving vehicles rather than ferocious beasts). Motorcycles and bicycles were the most common forms of transport, although walking was often the only option and horses were occasionally provided. In his memoirs, Bishop Thomas McGettrick gives an account of a cycling priest who 'lost his bearing and free-wheeled a gentle incline into the river', where he drowned. McGettrick comments: 'He was a very fine priest, but as I said earlier, he had a terrible sense of direction.'[60]

Buildings

Changing the landscape was a means of taking control of a hostile and foreign space and the construction of buildings was a project which occupied both missionaries and fundraisers. A substantial proportion of funds raised was for new churches or houses for the missionaries. In Africa, in common with missionaries of other nationalities, the Irish rejected local building materials, traditions and methods, insisting on building with brick. Their rejection of the roundhouse, a structure which was easily built with natural, local materials demonstrates a lack of consideration for the relevant issues and an insistence on imposing European standards. (Even if they had adopted a local design, building a roundhouse out of brick would have been unfeasibly difficult.) Bricks were not an easily available raw material and had to be handmade; in some cases, each was stamped with a shamrock.[61] The flaw in the construction of such permanent buildings was demonstrated in one instance when, immediately following the building of a school at Momba, the chief and the entire village had to move 20 miles away due to a 'plague' of locusts which had stripped the area and introduced a threat of starvation.[62] Another imposition of Irish ideas of design on buildings in Africa is apparent from the existence of 'a shamrock [shaped] church and school at Matroosfontein'[63] and a round tower in Sierra Leone.[64] Kiggins mentions the priests being ill-equipped for building work and cites the example of a two-storey house which was built with no stairs.[65] In his memoirs, Fr Kevin Doheny comments on the building of 'Irish-style churches in a climate which called for greater ventilation and an African design',[66] but notes that, in later years, Irish architects Niall Meagher and Pierce McKenna introduced a new architecture that took account of climate and culture.

Buildings for religious use were a focus for criticism too. There was a common assumption that the quality of one's church buildings was an index of the church's strength and permanence, and of the status of the Irish missionaries in their communities. A missionary comments that he cannot but make a comparison between the beautiful edifices in Cork city 'and the tumbledown mud huts which act as shelters for our Divine Lord in this poor pagan land'.[67] Another writer notes: 'Judging by appearances, Our Divine Lord is the biggest tramp and the worst housed beggar that ever lived in a Bohaneen in Ireland' and that 'the worst criminals in the Government prison here beside us have a decent dwelling in comparison'.[68] These descriptions are followed by appeals for funds to build churches in Africa. A commonly used illustration accompanying such appeals is of the 'church', depicted as a dilapidated shack or, alternatively, half-built in brick and waiting for funds for its completion. A sketch of a rickety grass-roofed mud hut on a 'children's page', is captioned: 'Is this a fitting home for Jesus? If you saw it you would call it a cattle shed it looks so little like a church.'[69] There was another point of view: in an article entitled 'Sticks and Stones', Lupton describes pagan altars and jujus as insignificant in size and appearance but standing 'for mighty ideas' and continues, 'why build big temples for worship, or have elaborate sacrifices when a square yard of bush railed in by palm branches can serve as a holy place'?[70] This seems to be a rhetorical question or at least a minority view as, in the following pages, a drawing of an incomplete church foundation is featured with a note that 70,000 bricks are needed. Readers were invited to contribute by 'purchasing a brick'. Whether these large buildings increased the authority of the Catholic Church or merely raised the profile of the particular missionary order is debatable. In 1926, Pope Pius XI felt it necessary to draw attention to the issues regarding mission buildings:

> If it is necessary, Venerable Brothers and Beloved Sons, in the cities where you have your residences and in other more important centers, to erect large churches and other mission buildings, you must, however, avoid building churches or edifices that are too sumptuous and costly as if you were erecting cathedrals and episcopal palaces for future dioceses.[71]

According to a nun in Ethiopia, these large buildings were seen as belittling the local efforts. She asked Doheny, 'Why do you want to put up such fine buildings? Is it to show us up?'[72] A Capuchin missionary, Michael Glynn, explains that if a station 'is "strong", there is a small bell – cast in Dublin – mounted high on four slim pillars'.[73] The bell, which called the people to prayer, exerted another control over the territory. The sound which disturbed the silence or the hum of indigenous life is one noted by African writers in their accounts of the missionary presence; the noise travelled over the land and through the sky, alerting all the inhabitants 'to its loud and alien voice'.[74]

The naming of these new churches provided opportunities for an exercise of cultural authority. The many churches all over the world named for St Patrick were the subject of much pride. In his message to the Irish people, read by Cardinal Lauri, at the Eucharistic Congress in Dublin in 1932, Pope Pius XI drew attention to the dedication to St Patrick of almost 800 churches, as one of the most conspicuous signs of Ireland's worldwide influence in the Church.[75] Hence, the list of schools and churches dedicated to Irish saints was a message to the world demonstrating Ireland's influence internationally; these dedications had more than a local significance. Naming is traditionally a family's responsibility; by undertaking the naming of buildings, the idea of a 'spiritual mother country' was being evoked. Children too were named for Irish saints, and other historical figures. Padraig Ó Máille, who worked as a missionary in Nigeria and Malawi, mentions meeting Brian Boru Davis Usanga, a bishop in Nigeria.[76]

Conclusion

It is evident that much of the Irish Catholic missionary discourse of this time is close in form and content to the imperial discourse of an earlier period in history and however unintentional these parallels are striking. Empire building, whatever its motivation, seems to draw on similar ideologies and representational strategies, probably due to the need to engage the support and approval of the public, in order to ensure the continuance and success of the project.

The negative connotations of imperialism were countered by the missionary presumption that a Spiritual Empire was in many ways the antithesis of an imperialism driven by mercenary greed and a hunger for power. It seems never to have been considered that elements of that impulse to hold sway over territories and influence populations might also be present in the spiritual empire. Even when different missionary orders clashed over control of territories, or individuals fought for control of their missionary organizations, this similarity was not recognized.

It is evident that missionary material shares many elements with other non-missionary, colonial texts and that missionary methods of defining the 'Other' relate closely to definitions of 'Other' to be found in the various texts of empire including fiction, adventure writing, travel writing and biography. 'Otherness' is an essential element of such a discourse as it functions to establish identity, justify intervention and provides motivation for action. Duty is established as the overriding principle – there is no choice other than to engage with the situation as it is presented. Edward Said has remarked that every empire 'tells itself and the world that it is unlike all other empires, that its mission is not to plunder and control but to educate and liberate'.[77] In the case of the Irish missionaries, the function of their discourse was not so much to strengthen the functioning of their 'colonial' power, but to strengthen Irish identity as Catholic and civilized.

It would seem that the disadvantages of the association of empire with the exploitation of people and resources, and its negative mercenary connotations were easily outweighed by the advantages and convenience of a discourse which provided adaptable elements/genres and enabled the focus to be firmly on the civilizing aspects of the missionary project, highlighting the benevolent motivation of this new and different empire.

Notes

1. For example, T. P. Coogan, *Wherever Green is Worn: The Study of the Irish Diaspora* (London: Palgrave, 2000); T. Keneally, *The Great Shame: A Story of the Irish in the Old World and the New* (London: Chatto and Windus, 1998).
2. S. Neill. *A History of Christian Missions*, 2nd edn (Harmondsworth: Penguin, 1986), p. 50.
3. Ibid., p. 86.
4. E. M. Hogan, *The Irish Missionary Movement: A Historical Survey, 1830–1980* (Dublin: Gill and Macmillan, 1990), p. 6.
5. Cardinal Newman's sermon of the same name, referring to the reinvigoration of the Catholic Church in England, seems to have been the source of this phrase. 'The Second Spring' was preached on 13 July 1852 in St Mary's, Oscott, in the first Provincial Synod of Westminster. Text in *Sermons Preached on Various Occasions*, 3rd edn (London, 1870).
6. J. P. Mullen, 'Ireland's Missionaries Ancient and Modern', *Student Missionary* (1931) 32.
7. J. Lee, *Ireland 1912–1985: Politics and Society* (Cambridge: Cambridge University Press, 1989), p. 396.
8. 'Editorial', *African Missionary*, no. 86 (July–August 1928) 182.
9. See, for example, L. Gibbons, *Transformations in Irish Culture* (Cork: Cork University Press in association with Field Day, 1996) and D. Lloyd, *Anomalous States: Irish Writing and the Post-Colonial Moment* (Dublin: Lilliput Press, 1993).
10. For further discussion of 'missionary discourse' see Esther Breitenbach's chapter (Chapter 5) in this volume and see also my own unpublished PhD thesis: 'The Spiritual Empire: Irish Catholic Missionary Discourse in the Twentieth Century', (NUI Galway, 2003).
11. L. P. Curtis, *Apes and Angels: The Irishman in Victorian Caricature* (Newton Abbott: David & Charles, 1971), p. 1. See also A. McClintock, *Imperial Leather: Race, Gender and Sexuality in the Colonial Contest* (New York and London: Routledge, 1995).
12. Charles Kingsley, Letter to his wife, 4 July 1860, in *Charles Kingsley: His Letters and Memories of His Life*, Francis E. Kingsley, ed. (London: Henry S. King and Co., 1877), p. 107, cited in Anne McClintock, ibid., p. 216.
13. *United Irishman*, Supplement, 17 March 1900.
14. R. Kipling, 'The White Man's Burden' (advice to the United States in their efforts to colonize the Philippines) originally published in 1899, may be found in *The Works of Rudyard Kipling* (London: The Wordsworth Poetry Library, 1994), p. 323. In its original context, Kipling's phrase would have included the Irish as a part of that burden: the 'new-caught, sullen peoples, Half devil and half child', who resisted civilization and made the mission difficult through their 'Sloth and heathen folly'.
15. E. Said, *Orientalism* (London: Vintage Books, 1978), p. 77.
16. T. Kiggins, *Maynooth Mission to Africa: The Story of St Patrick's, Kiltegan* (Dublin: Gill and Macmillan, 1991), p. 49.
17. For example, see Children's Page, *African Missionary*, no. 80 (July–August 1927) 79.
18. 'The Place of the Church in Africa', *St Patrick's Missionary Bulletin* (August 1935) 12.

19. 'Dympna's Corner', ibid. (January 1935) 4.
20. M. Slattery, 'The Provincial's Page: Bathurst, Freetown and Kruland', *The African Missionary* (March–April 1928) 148.
21. Rev. H. McGrath, 'Anua Mission Station', *St Patrick's Missionary Bulletin* (June 1934) 11.
22. Fr P. McDaid, 'A New Opening by the Cross River', ibid. (September 1935) 9.
23. S. Brown, 'Foreign Missions: A Survey', *Studies*, XV (March 1926) 105–20, 113.
24. Fr James Mellett CSSp, *If Any Man Dare: Missionary Memoirs* (Dublin: Fallons, 1963), p. 62.
25. Mary Louise Pratt, *Imperial Eyes: Travel Writing and Transculturation* (New York and London: Routledge, 1992), p. 51.
26. Thomas Gavan Duffy, *Let's Go* (London: Propagation of the Faith Office, 1928), p. 26.
27. Fr James [O'Mahony], *African Adventure* (Dublin: The Father Mathew Record Office, 1936), p. 132.
28. Rev. J. Lupton, 'Visiting Outstations', *African Missionary* (July–August 1927), 74–5.
29. Pratt, *Imperial Eyes*, p. 151.
30. Rev. E. Leen, 'In the Heart of Nigeria', *Pagan Missions*, V, no. 1 (March 1922) 16.
31. John Eliot Howard, *The Island of Saints or Ireland in 1855* (London: Seeleys, 1855), p. 175.
32. Ibid., p. 162.
33. Ibid., pp. 171, 190.
34. *King James Bible*, Genesis, 6: 11.
35. *The New Africa: The Holy Ghost Fathers and Their Missions* by a Member of the Congregation of the Holy Ghost, CTSI, Pamphlet 1280 (Dublin, 1934), p. 13.
36. Rt Rev. B. J. Wilson, 'A Native Ordination: Birth of a Native Clergy', *Missionary Annals*, X, no. 1 (January 1928) 11.
37. Fr James, *African Adventure*, p. 18.
38. John P. Jordan CSSP, *Bishop Shanahan of Southern Nigeria: The Life of the Great Apostle to the Igbo Tribe and Founder of the Missionary Sisters, Killeshandra 1871–1943* (Dublin: Clonmore & Reynolds, 1949), p. 205.
39. Rev. C. Plunkett, 'Missionary Life', *St Patrick's Missionary Bulletin* (June 1935), 9.
40. P. Birmingham, 'Missionary Reports', *African Missionary*, no. 94 (May 1929), 111.
41. Rt Rev. P. Rogan, 'Numbers! What They Mean', *Catholic Missions*, XCIII, no. 1 (January–February 1930) 127.
42. Joseph Conrad, *Heart of Darkness* (London: Everyman's Library, 1993), p. 19. First published in 1902.
43. Cover illustration, *The African Missionary*, VII, no. 1 (January 1929).
44. 'Editorial', ibid., no. 86 (July–August 1928) 182.
45. Brown, 'Foreign Missions: A Survey', 106.
46. Rev. John O'Leary, 'Vocations', *Pagan Missions*, III (June 1924) 78–9.
47. Fr James, *African Adventure*, p. 61.
48. Ibid.
49. H. A. Gogarty, *In the Land of the Kikuyus* (Dublin: M.H. Gill & Son, 1920), p. 102.

50. A typical feature of travel writing according to Pratt, *Imperial Eyes*, p. 148.
51. Gavan Duffy, *Let's Go*, p. 50.
52. Fr. James, *African Adventure*, p. 156.
53. G. Dawson, 'The Imperial Adventure Hero and British Masculinity: The Imagining of Sir Henry Havelock', *Gender and Colonialism*, ed. T. P. Foley, L. Pilkington, S. Ryder and E. Tilley (Galway, 1995), pp. 46–59, 47. For 'imaginative geography' see Said, *Orientalism*, pp. 54–5.
54. Mellett, *If Any Man Dare*.
55. 'Editorial', *African Missionary* (March–April 1928), 141.
56. Fr James, *African Adventure*, 15.
57. Ibid.
58. *Diary of a Medical Missionary of Mary*, I (Dublin, 1957), p. 12.
59. 'From the Archives: Our First Christmas in Nigeria', *The African Rosary*, I, no. 2 (November–January 1936–1937) 17.
60. Bishop Thomas McGettrick, *Memoirs of Bishop T. McGettrick* (Sligo: n.p., 1988), pp. 99–100.
61. Jordan, *Bishop Shanahan*, p. 146.
62. Fr James, *African Adventure*, p. 135.
63. Ibid., pp. 89–90.
64. Illustration in Joseph Dunn, *No Tigers in Africa: Recollections and Reflections on 25 Years of Radharc* (Dublin: The Columba Press, 1986), p. 158.
65. Kiggins, *Maynooth Mission to Africa*, p. 156.
66. Kevin Doheny, *No Hands but Yours: Memoirs of a Missionary* (Dublin: Veritas, 1997), p. 53.
67. E. McSweeney, 'Church Needs in Africa', *The African Missionary* (March–April 1928) 156.
68. 'Auntie's Playground', ibid. (March 1929) 58.
69. 'Auntie's Playground', ibid., no. 86 (July–August 1928) 197.
70. Lupton, 'Sticks and Stones', ibid., no. 85 (May–June 1928) 176.
71. Pius XI, *Rerum Ecclesiae*, 8 February 1926.
72. Doheny, *No Hands But Yours*, p. 54.
73. Fr M. Glynn, 'Saint Patrick's Missions', *Capuchin Annual* (1955) 377–80, 378.
74. T. O. Echewa, *The Land's Lord* (London: Heinemann, 1976), p. 116.
75. Papal message, read at the Pro-Cathedral Dublin at the opening of the Eucharistic Congress, 'Ireland, Mother of Heroes', reported in the *Irish Press*, 23 June 1932.
76. P. Ó Máille, *Dúdhúchas* (Baile Átha Cliath: Sairseal agus Dill, 1972), p. 127.
77. E. Said, 'The Blind Arrogance of the Imperial Gaze', *Irish Times*, 25 July 2003.

Further reading

Bateman, F. 'The Spiritual Empire: Irish Catholic Missionary Discourse in the Twentieth Century' (PhD Thesis: NUI Galway, 2003).

Coogan, T. P. *Wherever Green is Worn: The Study of the Irish Diaspora* (London: Palgrave, 2000).

Gibbons, L. *Transformations in Irish Culture* (Cork: Cork University Press in association with Field Day, 1996).

Hogan, E. M. *The Irish Missionary Movement: A Historical Survey, 1830–1980* (Dublin: Gill and Macmillan, 1990).
Kiggins, T. *Maynooth Mission to Africa: The Story of St Patrick's, Kiltegan* (Dublin: Gill and Macmillan, 1991).
Lloyd, D. *Anomalous States: Irish Writing and the Post-Colonial Moment* (Dublin: Lilliput Press, 1993).
Pratt, M. L. *Imperial Eyes: Travel Writing and Transculturation* (New York and London: Routledge, 1992).
Said, E. 'The Blind Arrogance of the Imperial Gaze', *Irish Times*, 25 July 2003.

14
Canadian Protestant Overseas Missions to the Mid-Twentieth Century: American Influences, Interwar Changes, Long-Term Legacies

Ruth Compton Brouwer

Popular stereotypes of colonial-era missionaries have not readily faded even in the face of research that has directly challenged those stereotypes. As iconic images of missionary heroes began to disappear in the 1960s along with hackneyed missionary jokes and cartoons, a new stereotype emerged, one that presented the missionary as 'part and parcel of the imperial project'.[1] The resilience of this 'new' image, not only in the popular mind but also in the broad scholarly community, probably owes a good deal to the published discourses of the mission churches themselves. Proponents of missions showed a pragmatic tendency to speak favourably of imperial rule and sometimes even to urge its extension, notwithstanding individual missionaries' private concerns about self-serving imperial policies and official or random acts of violence against colonized peoples. Several contributions in the present volume illustrate this pattern.[2] Other contributors deal with missionaries who, while they worked in British imperial terrain, had ethnic or faith backgrounds that differentiated them from, and sometimes put them at odds with, the mainly elite Protestant men who determined colonial policies.[3]

Writing from a Canadian perspective, one could readily substitute French-Canadian Catholic missionaries for the Irish Catholics who appear in this volume in the role of outsider missionaries within the British Empire. Although France's colonial empire in what is now Canada fell to Britain by conquest in 1763, French-Catholic culture survived as a central element in French-Canadian national identity and, in the twentieth century, supported a vigorous overseas missionary enterprise.[4] My concern in this chapter, however, is with a group who were very much insiders: mainline Canadian Protestants of Anglo-Saxon and Anglo-Celtic backgrounds. Focusing particularly on Presbyterians and, for the period after 1925, on missionaries of the United Church of Canada, I present some of the more salient features of mainline Canadian Protestants' overseas missionary activity in Asia and Africa to approximately the mid-twentieth century. In doing so I provide evidence that Canadian Protestants both supported *and* challenged Britain's secular imperial agendas. The chapter first highlights the importance of American models and linkages for the pioneer period and then turns to Canadians' involvement with international and ecumenical institutions and bureaucracies, particularly in the interwar years. This was a period when important shifts in the meanings attached to gender, race and professionalism in mission work were also taking place. Here I foreground two Canadian professional women, Dr Belle Choné Oliver (1875–1947) and Margaret Christian Wrong (1887–1948). Both women headed mid-level international and ecumenical organizations, one function of which was to assist and employ indigenous Christians of talent. The third part of the chapter continues to position the Canadian experience within larger patterns of internationalism and ecumenism and makes some tentative observations about mainline missions in the interwar period as agents in the emergence of an imagined global Christian community.[5] While a consideration of the long-term significance for Canada of its role in overseas missions is outside the scope of this chapter, I touch briefly in the final pages on the missionary legacy in development work in the 1960s.

American precedents and Canadian initiatives

The mainline Canadian Protestant denominations – Anglicans, Baptists, Methodists and Presbyterians – established the machinery for sending missionaries abroad in the late 1800s, initiating work in India, China, Taiwan, Japan, Korea and the Caribbean. Canada's tiny Congregationalist community also began overseas work in this period, in the Portuguese colony of Angola.[6] Presbyterians (marginally ahead of Methodists as Canada's largest Protestant denomination by the time of the 1911 census)[7] had begun missions in India and China in the 1870s. From the outset, three interrelated patterns emerged in this work that did not fit widespread, and still-common, assumptions about overseas missions.

First, women missionaries were numerically dominant. Second, US models and connections were more significant than British leadership and example. And third, like the Americans, Canadians leaned heavily towards institutional and social service work rather than direct evangelism. In the Canadian Presbyterians' India mission these patterns were particularly pronounced. Two single women were the pioneers, serving first in American Presbyterian missions, where the more forceful of the two worked with ordained missionaries to point the Canadian church to a site for its own 'field' and then urged the potential for making conversions through service-oriented mission work. More or less similar patterns would eventually emerge in the missions established by the other mainline denominations.

The tendency to look to the United States rather than Britain in late-nineteenth-century missionary matters did not reflect an absence of strong ties to the Mother Country. Many Canadian missionaries and home-base officials were only a generation or two removed from England, Ireland or – especially in the case of Presbyterians – Scotland. Sentimental ties of blood and belonging were still very real, nor was support lukewarm for Britain's imperial agenda. Indeed, English-speaking Canadians were ardent imperialists in the late Victorian and Edwardian years. They had no difficulty in combining an emerging sense of Canadian national identity with belief in and commitment to a British Empire that they saw as, whatever its flaws, a uniquely powerful force for good in the world.[8] A more distinctly Canadian sense of identity began to emerge after the First World War, but during the war, even Irish Catholics could be found in the ranks of loyal imperialists in English Canada.[9]

Unquestioned assumptions about the essentially benevolent and Christian intent of the British raj may have been responsible for a serious diplomatic and strategic gaffe on the part of the Canadian Presbyterians when they began their mission in India. The ordained men who in 1877 officially established the church's first station, at Indore, Central India, did so without consulting either the local colonial authority or the reigning maharajah. Whatever the reason for the missionaries' undiplomatic action, the result was immediate ill will towards the mission, a situation that improved only after 1885 when a new viceroy, Lord Dufferin, a former governor-general of Canada, intervened on behalf of the missionaries, ending what one ordained worker would later describe as the era of 'freezing coldness'.[10] Even after the thaw, few of the Canadians developed truly close personal relationships with colonial officials in Central India. Moreover, some had grave concerns about such local imperial policies as the fostering of opium production. Generally, though, the missionaries kept such criticisms private, believing that, on balance, colonial rule was a blessing for India, and especially for its women.

What probably contributed most to the long-term improvements in relations between the Canadians and local imperial and princely officials was the educational and medical work undertaken by the women missionaries. Schools were begun, including what reportedly became the first high school for girls in Central India. Dispensaries and hospitals were also established. At the home base, both the church's all-male Foreign Missions Committee and the Woman's Foreign Missionary Society (WFMS), its powerful auxiliary, were especially enthusiastic about supplying staff and funds for medical work. For a time in the 1890s, more than a third of the single women in the mission were doctors. The women missionaries'medical work was also facilitated by local imperial authorities and Indian princes. Here, unquestionably, the model of institution building and social service activism being followed by the Canadians in India was primarily an American model: even in this most valued of Britain's colonies, US women missionaries were ahead of their British counterparts in establishing institutionalized social services, both as tools for evangelism and as tangible expressions of Christian witness.[11]

Meanwhile, back in Canada the Presbyterian WFMS and the other women's missionary societies that were formed in the late nineteenth century to support their respective denominations' mission work among women and children drew upon their US denominational counterparts' older societies in drafting their own constitutions. They also sent delegates to their national or regional conventions and welcomed their American sisters to their own annual gatherings. Intending Canadian missionaries of both sexes sometimes went to the United States for preparatory training, and not a few married American fellow-workers. Canadian mission personnel, particularly men, sometimes served under a US mission board if it meant getting an overseas appointment more to their liking than the opportunities available through their own board. Some of these men, such as Dr William Wanless (1865–1933), would later become much celebrated 'American' missionary heroes.[12] In summary, it was primarily US precedents and cross-border linkages rather than examples from Britain that influenced early planning and policies in Canadian Protestants' overseas mission work. Factors related to geography and culture inevitably gave primacy to the cross-border ties.

Did the same hold true for discourses? Did their orientation towards mission metropoles in the United States rather than Britain and the fact that Americans and Canadians sometimes cooperated in producing mission study materials make much difference in terms of Canadian Protestants' discursive representations of the indigenous populations of 'their' mission lands? Perhaps not a great deal. As Nicholas Thomas and others have pointed out, there were significant differences between the ways that missionaries, on the one hand, and travellers and secular imperialists, on the other, depicted 'natives'.[13] And within mission discourses there were certainly important differences between what was represented about, for instance, Africans and native North Americans, on the one hand, and the peoples of South and East Asia, on the other. In the Asian fields, literate, complex and ancient cultures were undeniable phenomena, and perceived as barriers to conversion. It was in these fields that Canadian Protestants had most of their early overseas work, and in speaking of its needs to their home-base constituencies they could draw on a stock of metaphors and generalizations shared by their English-speaking counterparts on both sides of the Atlantic. References to the 'subtle Hindu' or the 'stubborn Confucian' – they were typically gendered male – or to 'benighted heathen women' could thus have originated in London, New York or wherever else early missionary presses had begun giving currency to such stereotypes.

Ecumenism, gender and race in interwar mission work

The Dominion of Canada in 1900 was a physically vast country, still mainly rural, its population tensely divided between a French-speaking Catholic majority in Quebec and an English-speaking Protestant majority in the other provinces. In the years leading up to the First World War unprecedentedly high rates of immigration added to the challenges facing the country's faith communities. Many of the so-called 'new' immigrants, those from southern and eastern Europe and Asia, were settling in the still-sparsely-serviced western provinces. Between 1901 and 1921 the Dominion's population rose from just over five to just over ten million, and in 1913 a record 400,000 immigrants arrived.[14] Given these circumstances, ecumenism made compelling sense to mainstream Protestants. In 1925, Canadian Methodists, Congregationalists and most Presbyterians came together to create the nation's largest and most liberal Protestant church, the United Church of Canada. As '[t]he first union across confessional lines', it was a unique development in world Protestantism.[15] A powerful motive was the desire for a strong and united approach to assimilate 'new' immigrants into English-Canada's cultural and Protestant Christian traditions. Proponents of the social gospel also favoured church union, seeing it as the best way to respond to social problems arising from industrialization as well as immigration.[16]

Support for union was even stronger among the uniting denominations' overseas missions constituencies. Virtually all overseas work of the three uniting denominations became the responsibility of the United Church of Canada.[17] This level of mission support for church union was not surprising. By 1925 Canada's Protestant foreign missionaries already had a good deal of involvement with institutions founded as an outcome of ecumenical or international cooperation, especially those providing advanced academic education and medical training. In India, for instance, Canadian Presbyterians contributed to the Women's Christian Medical College at Ludhiana, in the Punjab, established in the 1890s, while, further south, Canadian Baptists supported the Missionary Medical School for Women at Vellore in the Madras Presidency, begun in 1918 by a high-profile American Reformed Church missionary, Dr Ida Sophia Scudder.[18] Economic pressures and pressures related to missions' role (and prestige) as agents of modernization made participation in international and ecumenical mission work an increasingly necessary and attractive approach. At the same time, by the second decade of the twentieth century there was growing pressure on mission-support committees within Canada to commit more fully to home-missions work. The resulting competition for scarce mission funds was especially difficult for the overseas constituencies to deal with, since they had opted early on to follow the costly US model of institutionally oriented mission work. Clearly, they could not afford to open and operate more such institutions on a strictly denominational basis, much less raise their professional standards. Yet, pressures grew in the interwar period for higher standards of professionalism. Of the various factors responsible for this phenomenon, two are worth noting here, one related to supply and the other to demand. On the 'supply' side, the overseas fields had traditionally been viewed as the most appropriate destination for the Canadian churches' best and brightest mission volunteers.[19] In the interwar period this was still the case, but such volunteers, especially the doctors, were often better prepared than in the pioneering era, and they wanted to serve in settings where there was scope for good professional work. On the 'demand' side were indigenous Christian and nationalist youth seeking institutions that could give them high-quality Western-style academic and medical training. Such students had little or no interest in the theological differences in which early denominational institutions had been rooted.

Meanwhile, new and important ecumenical initiatives were taking place nationally and internationally at organizational and bureaucratic levels. These had their roots in the World Missionary Conference at Edinburgh in 1910 and took full flight in the post-war period following the establishment in 1921 of the International Missionary Council (IMC), whose 'members' were regional or national organizations of mission boards or churches. By far the largest of these were the Conference of British Missionary Societies (CBMS) and the Foreign Missions Conference of North America (FMCNA). Early in the 1920s, national Christian councils were established in India, China and Japan, their memberships made up of missionaries and indigenous Christian leaders. During the next three decades, the number of national councils grew, becoming the majority, if still not the most influential, of IMC member organizations. As well, as particular mission concerns came to the fore new regional or task-specific permanent ecumenical committees were established from the 1920s onwards to promote such facets of mission work as evangelization, literacy and literature and medical work. The new internationalism in missions was showcased by two much publicized IMC conferences in the interwar years, the first held at Jerusalem in 1928 and the second at Tambaram, near Madras (Chennai) in 1938.[20]

Canada appears to have contributed personnel to overseas missions in numbers disproportionately high for its small population. Still, the total was small by comparison with the number of missionaries deployed from the United States and Britain. That reality was reflected in Canadians' lack of leadership roles at the top of the bureaucratic structures that proliferated in the interwar period for international and ecumenical work. The towering figures were the Scot Joseph Oldham (1874–1969), the first permanent secretary of the IMC and American Methodist layman John R. Mott (1865–1955), its peripatetic chairman.[21] Yet if Canadians did not have the numbers and clout to find room at the top of these organizations, it was nonetheless in the interest of the American and British male leaders who headed them in these interwar years – sometimes in considerable tension with one another – to have *some* Canadians involved. Canadians typically participated in mid-level mission bureaucracies and as limited-term or permanent heads of committees, and here there was some space for women.

As Barbara Bush has observed, the interwar period was a time when some well-educated women were beginning to gain access to high-profile or influential roles in the world of secular British imperialism, as academics and as supporters or critics of empire. Bush cites as examples women as different as journalist and educator Margery Perham and literary radical Nancy Cunard.[22] The same phenomenon emerged in the IMC and some of its ancillary organizations. Two Canadian women who played influential roles in this world of interwar liberal Protestant ecumenism were Margaret Wrong, as secretary of the International Committee on Christian Literature for Africa (ICCLA), a sub-committee of the IMC, and Dr Belle Choné Oliver, as secretary of the Christian Medical Association of India.

Margaret Wrong remained as secretary of the London-based ICCLA from the time of its establishment in 1929 until her sudden death in Gulu, Uganda, in 1948 just as she was beginning her fifth major tour of Africa for the Committee.[23] Educated at Oxford like her three brothers and at the University of Toronto, where her father taught history and promoted all things imperial, Wrong came to the ICCLA secretaryship without any missionary experience. She had, however, established a reputation for effective ecumenical work, first, during her years as a travelling secretary for the Geneva-based World Student Christian Federation (WSCF), and then, from 1926 to 1929, in the same role for the British Student Christian Movement. Her role as a catalyst for the British SCM's 'Africa Group' reflected the fact that that continent had become an abiding interest following a six-month tour of sub-Saharan Africa with Columbia University professor Mabel Carney in 1926. On the 1926 tour, as in her earlier work with the WSCF, Wrong had demonstrated adaptability and enthusiasm, and liberal tendencies on 'race' questions, qualities that made her an attractive candidate to the Africanists in the IMC who were engaged during the 1920s in efforts at restructuring the educational work of missions in Africa. Joseph Oldham, the leading figure in the restructuring process and the person most responsible for recruiting Wrong, also believed that her Canadian identity would make her a congenial appointment for IMC and FMCNA officials in New York.

Working out of Edinburgh House, the London headquarters of the IMC and various other ecumenical groups, Wrong had the opportunity as the ICCLA's first secretary to put her own stamp on its mandate to foster 'the preparation, publication, and distribution of literature for use in connection with missionary work in Africa'.[24] Promoting the creation and distribution of 'Christian literature' was only one aspect of her work. Over the years, in the ICCLA's two periodicals and in the numerous books and booklets it disseminated, there were literary lessons for

Africans on everything from health matters to homemaking, geography and agriculture. At the time of Wrong's death most of this literature was still didactic and written by Westerners. But, from the beginning, there had been efforts to publicize writings about Africans and African Americans. And from the time of her 1936 tour, when she was invited to be present at what was later described as the first conference of black South African writers, Wrong increasingly interpreted the ICCLA's mandate as the development of literature *by* as well as *for* Africans. Following her unexpected death an article in the *Times* of London described Wrong's knowledge of Africa as 'unique in its depth, range, and sympathy' and called for the establishment of a prize in her name 'to encourage literary productions from Africa, either in English, French, or any other suitable language, and, if possible, to subsidize their production'. Future winners of the Margaret Wrong prize included such luminaries of African creative writing as Nigerians Chinua Achebe and Wole Soyinka.

The intensely spiritual, small-town, Presbyterian-born Choné Oliver was a very different figure in many ways from the cosmopolitan Anglican Margaret Wrong. Yet there are some significant parallels in their interwar careers.[25] In 1929, the year that Wrong began her career with the ICCLA, Oliver left almost three decades of work as a medical missionary in Central India to take up the secretaryship of the Christian Medical Association of India (CMAI) and simultaneously to serve as secretary for medical work for India's National Christian Council (NCC). In 1933, she settled in Nagpur, the bureaucratic centre of Christian ecumenism in India. The break with her old life in a remote denominational mission among a mainly aboriginal population was not as dramatic as it might appear at first. Even as a child Oliver had been open to religious diversity.[26] During the 1920s she had read systematically and attended conferences in an effort to keep abreast of actual and recommended changes in missions and missionary medicine. With a US male colleague, she had been asked to undertake a survey of medical missions in India and Ceylon to assess the challenges facing such work. The survey report, issued in 1928, called for various improvements to vivify the quality and reputation of medical missions in the rapidly changing professional and political environment of India. In the same year, Oliver was one of just two medical missionaries sent from India as the NCC's delegates to the IMC conference in Jerusalem. Oliver and colleague Christian Frimodt-Möller drew the results of the survey report to an international audience. They also presented what was, in effect, a new theology of medical missions, one that officially abandoned the old view that the function of medical missionary work was simply to open doors for the preaching of Christianity. A corollary was that if medical missions were to serve as an effective Christian witness and withstand comparison with government and private medicine they would need to demonstrate high professional standards as well as good intentions.

In the years between her appointment to the secretaryship of the CMAI and her permanent retirement from that position just a few months before her death in 1947, Oliver played a leadership role on a number of initiatives, among them cautious support for birth control.[27] The project about which she was most passionate, however, was the establishment of a co-educational, fully professional Christian medical college, a college that would enable Indian Christian men and women to become, in her words, missionaries' 'colleagues and eventually successors', rather than just their assistants. The project came to fruition in 1950 as Christian Medical College (CMC) Vellore, affiliated with the University of Madras and established through the transformation and upgrading of Dr Ida Sophia Scudder's Missionary Medical School for Women.

The protracted and contested process of establishing CMC reveals a good deal about the interwar period as a time of significant change in several aspects of mainstream Protestant missions. The project depended heavily for financial support upon contacts available through the IMC, the CBMS and the FMCNA. After initially having good hopes of such support through William Paton (1886–1943), former secretary of the NCC and now the IMC's Associate Secretary in London, Oliver saw the medical college project fall from favour as the Great Depression worsened and as Paton's support turned to resistance. And while the proposed Christian medical college had Indian Christian doctors who argued for its value on both professional and nationalist grounds, it had critics among other educated Indian Christians who, like Paton, had their own favoured mission projects, for which they wanted IMC support. Meanwhile, until well into the 1940s the co-educational medical college project failed to win over Dr Scudder and her supporters, who sought to maintain the medical school at Vellore as an exclusively female institution while upgrading it professionally to meet new state requirements. Yet if Oliver and Scudder were working during the 1930s towards different institutional goals, there was one matter on which they were like-minded: each of them identified Dr Hilda Lazarus (1890–1978), a devout Christian, a CMAI member, and the most senior Indian doctor in the Government of India's Women's Medical Service, as the ideal person to head up a new, fully professional Christian medical college. Their shared view on this matter made it clear that in late-colonial India race mattered in mission projects in ways that bore little resemblance to its attributed meanings half a century earlier.

In taking up international and ecumenical work for missions in the interwar era, both Choné Oliver and Margaret Wrong moved well beyond the 'women's work for women' paradigm that had dominated the era before the First World War. Close female friendships remained intensely important in their private lives and sometimes in their professional struggles, but both moved in work worlds where their colleagues, whether Western or indigenous, were more likely to be men rather than women.

What of their relationship to the larger secular world within which their respective projects were located, the world of the late British Empire? What, if any, were their links to colonial establishments and officials? Oliver appears to have had few personal contacts of this sort. There are remarkably few comments in her writings about the politics of late-colonial India, though the political context was obviously of great significance for her work. Wrong, by contrast, had numerous contacts with imperial structures and personalities, both on the ground in Africa in the course of her tours and, back in London, at the Colonial Office and in such state-linked agencies as the British Council and the BBC. Such contacts and involvements were vitally important in the development of her literature work, and at the same time presumably were symbiotic in the sense of serving the agendas of her non-mission associates. In 1936, for instance, meetings with the Acting Director of Education in Lusaka led to what would become the Northern Rhodesia and Nyasaland Joint Publications Bureau, one of the first publishing houses in south-central Africa. Wrong's connections with the Colonial Office and with various state-linked agencies increased during the years of the Second World War. From 1941, she worked half time on loan from the ICCLA as a section officer for the African Division of the BBC's London Transcription Service. And in 1944 she served on the sub-committee that prepared *Mass Education in African Society* for the Colonial Office's adult literacy work. Interestingly, it was also during this period that Wrong seems to have become increasingly concerned about evidence of officially sanctioned racism (policies, for instance, restricting Africans' opportunities in wartime employment) and conscious of the need to prepare for decolonization. As a member of the Fabian Colonial Bureau and a contributor to *Fabian Colonial Essays*, she was part of a group that operated both as a left-wing pressure group *and* in cooperative relationships with liberal imperialists in the Colonial Office to facilitate what historian J. D. Hargreaves termed 'progressive social engineering'.[28]

An emphasis on Wrong as a well-connected 'imperial woman' should not obscure the importance of her US contacts. Indeed, the latter were probably vitally important to the former. Tours of Africa were typically followed up by tours of North America. The latter were a means of cultivating financial and other support for the work of the ICCLA. This meant contacts not only with supportive groups within Canada but also with key figures on the New York-based FMCNA, with African Americans, and, importantly, with American foundations. The Phelps-Stokes Fund, the Carnegie Corporation and the Rockefeller Foundation, especially through its General Education Board (GEB), played a key role in helping to finance Wrong's tours of Africa and in aiding agencies with which she was associated. One of the latter was the International African Institute (IAI) and its journal, *Africa*, which in the 1930s had become an important vehicle for promoting writing in African vernaculars. When financial difficulties during the Second World War interrupted the publication of *Africa*, Wrong's intervention on its behalf with the GEB helped secure needed funds.

British mission leaders, Africanists like those in the IAI, and colonial officials may at times have chafed at their need to rely on American financial help. But by the war years US hegemony was increasingly a fact of life in all such international relations. Moreover, such groups recognized that their American benefactors shared their interest in using education and economic development to promote the conditions that would lead to peaceful decolonization and put future political decision-making in the hands of 'responsible' new leaders. For Jackson Davis, the associate head of the GEB and a close personal friend of Wrong, 'responsible' in the case of Africa seemed also to mean 'Christian'.

Towards an imagined global Christian community?

For more than two decades, Africa was the focus of Wrong's personal and professional interest. The result was a triangle of travels and contacts linking the African sub-continent to Britain and North America. Yet Wrong's outlook is perhaps best understood as that of a Christian humanitarian with a global vision, or, to borrow a phrase used by Jeffrey Cox, a 'Christian universalist'.[29] The contacts and perspectives of Wrong, Oliver and their colleagues in international and ecumenical work reflected and contributed to an increasing consciousness of an interlinked world of Christian believers, an 'imagined community' of global dimensions.[30] Another borrowed phrase, this one used in reference to cooperation between German-speaking and British missionaries in the early Church Missionary Society, also seems apt for the men and women associated with the IMC and its various offshoots: 'an *International* of Protestant activists'.[31]

A perception of being part of a global Christian community was not confined solely to those engaged directly in ecumenical work, nor wholly a phenomenon of the interwar era. Even for missionaries mainly engaged in local denominational work (they remained, of course, the majority) a larger identity could come about as a result of reading mission journals, participating in conferences, holidaying in ecumenical mission sites or travelling during furloughs to missions in other lands. This last was perhaps especially important. Single Canadian women missionaries were presumably not unusual in the way they organized their furlough agendas to allow time for visits with friends serving in other missions, travelling, for instance, from India to China or from Korea to India or Africa. Directly experiencing the geography of missions, they could later convey to those at home their perception of far-flung, diverse communities united in world Christendom. Missionaries also conveyed to their home churches an understanding that they had particular obligations to help besieged Christian communities, whether in the face of natural disasters or man-made tragedies. The Armenian 'genocide' was one such event that resonated with Canadians. While no Canadian church had its own work in Turkey, Canadian Presbyterian and Congregational missionaries serving there under other boards took a lead in alerting the Canadian public to the persecution of Armenian Christians, first in the late 1890s and again from 1915. Responding to wartime reports of mass persecution of the Armenians, the Anglican Primate of Canada joined other churchmen in calling for relief drives and, later, assistance to refugees.[32]

A sense of being part of an international Christian community was also a possibility for the handful of elite non-Western Christians who began to attain leadership roles in the 1930s and 1940s and broaden their contacts through the international mission bureaucracies. The Ghanaian C. G. Baëta (1908–1994), for instance, was a delegate from the Gold Coast in 1938 to the IMC conference at Tambaram, where non-Westerners were present in significant numbers and by no means a silent subaltern group. Baëta would later become vice-chairman of the IMC and in that role help orchestrate its merger into the World Council of Churches (WCC) in 1961. Also present from Africa at Tambaram was Thompson Samkange, whose nationalist- and Christian-inflected family history has been presented in a riveting account by Terence Ranger. Back home in Southern Rhodesia (Zimbabwe), Samkange experienced many disappointments and humiliations at the hands of white churchmen untouched by the Tambaram spirit. Nonetheless, he had come away from the conference with a sense of belonging to, and acceptance in, a larger Christian community. Naming his farm 'Tambaram' signified the importance of the experience.[33]

But what, if anything, did being part of a global faith community mean to the millions of colonized Christians in Asia and Africa whose lives remained bounded by their town or village and by deeply rooted traditions and identities? In the case of late-colonial and newly independent India, Indian Christians' adherence to the colonizers' religion constituted a significant barrier to their acceptance by the Hindu majority as authentically Indian and genuinely national. This was deeply ironic, since, in India as elsewhere, indigenous Christians' understanding and practice of their faith was often dramatically different from that of either colonial officials or the missionaries themselves, grafted as it was onto pre-existing concepts and local agendas.[34] Given such challenges, could any but the most privileged and cosmopolitan of Asian and African Christians develop a sense of Christian community that was not wholly parochial? In Chapter 12, in this volume, Elizabeth Prevost makes a compelling case for the Anglican Mothers' Union as an organization that operated at a grass-roots level to foster extra-local Christian identities. Focusing on the participation of Malagasy women in the MU, she argues that, especially after the First World War, the organization facilitated the growth of a sense of international Christian sisterhood based on vibrant spiritual links as well as women's shared concerns as mothers.[35]

This was precisely the kind of message that Margaret Wrong was seeking to convey in her *Five Points for Africa*, published in 1942. 'The spread of Christianity', she wrote, 'brings to Africans even in very remote parts some sense of being part of a world community'.[36] Wrong's comment and her illustrative anecdotes (a reference, for instance, to African schoolchildren contributing to China relief) need to be understood in the context of her concern to portray Africans as loyal contributors to the war effort justifiably anticipating a more prosperous and racially fairer post-war Africa than had followed the previous world war. At a time when rival European powers were once again carrying their conflicts into parts of the world where they had previously established fraternal Christian missions there were clearly many challenges to presenting a convincing vision of a global Christian community. Yet in the context of wartime Africa, it evidently seemed to Wrong more important than ever to proffer such a vision and to consider what it meant in terms of Western Christians' understandings of, and future actions on, vital questions of race, economics and colonial rule. Likewise, at his enthronement in 1942 as Archbishop of Canterbury, William Temple spoke of a new world fellowship of Christian peoples as 'the great new fact of our era'. For Temple, this 'great new fact' was a compelling reason for the establishment of a world council of churches, the first assembly of which would take place at Amsterdam in 1948.[37]

Attempts to vivify a concept of global Christianity probably contributed to a gradual decline in the most egregious forms of racism in missionary discourse and to positive changes in interpersonal and interracial relationships. Kevin Ward is undoubtedly correct in stating that, in the case of Africa, it was only in the 1960s that the 'colour bar' in social relations between Africans and Europeans, including missionaries, 'eroded beyond all recognition'.[38] Yet it seems important to recognize the significance of earlier steps in that direction and to consider the international and ecumenical institutions, organizations and conferences on which I have been focusing as a likely milieu for such change. Missionary paternalism undoubtedly persisted even within these sites. But the Asian and African elites who became involved had a level of education and a familiarity with a world beyond their own town or village that made it difficult to regard them as backward or primitive or in need of indefinite mentoring. The kind of infantilizing of the missionized other even when he or she was an adult that scholars such as Nicholas Thomas and, for Northern Canada, Myra Rutherdale, have noted for the nineteenth and early twentieth centuries[39] was becoming increasingly unthinkable in these newer and more cosmopolitan contexts.

With regard to Margaret Wrong and Choné Oliver, there appeared to be a genuine desire to promote cross-race friendships, both as the 'right' thing to do and for personal reasons. Thus, in the London home that Wrong shared with her companion, the anthropologist Margaret Read (1889–1991), the two kept a kind of open house for African and Asian visitors. Much of their hosting activity particularly with students, perhaps fit into the 'good works' category. But Wrong's relationships with the non-Westerners she met through her committee work seem to have been more collegial. And when such men – they were usually men – had occasion to travel to North America, she sometimes directed them to her father or her diplomat brother, Hume, expecting them to play the role of host and good-will agent on their side of the Atlantic.[40] In India, meanwhile, Oliver had a professional interest in obtaining Dr Hilda Lazarus for the headship of the planned Christian medical college. But she also knew Lazarus as, like herself, an earnest Christian and an enthusiastic gardener. For Oliver, these were appropriate bases on which to seek a personal as well as a professional bond.

Beyond the missionary era: Legacies in development

Writing about Christianity in post-colonial Africa in his contribution to *World Christianities, c.1914–c.2000,* David Maxwell focuses on the phenomenal growth in the number of Christian adherents since the 1960s, especially as a result of the activities of 'born again' African and North American missionaries. But he also deals with the historic mission churches and the challenges they faced during decolonization and the immediate post-colonial period. Many of them, he writes, 'NGOised', acting as agents of modernization as in the past, but now often 'adopting the same language and priorities of development as the states themselves'.[41]

Certainly, the United Church of Canada can be said to have 'NGOised' in the 1960s in terms of its role in the developing world. New terminology was the outward sign of changes that indigenous Christian leaders and forward-thinking missionaries were seeking, usually with the support of home-base officials but, perhaps inevitably, some resistance from traditionalists. In 1962, the Board of Overseas Missions became the Board of World Mission (BWM). The change in title was accompanied by another modernizing step, the integration of the formerly separate Woman's Missionary Society into the new BWM.[42] A decade later, the term 'mission' was itself jettisoned as the BWM became the Division of World Outreach (DWO).

Aspects of the new order were evident from the beginning of the 1960s: in a policy, for instance, of sending staff to overseas churches only on the basis of specific requests, typically with technical or professional expertise and as 'partners' or 'overseas personnel'. The latter term officially replaced the word missionary in the DWO lexicon only in 1989. As time passed, these would increasingly be short-term workers, and their numbers would steadily decline.[43] Some Asian countries and their churches, it was recognized, might no longer want missionaries at all. In East Africa and Angola, meanwhile, high-profile missionaries were speaking out in support of non-violent struggles for independence from colonial rule and calling on the BWM to do likewise.[44] By the time that the church's Commission on World Mission, appointed in 1962, issued its report four years later, its recommendations were to some extent already being put into practice. They included a call for a broader, more flexible and more ecumenical approach to the church's 'task in mission'; repentance for 'all arrogance, whether racial, cultural or ecclesiastical'; and 'dialogue with people of other faiths'. As well as partnership with autonomous churches, there was to be cooperation with governments and international aid organizations in a mission field now understood to be of global dimensions.[45]

To Donald K. Faris, writing to his son in 1951, the kinds of transformations that took place in the United Church's understanding and practice of missions during the 1960s were, if not inconceivable, frustratingly distant. An ordained missionary in China from 1925 to 1942, Faris had quickly shifted his focus from evangelism to rural development. From 1945 to 1949 he had served on UN rehabilitation and relief projects in China, and in the next two decades he would administer UN-sponsored development projects elsewhere in Asia. But his clear preference in 1950 was to return to China under United Church auspices to do more work in rural development. When that door closed with the Communists' victory, he sought similar work in India. His vision of grass-roots agricultural initiatives linked to government goals and undertaken in a consultative but unconfining relationship with the church's India mission was not supported. Facing this reality, he concluded that the kind of work he wanted to do and for which his experience had prepared him was

> something that the church has no machinery or thinking prepared to use. It forces me to a conclusion that ... my best contribution to the world now, perhaps is outside the church organization. ... The exodus from China and growing sentiment in all countries of the world ... make the older concept of missions a complete impossibility.[46]

In making his personal journey from missionary to development worker Faris also influenced a younger generation of globally minded Canadians. His 1958 book, *To Plow with Hope*, undertaken initially to enlarge his own understanding of development, contained a brief passage that served as a call to action for an idealistic University of Toronto graduate student with an interest in the developing world. Keith Spicer's subsequent organizational efforts at the university resulted in the formation of the first contingent of overseas volunteers in what would become one of Canada's first and most successful secular NGOs, Canadian University Service Overseas (CUSO). Established in 1961 under the umbrella of a national organization of university administrators, CUSO sent university graduates and professionally trained volunteers to developing countries for two-year terms, typically to work in educational, medical or community development roles at local rates of pay.[47]

Anxious to ensure its secular identity, CUSO's organizers kept church and mission leaders at arms length during their planning meetings. Nonetheless, in the first decade mission-linked advisers and mission institutions, Protestant and Catholic, played an important role the selection, orientation and deployment of CUSO volunteers. Although placement decisions were made on the basis of requests from the host country, volunteers sometimes wound up serving in institutions founded by missions. The fact that such positions generally came with no religious strings attached made them acceptable to most volunteers. A large number were themselves the products of a church upbringing. Yet it seems clear that few would have wanted a mission appointment even if they had realized that the churches were in the process of significantly broadening their understanding of mission, for by the time they graduated and joined CUSO many volunteers had loosened or altogether abandoned their church ties.[48]

As secularization and other social changes proceeded in Canada in the late 1960s and as returned volunteers brought the ferment of decolonizing societies back home, CUSO became a more politicized organization. Its Ottawa office increasingly staffed by returned volunteers, it was critical of the federal government's limited response to problems of global underdevelopment and just as critical of its own idealism and early naiveté. Still, traces of the missionary legacy lingered. Indeed, what was in some respects CUSO's most radical phase took place in the mid-1970s under an executive director who was a mish kid.[49] Moreover, there was now a significant degree of convergence in the overseas concerns of Canadian churches and secular NGOs. The new understanding of world mission that shaped the United Church of Canada and other mainline Canadian churches from the 1960s enabled them to make common cause with organizations like CUSO on such matters as foreign aid, anti-apartheid activism and peace issues.[50]

By the late twentieth century, most Canadians had little or no awareness of the period when the mainstream Protestant churches had proudly participated in overseas missions. On the rare occasions when the era was recalled to public consciousness, it was likely to be in the context of an anxious reconsideration of Canada's role in British imperialism, most notably in 1989–1990 in connection with the Royal Ontario Museum's controversial 'Into the Heart of Africa' exhibition.[51] Even within the churches, there seemed to be little institutional memory of the American rather than the British complexion of Canada's early foreign missionary endeavours; of their gendered and ecumenical dimensions; and of missionaries' increased attentiveness, from the First World War onwards, to the practical needs, capabilities and aspirations of Asian and African peoples. That this attentiveness did not go far enough or come soon enough was signalled by Faris in 1951 and, a decade later, by the young, secular idealists who joined CUSO and by many missionaries themselves. Working from within the churches and affected by pragmatism as well as the *zeitgeist* of the 1960s, the liberal missionaries of this decade sought what were unquestionably dramatic changes in the discourse and practice of mission. But as this chapter has shown, they were not so much breaking a previously static mould as drawing on a legacy of evolutionary change. As for CUSO, despite its concern to keep the missionary image at a distance, it found serviceable elements in the legacy. Meanwhile, for those at the receiving end a century after it all began there was perhaps not much discernable difference between Christian and secular development work.

Notes

1. Kevin Ward, 'Christianity, Colonialism and Missions', in *Cambridge History of Christianity, Vol. 9, World Christianities, c.1914–c.2000*, ed. Hugh McLeod (Cambridge: Cambridge University Press, 2006), pp. 71–88, quotation at 86. The assumption is most fully analysed in Andrew Porter, *Religion versus Empire? British Protestant Missionaries and Overseas Expansion, 1700–1914* (Manchester: Manchester University Press, 2004). See also Jeffrey Cox, 'Master Narratives of Imperial Missions', in *Mixed Messages: Materiality, Textuality, Missions*, eds. Jamie S. Scott and Gareth Griffiths (New York: Palgrave MacMillan, 2005), pp. 3–18.
2. See Chapters 5 and 9 by Breitenbach and O'Brien respectively.
3. See especially Chapters 13 and 6 by Bateman and MacKenzie respectively.
4. Roberto Perin, 'French-Speaking Canada from 1840,' in *A Concise History of Christianity in Canada*, ed. Terrence Murphy and Roberto Perin (Toronto: Oxford University Press, 1996), 190–259.
5. For detailed references to sources referred to in this chapter, see Ruth Compton Brouwer, *New Women for God: Canadian Presbyterian Women and India Missions, 1876–1914* (Toronto: University of Toronto Press, 1990); *Modern Women Modernizing Men: The Changing Missions of Three Professional Women in Asia and Africa, 1902–69* (Vancouver: UBC Press, 2002); 'Shifts in the Salience of Gender in the International Missionary Enterprise during

the Interwar Years', in *Canadian Missionaries, Indigenous Peoples: Representing Religion at Home and Abroad*, ed. Alvyn Austin and Jamie S. Scott (Toronto: University of Toronto Press, 2005), pp. 152–76.

6. Alvyn J. Austin, *Saving China: Canadian Missionaries in the Middle Kingdom, 1888–1950* (Toronto: University of Toronto Press, 1986); Brouwer, *New Women;* Rosemary R. Gagan, *A Sensitive Independence: Methodist Women Missionaries in Canada and the Orient, 1881–1925* (Montreal: McGill-Queen's University Press, 1992); A. Hamish Ion, *The Cross and the Rising Sun: The Canadian Protestant Missionary Movement in the Japanese Empire, 1872–1931* (Waterloo: Wilfrid Laurier University Press, 1990). For a useful overview produced by the churches after the First World War see H. C. Priest, ed., *Canada's Share in World Tasks* ([Toronto]: Canadian Council of the Missionary Educational Movement, 1920).
7. Brian Clarke, 'English-Speaking Canada from 1854,' in *Concise History,* 263.
8. Carl Berger, *The Sense of Power: Studies in the Ideas of Canadian Imperialism, 1867–1914* (Toronto: University of Toronto Press, 1970).
9. See Mark McGowan, *The Waning of the Green: Catholics, the Irish, and Identity in Toronto, 1887–1922* (Montreal: McGill-Queen's University Press, 1999).
10. Norman Russell, 'Shall We Retreat?' *Presbyterian Record,* 22 (August 1897) 202.
11. Torben Christensen and William R. Hutchison, eds, *Missionary Ideologies in the Imperialist Era, 1880–1920* (Aarhuis, Denmark: Farlaget Aros, 1982), Introduction.
12. Lillian Emery Wanless, *Wanless of India: Lancet of the Lord* (Boston, MA: W. A. Wilde, 1944).
13. Nicholas Thomas, 'Colonial Conversions: Difference, Hierarchy, and History in Early Twentieth-Century Evangelical Propaganda', in *Cultures of Empire: A Reader,* ed. Catherine Hall (Manchester: Manchester University Press, 2000), pp. 298–328.
14. R. Craig Brown and G. R. Cook, *Canada, 1896–1921: A Nation Transformed* (Toronto: McClelland and Stewart, 1974), pp. 50, 79.
15. David M. Thompson, 'Ecumenism', in *World Christianities,* p. 63. Some one-third of Presbyterians remained in the Presbyterian Church in Canada, now generally a more conservative denomination; John Webster Grant, *The Church in the Canadian Era* (Burlington: Welch Publishing, 1988), p. 128.
16. Neil Semple, *The Lord's Dominion: The History of Canadian Methodism* (Montreal: McGill-Queen's University Press, 1996), ch. 16.
17. Semple, *The Lord's Dominion,* p. 439; John S. Moir, *Enduring Witness: The Presbyterian Church in Canada,* new edn. (np: Eagle Press Printers for Presbyterian Church in Canada, nd), pp. 229–31.
18. Brouwer, *Modern Women,* ch. 2; Austin, *Saving China; Canada's Share in World Tasks,* pp. 52–74.
19. By contrast, missions to Canada's aboriginal peoples were viewed as appropriate destinations for candidates considered too old or ill prepared to be sent overseas. See, for instance, Gagan, *Sensitive Independence,* ch. 5.
20. William Richey Hogg, *Ecumenical Foundations: A History of the International Missionary Council and Its Nineteenth-Century Background* (New York: Harper & Brothers, 1952); Ruth Rouse and Stephen Charles Neill, eds, *A History of the Ecumenical Movement, 1517–1948* (London: Society for the Promotion of Christian Knowledge, 1954), especially ch. 8.

21. There is no recent scholarly biography of Mott, but the two men's distinctive approaches are discussed in Keith Clements, *Faith on the Frontier: A Life of J. H. Oldham* (Edinburgh: T & T Clark, 1999).
22. Barbara Bush, 'Gender and Empire: The Twentieth Century', in *Gender and Empire*/Oxford History of the British Empire Companion Series, ed. Philippa Levine (Oxford: Oxford University Press, 2004), especially pp. 89–90, and '"Britain's Conscience on Africa": White Women, Race and Imperial Politics in Inter-War Britain', in *Gender and Imperialism*, ed. Clare Midgley (Manchester: Manchester University Press, 1998), pp. 200–23.
23. What follows on Wrong is drawn mainly from Brouwer, *Modern Women*, ch. 4.
24. *Books for Africa* (January 1931), p. 1.
25. What follows on Oliver is drawn mainly from my *Modern Women*, ch. 2.
26. Ruth Brouwer, 'The Varieties of Religious Experience in an India Medical Missionary: Belle Choné Oliver', *Touchstone*, XXIII, no. 2 (May 2005) 41–51.
27. Ruth Compton Brouwer, 'Learning and Teaching about Birth Control: The Cautious Activism of Medical Missionaries in India', in *Rhetoric and Reality: Gender and the Colonial Experience in South Asia*, ed. Avril Powell and Siobhan Lambert-Hurley (New Delhi: Oxford University Press, 2005), pp. 154–84.
28. J. D. Hargreaves, *Decolonization in Africa* (London: Longman, 1988), pp. 61–2.
29. Review of Catherine Hall's *Civilising Subjects: Metropole and Colony in the English Imagination, 1830–1867*, in *Journal of Imperial and Commonwealth History*, XXXII, no. 1 (January 2004) 122–5.
30. The phrase 'imagined community' is, of course, borrowed from Benedict Anderson's *Imagined Communities: Reflections on the Origin and Spread of Nationalism* (London: Verso, 1983).
31. Paul Jenkins, 'The Church Missionary Society and the Basel Mission: An Early Experiment in Inter-European Cooperation', in *The Church Missionary Society and World Christianity, 1799–1999*, ed. Kevin Ward and Brian Stanley (Grand Rapids, MI: William B. Eerdmans, 2000), pp. 43–65, quotation at p. 50; drawn to my attention through Andrew Porter's review in *Journal of Imperial and Commonwealth History*, XXIX, no. 1 (January 2001) 129–30.
32. Isabel Kaprielian-Churchill, *Like Our Mountains: A History of Armenians in Canada* (Montreal: McGill-Queen's University Press, 2005), pp. 141–9.
33. *The World Mission of the Church: Findings and Recommendations of the International Missionary Council*, Tambaram, Madras, India, December 12–29, 1938 (London: International Missionary Council [1939]); J. S. Pobee, ed., *Religion in a Pluralistic Society: Essays Presented to Professor C. G. Baëta* (Leiden: Brill, 1976), preface and ch. 10; Terence O. Ranger, *Are We Not Also Men? The Samkange Family and African Politics in Zimbabwe, 1920–64* (Harare: Baobab, 1995), ch. 3, entitled 'Tambaram: A Re-making'.
34. Recent works that deal with the indigenization and hybridization of Christianity in mission settings are Jeffrey Cox, *Imperial Fault Lines : Christianity and Colonial Power in India, 1818–1914* (Stanford: Stanford University Press, 2002) and Elizabeth Elbourne, *Blood Ground: Colonialism, Missions, and the Contest for Christianity in the Cape Colony and Britain, 1799–1853* (Montreal: McGill-Queen's University Press, 2002).
35. See Chapter 12 by Elizabeth Prevost in this volume.

36. *Five Points for Africa* (London: Edinburgh House Press, 1942), p. 18.
37. Thompson, 'Ecumenism', quotation at 51.
38. Ward, 'Christianity, Colonialism and Missions', p. 86.
39. Thomas, 'Colonial Conversions'; Myra Rutherdale, 'Mothers of the Empire: Maternal Metaphors in the Northern Canadian Mission Field', in *Canadian Missionaries*, pp. 46–66.
40. Brouwer, *Modern Women*, pp. 100, 106, 166–7, n. 28. Hume Wrong served in Washington for many years, becoming ambassador in 1946.
41. 'Post-colonial Christianity in Africa', *World Christianities*, pp. 401–21, quotation at p. 411.
42. United Church of Canada/Victoria University Archives (UCA), acc. no. 83.010C, Board of World Mission, United Church of Canada, 1962, box 1, file 1, Minutes of BWM Interim Executive Committee, 18 January 1962 and of General Meeting, 9 April 1962; Donna Sinclair, *Crossing Worlds: The Story of the Woman's Missionary Society of the United Church of Canada* (Toronto: United Church Publishing House, 1992).
43. Rebekah Chevalier, 'Where Have All the Missionaries Gone?' in *Fire & Grace: Stories of History and Vision*, ed. Jim Taylor (Toronto: United Church Publishing House, 1999), pp. 37–42. The number of United Church missionaries declined from 602 in 1927 to 261 in 1964. By 1998 the total number of overseas personnel was down to 35. Similar changes characterized the continuing Presbyterian Church in Canada, though the language of mission and gendered structures lasted longer. See Moir, *Enduring Witness*, pp. 272–3, and Lois Klempa and Rosemary Doran, *Certain Women Amazed Us: The Women's Missionary Society/Their Story/1864–2002* (Toronto: Women's Missionary Society [WD], 2002).
44. See UCA, BWM Minutes, box 1, file 1, 1962, interim, executive, and general board minutes, for the range of issues being considered in that year. See also *The United Church Observer*, Vol. 24, which provides a useful window on actual and proposed changes in overseas mission policies and the reactions of *Observer* readers.
45. *World Mission/Report of the Commission on World Mission* (Toronto: General Council, United Church of Canada, 1966), 'Recommendations', pp. 135–9.
46. Donald K. Faris papers, uncatalogued, privately held, Faris to son Ken, 5 February 1951, from Hong Kong. Other indications of Faris's frustrations with the church and with his own uncertain future in the wake of the Communist victory in China can also be found in this collection.
47. *To Plow with Hope* (New York: Harper & Brothers, 1958), pp. 202–3; Keith Spicer, *Life Sentences: Memoirs of an Incorrigible Canadian* (Toronto: McClelland and Stewart, 2004), ch. 5; Ian Smillie, *The Land of Lost Content: A History of CUSO* (Toronto: Deneau , 1985).
48. Information about volunteers is drawn from CUSO fonds, MG28 I 323, at Library and Archives Canada, Ottawa, and from my personal or electronic interviews with CUSO alumni.
49. Smillie, *Lost Content*, ch. 7, 'The Mahatma'.
50. See, for instance, Bonnie Green, ed., *Canadian Churches and Foreign Policy* (Toronto: James Lorimer, 1990), especially Introduction, and 'The World Church and the Search for a Just Peace'.

51. Jeanne Cannizzo, *Into the Heart of Africa* (Toronto: Royal Ontario Museum, 1989); Linda Hutcheon, 'The Post Always Rings Twice: The Postmodern and the Postcolonial', *Material History Review*, 41 (Spring 1995) 4–23; Barbara Lawson, 'Collecting Cultures: Canadian Museums, Pacific Islanders, and Museums', in *Canadian Missionaries*, pp. 235–61, Introduction.

Further reading

Austin, A. and J. S. Scott, eds. *Canadian Missionaries, Indigenous Peoples: Representing Religion at Home and Abroad* (Toronto: University of Toronto Press, 2005).

Austin, A. J. *Saving China: Canadian Missionaries in the Middle Kingdom, 1888–1950* (Toronto: University of Toronto Press, 1986).

Berger, C. *The Sense of Power: Studies in the Ideas of Canadian Imperialism, 1867–1914* (Toronto: University of Toronto Press, 1970).

Brouwer, R. C. *Modern Women Modernizing Men: The Changing Missions of Three Professional Women in Asia and Africa, 1902–69* (Vancouver: UBC Press, 2002).

Brouwer, R. C. *New Women for God: Canadian Presbyterian Women and India Missions, 1876–1914* (Toronto: University of Toronto Press, 1990).

Christensen, T. and W. R. Hutchison, eds. *Missionary Ideologies in the Imperialist Era, 1880–1920* (Aarhuis, Denmark: Farlaget Aros, 1982).

Ion, A. H. *The Cross and the Rising Sun: The Canadian Protestant Missionary Movement in the Japanese Empire, 1872–1931* (Waterloo: Wilfrid Laurier University Press, 1990).

McGowan, M. *The Waning of the Green: Catholics, the Irish, and Identity in Toronto, 1887–1922* (Montreal: McGill-Queen's University Press, 1999).

Murphy, T. and R. Perin, eds. *A Concise History of Christianity in Canada* (Toronto: Oxford University Press, 1996).

Semple, N. *The Lord's Dominion: The History of Canadian Methodism* (Montreal: McGill-Queen's University Press, 1996).

Thomas, N. 'Colonial conversions: difference, hierarchy, and history in early twentieth-century evangelical propaganda'. In *Cultures of Empire: A Reader*, ed. Catherine Hall (Manchester: Manchester University Press, 2000), pp. 298–328.

15

Empire and Religion in Colonial Botswana: The Seretse Khama Controversy, 1948–1956

John Stuart

Introduction

Western enthusiasm for 'planting' Christianity in Africa was often matched by that of indigenous peoples, in appropriating Christianity for their own ends.[1] In the case of the Ngwato kingdom, one of eight such polities in what is now the Republic of Botswana; African appropriation of Christianity in the late nineteenth century had important, long-term implications for religion and for empire in southern Africa. Here religion undoubtedly served, thwarted, transformed, mitigated and even at times reinforced the bonds of empire. In a major study of these complex processes, Paul Stuart Landau has described how 'an originally tiny "Ngwato" polity' wrested 'a form of ecclesiastical statehood from the expressions and habits propounded by a missionary society, expanded its own tenuous loyalties into a kingdom, and flourished for decades in the environment of British imperialism'.[2] Empire, along with Christianity, undoubtedly helped sustain this kingdom, and Ngwato kings were adept at playing off Western interests – imperial, commercial and religious – against each other. But changing imperial circumstances in the aftermath of the Second World War also helped to bring about the kingdom's demise. A crucial factor in this occurrence was the highly controversial marriage that took place in London in 1948, between Seretse Khama, heir to the Ngwato kingdom, and Ruth Williams, a white English woman. The marriage precipitated a crisis in the Ngwato kingdom and in British imperial affairs, details of which Susan Williams has recently documented in her book on Seretse and Ruth.[3] However, the marriage was no less a crisis for the local church and for that British Protestant mission organization from which the people of the Ngwato – the Bangwato – had originally wrested a form of ecclesiastical statehood and with which they had since lived in uneasy harmony: the London Missionary Society (LMS), an avowedly interdenominational but essentially Congregationalist body. This chapter relates the story of a notable crisis of religion and empire in late colonial Africa.

The London Missionary Society and the Bangwato

For a long time, the history of British Protestant overseas missions was essentially a history of Western evangelistic endeavour. The LMS was well represented in this respect, by John Philip, David Livingstone and John MacKenzie amongst others. In the early 1860s, the Scottish-born MacKenzie befriended Khama (*c*.1835–1923), king (or kgosi) of the Bangwato and recent Christian convert. MacKenzie was keen to extend the reach of Christianity, and Christian education and medicine were of value to Khama as an African leader. Following his accession to power in 1875 Khama decreed that the LMS would become his kingdom's sole representative Christian institution, in effect a state church.[4] There was too an imperial dimension to Bangwato affairs. Worried about white settler intentions towards African territories, MacKenzie and the LMS successfully lobbied government in London for the creation of what became in 1885 (and remained until Botswana's independence in 1966) the Bechuanaland Protectorate. Ten years later and again with the assistance of the LMS, Khama, with two fellow-African leaders, journeyed to Britain and secured from imperial government and monarch a promise of continued protection.

The actions of the LMS have gone down in history, as an example of late Victorian 'missionary imperialism'.[5] Yet LMS and Congregationalist attitudes towards empire were nothing if not varied and prone to change over time.[6] Primarily committed to the building up of Christianity in the Bangwato Reserve (as that part of the Protectorate was from 1885, officially defined), missionaries continued to support its inhabitants against depredatory white settler interest particularly that of the neighbouring Union of South Africa, self-governing from 1910. Its government in Pretoria wished to incorporate the Protectorate within South Africa's frontiers, along with the two other High Commission Territories of Basutoland (now Lesotho) and Swaziland. The concern of the LMS about such an eventuality was echoed by that of other Protestant mission and ecumenical bodies.[7] Yet no missionary could ever be quite certain that Britain, under pressure from Pretoria, might *not* renege on its promise to protect African interests. The situation of the Bangwato was complicated. Imperial 'protection' never included adequate financial or economic provision. So dire was the situation in the Reserve in the 1930s that most able-bodied adult men had to leave, to seek work in the Union. Secure, at least, in its 'monopoly' religious status the LMS had meanwhile been slowly expanding its presence in the Reserve, mainly through the efforts of African evangelists: the first African clergy were ordained in 1910. The missionaries' relations with the politically astute Khama were never less than fraught.[8] Khama's caginess was born of expediency; Bangwato kings had ever to be mindful of their situation, for kingship was often a matter of dissension and even dispute. As it happened Khama died in February 1923 to be succeeded by his son, Sekgoma. But Sekgoma died in turn only 20 months later. His son and heir, Seretse Khama, was only four years old. The Bangwato appointed Sekgoma's half-brother, the 20-year-old Tshekedi Khama, to rule on his behalf. Representatives of the LMS, both in the Protectorate and in Britain, subsequently supported Tshekedi, in a campaign against mining concessions in the Protectorate and against continuing South African attempts at incorporation.[9] Tshekedi was devout, strong-willed and as capable as Khama had been of utilizing the LMS to his advantage. His intention was not personal aggrandizement as such; he was committed to the interests of all his people and keen to entrust a strong legacy to Seretse. By his imperious, sometimes overbearing manner, however, Tshekedi nevertheless antagonized elements of Bangwato society. This would prove immensely troublesome for him when his leadership came under challenge in the aftermath of Seretse's marriage to Ruth. The LMS would also be affected; having for so long associated closely with Tshekedi as acting king the Society risked being identified with his interests rather than with those of the Bangwato as a whole.

Mission, church and the Khamas in the aftermath of war

In 1945 the LMS was taking stock of the post-war situation. Several thousand men from the Bechuanaland Protectorate had served during the war with the African Auxiliary Pioneer Corps and the Society was working to expedite their demobilization and return home from North Africa where most were stationed.[10] Since 1939 wartime conditions had necessitated restrictions on overseas missionary work. With the ending of hostilities, LMS officials and supporters in Britain intended that it should resume evangelism at a greatly increased level. As in the past, mission work in Africa, India, China and the South Pacific would largely be funded by contributions from Congregational church members in the denomination's strongholds in England and Wales. The success of a special fundraising venture initiated in 1942 appeared to confirm strong Congregationalist interest in overseas mission: within three years contributions of almost £100,000 had been received, for a 'New Advance' in mission, which, it was hoped, could commence in 1945. During that year the LMS, which had been founded in 1795, would celebrate its triple jubilee and a century and a half's commitment to mission.

The overall mood within the LMS was one of expectation mixed with thankfulness. The sudden ending of the war in Asia in August 1945 meant that missionaries interned by the Japanese would be repatriated sooner than expected. A series of events celebrating the triple jubilee were held in London in September culminating in a service of thanksgiving in Westminster Abbey.

Seretse Khama, now aged 25 and still heir to a kingdom, was a guest at several of these events, during which he dutifully acknowledged the benefits that the LMS had brought to the Bangwato.[11] But his recent arrival in London had nothing to do with the Society's celebrations; he was intending to enrol as a student of law at Balliol College, Oxford, to enable him to complete his education and, thereby, his preparations for kingship. Taking a parental-like interest in Seretse's progress and hoping that the LMS would display similar concern, Tshekedi was extremely disappointed to learn that his nephew had apparently been used by the Society for 'propaganda' purposes.[12] Outwardly less displeased than Tshekedi, Seretse nonetheless appeared less than enamoured of the attentions of the LMS. Unlike his uncle, who espoused a somewhat puritanical Christian faith, Seretse was not overtly religious. Yet the Society's motives, even if questionable, seem understandable enough. For missionaries and supporters of the LMS Seretse represented a living link between the historic past and the anticipated future of the church in Africa. Arguably, no LMS member in Britain felt this more strongly than Harold Moody, the black, Jamaican-born medical doctor who as well as being a director of the LMS was also president and founder of the League of Coloured Peoples. A Congregationalist, Moody's political ideas were strongly influenced by his Christian faith.[13] He had high expectations of Seretse as a young, African Christian leader.[14] Moody hoped that Seretse would form strong personal links with a local Congregational church in London. Moody had done so to good effect at the time of his own arrival in Britain some forty years earlier. In the event Seretse stayed for a time with Moody and his family in southeast London, before leaving for Oxford. He made little attempt to keep in touch either with the LMS or Moody thereafter.

The Oxford experience was a fiasco. Although Seretse was a graduate of the South African Native College at Fort Hare, he had insufficient Latin to be able to undertake his studies. Balliol hardly knew what to do with him. He abandoned Oxford and in 1947 returned to London to study for the Bar, enrolling with the Inner Temple. Strangely, in the light of what was to follow, the LMS indirectly facilitated his initial meeting with Ruth Williams, who worked as a clerk at a London firm of underwriters. The meeting occurred at a function organized by the Society and held at a hostel for colonial students.

Such hostels were by no means a new idea. The London-based West African Students' Union had been promoting them since the 1920s.[15] But during the war missionary societies began taking a much more active interest in visiting students from Africa, with the aim of improving 'race relations'. The Empire's catastrophic military defeat at Singapore in February 1942 had undoubtedly damaged white prestige, a development of which missionaries even in Africa were very aware.[16] The subsequent arrival in Britain of thousands of black American servicemen was the cause of no little unease, in official and religious circles. By 1942 also, the view had formed within the LMS that the British government would reward South Africa for its participation in the war by allowing it, after the cessation of hostilities, to finally incorporate the High Commission Territories. Consequently, missionary societies strove to emphasize as strongly as possible their commitment to the indigenous peoples of Britain's colonies. In 1945 they published a 'manifesto' on racial discrimination.[17] They also gave financial and administrative support to student hostels, such as that at Nutford House in central London where in June 1947 Ruth Williams's sister, Muriel, introduced her to Seretse.[18] The attraction was mutual and apparently instantaneous. Things moved quickly thereafter. In 1948, despite unease on the part of Ruth's parents, the couple decided to marry.

It may well have been the case, as Susan Williams has suggested, that Seretse had become aware during his time in England of how African antagonism to empire was growing. The cautious, reforming approach typified by Moody was being supplanted by something more radical, as evidenced by the proceedings of the Pan-African Congress, held at Manchester in October 1945.[19] Yet Seretse does not at this time appear to have possessed any recognizably political outlook. He was, to be sure, continuing with his studies; but, he was arguably less concerned with his political and religious inheritance than were Tshekedi and the LMS. Both uncle and missionary society had a vested interest in Seretse living up to their differing, yet interrelated expectations. Those expectations made no allowance for Seretse's marriage to anyone other than someone of his own race.

The Bangwato too had high hopes of Seretse not least as successor to Tshekedi, who had been at his most demanding during the war. The raising of a Bangwato detachment of the Pioneer Corps (accompanied by an LMS missionary and African clergy) had been his idea.[20] This display of African loyalty was intended to counteract a renewed post-war request from South Africa for control of the Protectorate. But South Africa also had designs on incorporating into the Union the former German colony and League of Nations mandated territory of South West Africa (now Namibia). Early in 1946 South African Prime Minister Jan Smuts petitioned the recently formed United Nations (UN) accordingly. Tshekedi immediately acted against Smuts, seeking the assistance of the LMS and other organizations (such as the Anti-Slavery Society) in Britain. The LMS headquarters, which had no knowledge of the South West Africa situation, found Tshekedi's request incomprehensible.[21] Its missionaries in the Protectorate, weary of his politicking tendencies, meanwhile hoped that the Ngwato succession would not be long delayed. Tshekedi subsequently formed an alliance with the radical Anglican priest Michael Scott, who personally took the case of the Herero people of South West Africa to the UN.[22]

A marriage and its consequences

Tshekedi's action was but one example of how, in the aftermath of the Second World War, African leaders were becoming more assertive and outward looking, sometimes to the discomfort of missionaries. For his part Seretse was in England ostensibly to study – not to find a partner. For the Bangwato no less than for other African societies marriage was a means of allying families, of creating and strengthening bases of authority and influence. It was an important matter, the form and timing of which Bangwato leaders were understandably much concerned about. In September 1948 Seretse wrote to Tshekedi (addressing him as 'Father'), to inform him for the first time of his intention to marry Ruth. Unprepared for this news, Tshekedi's reaction was one of disappointment and anger. Such a marriage would at the very least offend his racial sensibilities. Worse, it would also threaten the succession. Seretse had not consulted with his people, and they had not given their consent to his plans. Tshekedi made every effort to halt the marriage. So too did officials of the LMS, acting on his behalf. In so doing they strained further a relationship with Seretse that had never been less than uneasy since the time of his arrival in England. Being adults – Seretse was now 27 and Ruth was 24 – the couple were, however, entitled to marry. Having failed to obtain permission for a church wedding they finally married at Kensington register office, on 29 September 1948. Ruth defiantly wore black.[23]

For Tshekedi's legal representative, the Cape Town-based King's Counsel Douglas Buchanan, the problems presented by the marriage did not appear insuperable. Seretse, he thought, would shortly return to the Protectorate to explain himself to his people, at a kgotla, or council, which would be convened for the event. Buchanan contacted Rev. Ronald Orchard, the LMS secretary for Africa, informing him that the Bangwato would hardly accept any son of the marriage as a future king: 'when the parties come to a fuller realisation of what is involved', wrote Buchanan, 'a divorce by reasons of desertion might still be the best solution should Seretse return to South Africa and leave his wife in England'.[24] This was not at all an outcome wished by Orchard, although he personally doubted Seretse's judgement in marrying so hastily and without consultation. The LMS had no policy as such towards interracial marriages. But its officials took the view that these should be entered into with great care on the part of those involved, and only after having partaken of spiritual advice and counsel from clergy.

At this time Orchard had made but one official visit to Africa, in 1947–1948. He was heavily reliant for information about Bangwato affairs on two overseas colleagues. These were the veteran Rev. Alfred Haile, the LMS Board representative for the region, based near Bulawayo in Southern Rhodesia (now Zimbabwe), and the relative newcomer, Rev. Alan Seager, who, with his wife Ruth, was based at the mission in Serowe, the Ngwato capital.[25] Haile had taught Seretse at secondary school and was on close personal terms with Tshekedi. He was also a regular visitor to the Bangwato Reserve. The responsibility for coordinating any public response that the LMS might wish to make to the outcome of the kgotla would, however, fall to Orchard in London.

Seretse returned alone to Bechuanaland, and the kgotla, in which only adult males were allowed to participate, took place at Serowe over four days in mid-November. There, criticism was duly levelled at Seretse, not merely by Tshekedi but by representatives of other peoples in the Protectorate. He responded eloquently and emotionally, declaring his commitment both to Ruth and to his people. Tshekedi, conversely, was insistent that Seretse had neglected his duty as heir and his tone was equally emotional: 'I am your father...I have looked after the Tribe for you'[26] The outcome was indecisive, as was that of a further kgotla, held in December. Seretse travelled back to England (he was facing Bar examinations), intending to return for yet another kgotla in mid-1949. Seager had been in attendance at the proceedings, the kgotla customarily being opened and closed with prayer. He noted presciently 'for the next few months at least, probably years, there is going to be so much intrigue that any reasonable progress in Tribal affairs will cease'.[27] Seager now began to solicit local opinion, as did Haile further afield, in the Protectorate's other kingdoms. Kgari, king of the Bakwena, believed Seretse in the wrong. He had noticed nonetheless (he informed Haile) that at the second kgotla support for Tshekedi had been much less evident than expected. Further informal consultations with Bangwato confirmed this view and much else besides: many people, especially women (who were excluded from the kgotla) would apparently welcome Seretse back, even accompanied by Ruth. Haile's advice to Orchard was firm: on no account should the LMS evince open support for Tshekedi, to do so would risk dividing the local church.[28]

Orchard, however, was in a quandary, for Tshekedi, dissatisfied and frustrated with his inability to win over the kgotla, now decided to play on his long-standing connection with the LMS. Suspecting that Seretse might attempt to bring Ruth out to the Protectorate and that this might weaken his own support further, Tshekedi wrote to Orchard, asking him to intercede with the department of government directly responsible for Protectorate affairs, the Commonwealth Relations Office (CRO), to prevent such a visit. Tshekedi also suggested that the LMS liaise with other British organizations likely to be sympathetic to his cause, namely the Anti-Slavery Society and the Fabian Colonial Bureau. Wary of dealings with secular organizations, Orchard was positively dismayed by Tshekedi's recommendation to contact Michael Scott, his collaborator on the South West Africa question.[29] The outspoken Scott was the last person with whom Orchard wished to consult, given his by now well known penchant for 'stirring up...public agitation'.[30] Orchard did make a discreet approach to the CRO, and met informally with one of its officials in an attempt to discern the government's line on the Bangwato succession. He did not learn much; the CRO appeared interested only in avoiding undue controversy.[31] Orchard kept Tshekedi informed of his dealings with civil servants. He was determined to keep lines of contact open, fearful that Tshekedi might decide to rely exclusively on secular interlocutors, who might seek to make political capital out of the dispute.

Tshekedi had no qualms about asking for assistance from British organizations, either secular or religious. During the 1930s he had successfully attracted the support of those in Britain concerned not only about South African intentions towards the Protectorate but also about official overreaction to an incident in 1933 when he ordered the flogging of a white resident of Serowe.[32] In 1949, Tshekedi's situation was very different with his appeals for help apparently designed to entrench his position against that of Seretse rather than being for the benefit of the Bangwato.

The dispute over the Bangwato succession was meanwhile one of the many problems that the LMS now faced in 1949. The much-vaunted 'New Advance' of the post-war era had stalled. New recruits were not forthcoming, and the Society's finances were being seriously affected by monetary inflation in colonial territories. The devaluation of sterling in September 1949 would cause further financial difficulty. Worse, in its way, was the situation in China. The completion of the Communist takeover would presage the expulsion of all foreign missionaries whatever their denomination or nationality.

None of this could necessarily have been foreseen. And only gradually were British missionaries coming to terms with the theological challenges offered by 'separatist' indigenous churches, such as those in southern Africa noted by the Swedish missionary and theologian Bengt Sundkler in his influential 1948 book *Bantu Prophets in South Africa*. Orchard and his colleagues at LMS headquarters took note of Sundkler's work but they relied still on what they perceived as the old certainties, when missions, although primarily committed to evangelization, considered themselves by dint of their long-standing work amongst African peoples best placed to represent the interests of those peoples to colonial and imperial authority. In Africa, missionaries often adopted 'official' roles, as representatives of African interests in colonial legislatures. The situation in the Bechuanaland Protectorate, particularly in the Bangwato Reserve, was different. There, political and religious authority alike was mediated through complex, shifting arrangements between African leader, church and colonial administrators.[33] But the central political institution was the kgotla, at which the people were represented and to which the king was answerable. In June 1949 the kgotla finally came to a conclusive if unexpected decision on Seretse's fitness to succeed Tshekedi. By a decisive majority the kgotla accepted and acclaimed Seretse as king. Tshekedi, defeated, pledged to go into internal exile with his supporters.[34] The question of succession was now ostensibly settled but Seretse would also require confirmation by the imperial authorities of his right to rule. Would this be forthcoming?

The crisis widens

In Serowe, Seager moved quickly to improve relations between Seretse and the church, and offered to put Ruth up at his family house when she arrived in the Reserve. Seretse accepted the offer, although Seager and Ruth Khama do not appear to have warmed to one another.[35] Seager was more worried about the impact of Tshekedi's exile on the church, especially the loss to it of the services of Tshekedi's wife Ella and sister Bonyerile, both active Christians. Would it be politic at this time, he wondered, for the church to provide Tshekedi's party with a minister of its own? And what in any case would be Tshekedi's attitude to the church in future? There still appeared scope for dispute.[36] Seager's fear of this was soon realized. One evangelist, Seakgano Ncaga, accused him and his African colleague Rev. Joshua Danisa of collaborating against Seretse. In response Seager called a meeting of church members from Serowe and Tonota at which he and Haile vigorously asserted the church's neutrality in political matters.[37]

Church neutrality in the Ngwato kingdom was a fiction, as Seager well knew. But with the Bangwato still ostensibly divided it was important that the church not appear overtly supportive of Seretse. Tshekedi was after all still acting regent. However, the Labour government in London appeared in no hurry to accord recognition to Seretse; it had indeed decided in July 1949 to hold a judicial inquiry into the succession. This was a means by which to buy time. In the wake of the decision of the kgotla the government of South Africa had made representations to the CRO, arguing that it should refuse to recognize Seretse. The government of Southern Rhodesia also lobbied against recognition. Their reasons were varied, but the underlying rationale was Afrikaner and Rhodesian hatred and fear of miscegenation. Since coming to power in 1948 the ruling National Party in South Africa had begun legislating in support of apartheid. In 1949 it passed a law against 'mixed marriages'; in 1950, it would legislate against interracial sexual intercourse. Desperate to avoid offending South Africa for fear that it might threaten to leave the Commonwealth government in London hoped that the judicial inquiry might find against Seretse's right to rule. It duly did so, but it cited South African and Rhodesian hostility as important factors.[38] The government had no wish to make this known and did not publicize the report's findings, to the frustration of the LMS. Orchard had been lobbying the CRO discreetly in a futile attempt to seek clarification as to whether South Africa would be allowed to incorporate the High Commission Territories. As cagey in his way as any Bangwato leader, undersecretary of state Patrick Gordon Walker would say no more than that such an outcome was 'unlikely'.[39] In November 1949, after much deliberation Orchard organized an ecumenical deputation of Protestant church and mission representatives, headed by the Bishop of London, to the CRO. It requested that government increase colonial welfare and development funding to the High Commission Territories to improve the lives of their inhabitants.

The deputation had been organized for a particular reason: to demonstrate publicly the continuing commitment of British Protestant churches and missionary societies to the interests of African colonial peoples. It was an essentially defensive gesture. It achieved nothing. South Africa's formal instituting of apartheid had provoked strong public criticism from only a few British clergymen such as Michael Scott and Father Trevor Huddleston. Protestant churches and missionary societies remained publicly silent, partly in deference to the wishes of South African religious leaders such as Geoffrey Clayton, Archbishop of Cape Town, who insisted that South Africa's churches be allowed to fight their own battles with the state, free from outside interference.[40] In lobbying, via the Bishop of London's deputation, for the peoples of the Bechuanaland Protectorate rather than for Africans in South Africa British missionaries were on what they imagined to be less contentious ground. So it still seemed as 1949 drew to a close with the future of the High Commission Territories and of the Bangwato succession still unresolved.

But for Prime Minister Clement Attlee and his Labour government colleagues, exhausted after four and a half arduous years in government, the succession question had now become a nuisance. In February 1950, Seretse returned to London at the government's request. There, Secretary of State for Commonwealth Relations Philip Noel-Baker asked him to voluntarily relinquish his claim to the kingship. Government would then institute direct colonial rule over the Bangwato. Seretse refused to comply. A general election then took place in Britain, which Labour won narrowly. Gordon Walker replaced Noel-Baker as secretary of state, and he was determined to resolve the matter. Following an abortive meeting with Seretse Gordon Walker obtained cabinet approval for Seretse's banishment from the Protectorate. Tshekedi would be banished from the Bangwato Reserve. A district commissioner would rule in his stead as 'Native Authority'. Gordon Walker duly informed Seretse in person and two days later, on 8 March, made a statement in the House of Commons, denying, untruthfully, that South Africa had influenced the decision. Uproar ensued in Parliament and beyond.

'The lid has blown well and truly off', Orchard wrote to Haile in despair.[41] His own cautious attempts to keep the LMS in touch with developments had been exposed as useless. For some time now, events had been moving more swiftly than he or anyone else in the Society realized. Also, Seretse had pre-empted Gordon Walker by means of a statement to the press, expressing his disappointment at how he had been treated. To Orchard's further chagrin, Seretse was now besieged by offers of support and assistance from every quarter, some of it politically motivated. The cricketer Learie Constantine helped form a Seretse Khama Fighting Committee. The Labour MP Fenner Brockway became involved. Orchard was appalled; he firmly believed that Brockway and others were seeking to manipulate Seretse for their own ends. But Orchard was out of his depth. Seretse meanwhile accepted with alacrity assistance from secular organizations even to the extent of sharing a platform with Brockway at a London rally. Insistent meanwhile that LMS headquarters retain a 'neutral' position Orchard refrained from any attempts to contact either Seretse or Tshekedi directly. But the Society was now itself being subjected to innumerable requests from its supporters, from Congregational churches, and from the press to make known publicly its attitude to the government's decision. Orchard reluctantly drafted a statement. In this he emphasized that the LMS was concerned equally for all the Bangwato; and he requested that government reconsider its decision. The statement also formed the basis for a public pronouncement by the British Council of Churches (BCC) and a personal letter from Archbishop of Canterbury Geoffrey Fisher to Prime Minister Attlee.[42] 'I would have much preferred', Orchard admitted to Haile, 'to avoid any kind of public statement'.[43]

The Bangwato were overwhelmingly dismayed by news of the banishing. There was some disorder, and a boycott of official activities. This was a time when leadership from missions and church appeared vital. The growing controversy appeared to demonstrate instead the fragility of the church's hold on the Bangwato. For Haile, the LMS statement had given insufficient indication of the 'hurt' felt by Africans: the 'stock of British justice' in Africa, he reported, was 'about as low as it could be'.[44] Danisa, in Serowe, meanwhile reported that amongst church members the motives of the LMS were being openly questioned. This was causing particular difficulty for African clergy, who were being expected by church congregations to take sides, not merely on the succession issue but also on local disputes over land and livestock.[45] Bangwato women increasingly made their protests heard.[46] In June they wrote to Orchard imploring the LMS to give more of a lead 'in this sad state of affairs'.[47] Church evangelists with whom Haile was in regular contact reported from Serowe that the Bangwato were despondent and discouraged.[48] These were circumstances in which a dispirited Christian community might, he thought, become susceptible to entreaties from other 'rival' Zionist or Pentecostal churches.[49] The missions, as before, placed great reliance upon African clergy such as Danisa and his colleague Rev. Odirile Mogwe but training of clergy and lay evangelists alike had failed to keep pace with the needs of the church. Its situation appeared parlous. Other problems loomed also. Despite Gordon Walker's public assurances to the contrary, many people in Britain suspected South African involvement in the decision to banish Seretse. Missionaries also believed that the Dutch Reformed Church in South Africa had pressured its government to act. Pretoria was reviewing its African education policy and in 1949 the LMS received devastating news that non-Union students might no longer attend schools within South Africa's frontiers. Such a decision would inflict a grievous blow on the Society's prestigious secondary school and teacher-training facility at Tiger Kloof in northern Cape Province, where Seretse and many others from the Protectorate (and thus of course from outside the Union) had been pupils.[50]

By this time Seretse and Ruth with their infant daughter Jacqueline were in Britain, in exile. On furlough in England, Seager worked at reconciling Seretse and Tshekedi.[51] It is difficult to ascertain quite how effective were Seager's entreaties but by August 1950 Tshekedi was confirming his willingness to 'join hands' with Seretse, so that they might mount a joint argument for reconsideration of their case by the CRO.[52] The reconciliation did not last, and relations between the two men remained difficult.

Imperial politics and missionary dilemmas

Tshekedi was no longer acting regent of the Bangwato but he chafed against British government attempts to institute new forms of authority in the Reserve. In March 1951 he travelled to London to argue for his reinstatement and to be allowed to return to the Reserve. He did not attract support from third parties in Britain on anything like the scale of Seretse. Even Michael Scott, who was sympathetic to both men, found it difficult to effect a rapprochement.[53] Scott received no encouragement from the LMS.[54] Tshekedi, on his best behaviour, strove now to create a good impression. At a meeting with representatives of the Society in July he outlined his hopes of persuading government to provide for economic development of the Protectorate. He also addressed a meeting of Protestant missionaries of various denominations at their ecumenical headquarters, Edinburgh House, in central London. Following Tshekedi's visit to LMS headquarters Orchard considered the situation and wrote to Haile:

> I am just a little uneasy about our policy of neutrality... lest it be interpreted as a lack of courage on the part of the Mission... I am anxious to be sure that neither pressure of work nor fear of consequences lets us be false to the tradition of John Philip, John Mackenzie and all our other notable predecessors.[55]

Earlier that year Orchard had obtained from a CRO official private assurances that transfer of the High Commission Territories to South Africa was now 'politically unthinkable' and this may have induced him to feel more sanguine about the Protectorate's future.[56] Yet there is no evidence of an approach by the LMS to Seretse, and he had by now ceased to maintain contact with its head office. Tshekedi meanwhile continued to make progress with his representations; the government agreed to send to Bechuanaland a team of unofficial observers, who would report on the situation there.[57] But their visit became a farce, and the conclusions in the ensuing, unpublished report were overshadowed by yet another general election in Britain in October, won this time by the Conservative Party.

The Bangwato were optimistic, but uncertain. Then Tshekedi, having earlier returned to the Protectorate, flew back to London and successfully petitioned the new government for an end to his exile. This caused fresh unease in the Reserve: might Tshekedi now attempt to manipulate the post of 'Native Authority'? The situation was complicated further by unrest amongst district officers at Serowe.[58] In February 1952, a group of Bangwato decided that a deputation should visit London to request that Seretse be allowed to return to his people, even for a probationary period. Opinion in the Reserve as to this course of action was by no means unanimous. Lengthy debate ensued before it was finally decided that a deputation should indeed go, to petition the CRO. Seager provided its members with a message from the church to the LMS in London.[59] Before the arrangements could be finalized the Conservative government made in late March an unexpected decision of its own: to make Seretse's banishment permanent.

The decision came as a further blow, not least to the Bangwato. It was made, as had been that of the Labour government two years later, for geopolitical reasons, in essence to allow Britain to maintain good relations with South Africa and Southern Rhodesia. 'Isn't it stupid', Haile wrote, 'that the first people to object to this mixed marriage were the Ba[ma]ngwato themselves, and the Govt. might have acted at once on the strength of that support'. His opinion that 'Tshekedi is at the bottom of the Govt.'s decision' was widely shared in the Reserve.[60] The Bangwato deputation's departure acquired new urgency. In London the LMS Africa Committee had already decided to make discreet representations to the CRO. Seager was worried about this decision. He thought that it would be badly received by the Bangwato, if ever they learned of it. They might suspect collusion of some sort between mission and government.[61] But Orchard, like many British clergymen had by now concluded that in banishing Seretse the government was primarily motivated by a desire to appease South Africa. At a meeting of the BCC in Belfast, it was arranged that Fisher, on behalf of Protestant churches and missionary societies, should lead a deputation to meet with new Secretary of State for Commonwealth Relations, Lord Salisbury, to discuss the government's decision. This deputation, under the auspices of the BCC, would, Orchard hoped, usefully complement the more discreet representations being made to the CRO by the LMS.

What began in September 1948 with a marriage entailing difficulty for an African Christian kingdom and church in southern Africa had by early 1952 developed into an imperial crisis. Since 1949, with the decision of the kgotla in favour of Seretse and consequent lobbying by South Africa and Southern Rhodesia, matters relating to the Bangwato, once so susceptible to influence by African king and church and Western mission, had moved inexorably beyond the control of each. Now the CRO primarily dictated policy. By the time that Fisher, accompanied by Orchard and other church and mission representatives, met with Salisbury on 9 May the secretary of state had already held two meetings with the six-man Bangwato deputation, ultimately turning down their request that Seretse be allowed to return to the Reserve. The religious representatives' deputation proved no more fruitful. Salisbury merely reiterated the official line: the decision about Seretse had been taken in the interests of order and stability in the Reserve.[62]

The CRO had remained stubbornly immune to the private and public entreaties of churches and missionary societies. It was pointless, Orchard now concluded, for British religious organizations to press Seretse's case any further; there was no likelihood of further progress. He had lately been much exercised by government plans for a federation of colonial territories in central Africa with Southern Rhodesia at its head, which was intended to act as a counterweight to South African influence in the region. Far better, Orchard now thought, that British churches and missionary societies should focus instead on ensuring that African interests in those territories were not jeopardized by such plans. Representations could be made to the Colonial Office, and these were likely to be more productive than those that had just been made to the CRO.[63]

By this time the Bangwato deputation was about to return home. Orchard had met on three separate occasions at LMS headquarters with Moutlwatsi Mpotokwane, one of its members and its main spokesman. At the first meeting on 1 May, which other LMS officials also attended, Mpotokwane had been particularly keen to dispel any idea that the Bangwato were disunited: opinion was solidly behind Seretse, he reported. The government, he asserted, was guilty of 'hypocrisy'. Only at the two subsequent meetings, on 5 and 7 May, at which Orchard was the sole LMS representative present, did Mpotokwane speak more openly about mission and church affairs (the two men had previously met, during Orchard's visit to Africa). He conceded that the Bangwato were worried not merely about government attitudes but about those of the LMS also. Why, he now asked, had the Society been so reticent about supporting Seretse openly? Could it not have been more forthcoming, given its long-standing links with the Bangwato? Orchard could not have found this line of questioning anything but difficult. He had no wish to make known his own recent entreaties to the CRO. He attempted to reassure Mpotokwane that the LMS was committed equally to all the Bangwato. Mpotokwane did not appear convinced and suggested to Orchard that the Society risked its motives being misinterpreted.[64] A week later the deputation left London to return to Bechuanaland. They were angry and dissatisfied with the British government, and unsure of the attitude of the LMS towards Seretse.

What was the attitude of the LMS and Seretse towards each other by 1952? Essentially, their relationship had become estranged, a development that had been underway since Seretse first arrived in Britain seven years earlier but which had accelerated following his marriage to Ruth, an event of which the Society's senior officials disapproved and which they had tried to stop. It is likely that Orchard hoped privately that the marriage would not work out, but he could hardly admit as much openly. Perhaps most worrying from his point of view was the way in which the controversy surrounding the marriage and the Bangwato succession had become increasingly politicized. By 1952 the bipartisanship which is often held to have characterized post-war party political attitudes to empire in Britain had all but collapsed, in part a consequence of crises in colonies such as Kenya. Even as Orchard and his ecumenical colleagues planned their meeting with Salisbury at the CRO the Seretse Khama Campaign Committee (the epithet 'Fighting' having been dropped) was attempting to rouse public opinion on behalf of Seretse and against government. Orchard was suspicious of the motives of this organization and even Seretse appeared ambivalent. The Committee, Stephen Howe has suggested, may have been susceptible to communist influence.[65] In any case, Brockway set up another organization with a much broader remit. This Council for the Defence of Seretse Khama and the Protectorates appealed to a cross section of people in Britain beyond the world of party politics. It was a significant development, as was the formation also in 1952 of the Africa Bureau (in which Michael Scott was involved), again with the intention of facilitating closer British public engagement with colonial affairs.[66] Until the Second World War Protestant missionary societies had been able to lay fair claim to being the most knowledgeable and even influential unofficial representatives in Britain of African colonial interests. This was no longer the case, even by 1950. It was not the Bangwato alone who suspected that missions were too closely linked with empire to be capable of criticizing it openly: the failure of the LMS to vigorously challenge either apartheid in South Africa or official treatment of Ruth and Seretse seems to have been a cause of bewilderment and even dismay to its Christian constituency of supporters in Britain.[67] Yet there was little sign of regret in Orchard's answer to an enquiry from Seager about Seretse in early June 1952: 'I am not in touch with him', Orchard candidly admitted, 'and he has made no effort to get into touch with us'.[68]

Eleven days after the arrival of the Bangwato deputation back in the Reserve dissatisfaction with government, which until this point had mostly taken the form of non-co-operation, flared into violence. Three African policemen were killed.[69] Seager vainly attempted to intervene. The church was attracting criticism now from supporters of Tshekedi as well as of Seretse. From London Orchard wearily acknowledged that despite discomfort and even danger Seager had at least 'been successful in pursuing a policy of neutrality'. The unrest, he conceded, might at least force government to take more seriously its supposed commitment to order and stability in the Reserve.[70] Gradually, the tension eased a little. In May 1953, the government installed Rasebolai Kagamane as 'Native Authority'. He was not of the house of Khama and the decision did not meet with complete Bangwato approval. Yet Seager and his African colleagues were now able to focus more clearly on church affairs, which included revision of the constitution of the courts of the local church, thus permitting greater African representation and participation. Seager also renewed contact with Tshekedi, now settled back in the Protectorate, but outside the Reserve at Pilikwe. Seager wished to improve relations between Tshekedi (and his supporters) and the church. This proved difficult, and Tshekedi, keen to have the medical and educational benefits that other religious organizations might provide, would eventually invite Anglican and Roman Catholic missions into the territory.[71] The 1952 disturbances and continuing residual tensions affected the church in other ways. Independent African churches gradually increased their presence in the Reserve and throughout the Protectorate. The final ending of the Khamas' exile in 1956 – which necessitated Seretse renouncing his claim to leadership of the Bangwato – coincided with further independent church growth ultimately at the expense of the LMS, which for so long had been in effect the state church of an African kingdom.[72]

Conclusion

The fortunes of mission and church in the Bangwato Reserve were inextricably linked with those of the Ngwato kingdom. Ngwato rulers from Khama onwards proved capable of utilizing the church as well as colonial authority to their advantage. In 1945 both Tshekedi and the LMS expected things to continue largely as they had before. The marriage of Seretse to Ruth ensured that this would not be the case. The harshness of Tshekedi's authority was now turned against him not least by Bangwato women, many deprived during the difficult wartime period of husbands and sons serving in the Pioneer Corps. They were especially keen to have Seretse as their king. For Olufemi Vaughan, the controversy surrounding the marriage of Ruth and Seretse revealed the profundity of tensions 'between tradition and modernity' amongst the Bangwato.[73] There were tensions too between religion and empire. This had always been the case, since MacKenzie's time. 'Nothing happens here that is *not* political', one missionary had written from the Reserve in the early twentieth century.[74] As a result of the marriage and the succession question, however, the LMS found itself in an irreconcilable dilemma. Its head office staff attempted to resolve an imperial and religious problem by appealing, as so often in the past, directly to government. But after 1945, government was less receptive than in the past to religious concerns in general and to those of missionary societies in particular. The situation in colonial Africa meanwhile presented its own difficulties for missions. Within the Bangwato Reserve mission and church had been important influences and the LMS had benefited from its unusual status. Its missionaries tended to perceive themselves not only as evangelists first but also as guardians of African interests, in the missionary and 'imperial' traditions of Livingstone and MacKenzie. On several occasions they had been able to argue with apparent success against the incorporation of the Bechuanaland Protectorate by South Africa. Orchard and his ecumenical colleagues in Britain were keen to continue that argument in the 1940s and early 1950s. In reality, however, the High Commission Territories were never truly in danger of incorporation. It suited Britain to keep the question open, as a bargaining ploy in negotiations with South Africa. Seretse Khama was banished for reasons of expediency. The LMS petitioned government in an ultimately hopeless cause. Its unwillingness to openly criticize government for its treatment of Ruth and Seretse lost its support in Britain and, more importantly, in the Bangwato Reserve also.

Notes

1. R. Gray, *Black Christians and White Missionaries* (New Haven, CT: Yale University Press, 1990), pp. 59–78.
2. P. S. Landau, *The Realm of the Word: Language, Gender and Christianity in a Southern African Kingdom* (Portsmouth, NH and London: Heinemann, 1995), p. xvi.
3. S. Williams, *Colour Bar: The Triumph of Seretse Khama and His Nation* (London: Allen Lane, 2006). See also R. Hyam, 'The Political Consequences of Seretse Khama: Britain, the Bangwato and South Africa, 1948–52', *Historical Journal*, XXIX, no. 4 (1986) 921–47; M. Dutfield, *A Marriage of Inconvenience: The Persecution of Ruth and Seretse Khama* (London: Unwin Hyman, 1990).
4. Landau, *Realm*, p. 172.
5. B. Stanley, *The Bible and the Flag: Protestant Missions and British Imperialism in the Nineteenth and Twentieth Centuries* (Leicester: Apollos, 1990), pp. 116–21.
6. D. W. Bebbington, *The Nonconformist Conscience: Chapel and Politics, 1870–1914* (London: Allen & Unwin, 1982), pp. 112–15.
7. J. H. Oldham to Rev. A. M. Chirgwin, 3 May 1934, LMS file, Conference of British Missionary Societies/International Missionary Council Papers at School of Oriental and African Studies, London (SOAS), 1234.

8. N. Goodall, *A History of the London Missionary Society, 1895–1945* (London: Oxford University Press, 1954), p. 259.
9. M. Crowder, *The Flogging of Phinehas McIntosh: A Tale of Colonial Folly and Injustice, Bechuanaland 1933* (New Haven, CT and London: Yale University Press, 1988), pp. 18–22.
10. LMS to Secretary of State for War, 4 October 1945, Sandilands file, Council for World Mission Papers (CWM) at SOAS, AF13.
11. LMS Board of Directors meeting minutes, 19 September 1945, CWM BM/1.
12. Rev. A. J. Haile to Chirgwin, 16 October 1945, Haile file, CWM AF8.
13. A. S. Rush, 'Imperial Identity in Colonial Minds: Harold Moody and the League of Coloured Peoples, 1931–50', *Twentieth Century British History*, XIII, no. 4 (2002) 370–2.
14. Dr H. A. Moody to Rev. T. C. Brown, 27 July 1945, CWM AF37/79A.
15. G. O. Olusanya, *The West African Students' Union and the Politics of Decolonisation, 1925–58* (Ibadan: Daystar Press, 1982), pp. 16–45.
16. Haile, report to LMS Board of Directors, 1942, CWM AF/44.
17. Texts of archbishop's and other addresses to 'colour bar' public meeting, Westminster Central Hall, London, 12 June 1945, Statement and Meeting file, Conference of British Missionary Societies Papers at SOAS (CBMS) 260.
18. The Williams family was Anglican. Muriel, one year older than Ruth, had been evacuated to Wales during the war. She became a Congregationalist having stayed with a family of that denomination.
19. Williams, *Colour Bar*, pp. 11–13.
20. A. Jackson, *Botswana, 1939–45: An African Country at War* (Oxford: Oxford University Press, 1999), pp. 115–20.
21. T. Khama to LMS, 21 May 1946, CWM AF30/47A.
22. M. Crowder, 'Tshekedi Khama, Smuts and South West Africa', *The Journal of Modern African Studies*, XXV, no. 1 (1987) 25–42.
23. Williams, *Colour Bar*, pp. 16–26.
24. D. M. Buchanan to Rev. R. K. Orchard, 5 October 1948, CWM AF37/79A.
25. The other LMS Bechuanaland missions were at Kanye, Maun and Tonota. The LMS also maintained missions in Southern Rhodesia, Northern Rhodesia (now Zambia) and Madagascar.
26. Seager to Orchard, 22 November 1948, CWM AF37/79A.
27. Seager to Orchard, 30 December 1948, CWM AF30/47C.
28. Haile to Orchard, 17 May 1949, CWM AF37/79A.
29. T. Khama to Orchard, 25 April 1949, CWM AF30/47D.
30. Orchard to Haile, 13 May 1949, CWM AF37/79A.
31. Orchard, memo on meeting with H. N. Tait, 20 May 1949, CWM AF30/47D.
32. B. Bush, *Imperialism, Race and Resistance: Africa and Britain, 1919–45* (London: Routledge, 1999), pp. 192–4.
33. N. Parsons, 'Colonel Rey and the Colonial Rulers of Botswana: Mercenary and Missionary Traditions in Administration, 1884–1955', in *People and Empires in African History: Essays in Memory of Michael Crowder*, ed. J. F. A. Ajayi and J. D. Y. Peel (Harlow: Longman, 1992), pp. 197–216.
34. M. Benson, *Tshekedi Khama* (London: Faber, 1960), pp. 189–92.
35. Landau, *Realm*, p. 216; Williams, *Colour Bar*, p. 79.
36. Seager to Haile, 5 July 1949, CWM AF37/79A.

37. Haile, memo, 22 February 1950, CWM AF37/79B.
38. Hyam, 'Political Consequences', pp. 925–32.
39. Orchard, notes on visit to CRO, 5 July 1949, CWM AF40/82A.
40. Most Rev. G. H. Clayton to Most Rev. G. F. Fisher, 4 March 1949, Fisher Papers at Lambeth Palace Library, London, 65.
41. Orchard to Haile, 10 March 1950, CWM AF37/79B.
42. Fisher to C. R. Attlee, 21 April 1950, Fisher Papers 67.
43. Orchard to Haile, 28 March 1950, CWM AF37/79B.
44. Haile to Orchard, 14 March 1950, CWM AF37/79B.
45. Danisa, being an Ndebele from Southern Rhodesia, was less drawn into such disputes.
46. D. Wylie, *A Little God: The Twilight of Patriarchy in a Southern African Chiefdom* (Hanover, NH and London: Wesleyan University Press, 1990), pp. 200–1.
47. Martha Mokgwathi to LMS, n.d. but *c.*1 June 1950, CWM AF37/79C.
48. Haile to Orchard, 17 April 1950, CWM AF37/79C.
49. Rev. A. Sandilands to Orchard, n.d. but *c.*1 August 1950, CWM AF37/79D.
50. Haile to Orchard, 22 November 1950, CWM AF40/82A. Tiger Kloof closed in 1954, as a result of the Bantu Education Act.
51. Seager to T. Khama, 3 May, and to S. Khama, 4 May 1950, CWM AF37/79C.
52. T. Khama to Haile, 18 August 1950, CWM AF37/79D.
53. A. Yates and L. Chester, *The Troublemaker: Michael Scott and His Lonely Struggle against Injustice* (London: Aurum, 2006), pp. 141–2.
54. Orchard to Rev. G. M. Scott, 16 March 1951, CWM AF11L.
55. Orchard to Haile, 25 July 1951, CWM AF26/17A.
56. J. P. Gibson to Orchard, 10 March 1951, CWM AF26/17A.
57. M. Crowder, 'Professor Macmillan Goes on Safari: The British Government Observer Team and the Crisis over the Seretse Khama Marriage, 1951', in *Africa and Empire: W. M. Macmillan, Historian and Social Critic*, ed. H. Macmillan and S. Marks (Aldershot: Institute of Commonwealth Studies, 1989), pp. 254–78.
58. N. Parsons, W. Henderson, T. Tlou, *Seretse Khama, 1921–80* (Gaborone: Macmillan, 1995), pp. 116–18.
59. A. Seager, *The Shadow of a Great Rock* (Connah's Quay: ID Books, 2004), pp. 75–81.
60. Haile to Orchard, 29 March, and 3 April 1952, CWM AF 33/35D.
61. Seager to Orchard, 17 April 1952, CWM AF33/35D.
62. British Council of Churches press release, 9 May 1952, CBMS 571.
63. Orchard to Haile, 12 May 1952, CWM AF33/35e. The other territories were Northern Rhodesia and Nyasaland (now Malawi).
64. Orchard, notes on meetings with M. Mpotokwane, 1, 5, and 7 May 1952, CWM AF33/35D.
65. S. Howe, *Anticolonialism in British Politics: The Left and the End of Empire, 1918–64* (Oxford: Oxford University Press, 1993), pp. 196–7.
66. Yates and Chester, *The Troublemaker*, pp. 127–30.
67. See Orchard 1950 correspondence, justifying LMS silence, CWM AF40/83A.
68. Orchard to Seager, 6 June 1952, CWM AF33/35E.
69. Seager to Haile and Orchard, 2 June 1952, CWM AF33/35F.
70. Orchard to Seager, 3 July 1952, CWM AF33/35G.

71. Seager, *Shadow of a Great Rock*, pp. 116–19.
72. J. N. Amanze, *African Christianity in Botswana: The Case of African Independent Churches*, (Gweru: Mambo Press, 1998), pp. 72–86.
73. O. Vaughan, *Chiefs, Power and Social Change: Chiefship and Modern Politics in Botswana, 1880s–1980s* (Trenton, NJ: Africa World Press, 2003), pp. 49–52.
74. Cited in B. Head, *Serowe: Village of the Rain Wind* (London: Heinemann, 1981), p. 26.

Further reading

Amanze, J. N. *African Christianity in Botswana: The Case of African Independent Churches* (Gweru: Mambo Press, 1998).

Crowder, M. *The Flogging of Phinehas McIntosh: A Tale of Colonial Folly and Injustice, Bechuanaland, 1933* (New Haven, CT: Yale University Press, 1988).

Dutfield, M. *A Marriage of Inconvenience: The Persecution of Ruth and Seretse Khama* (London: Unwin Hyman, 1990).

Howe, S. *Anticolonialism in British Politics: The Left and the End of Empire* (Oxford: Oxford University Press, 1993).

Hyam, R. and P. Henshaw. *The Lion and the Springbok: Britain and South Africa since the Boer War* (Cambridge: Cambridge University Press, 2003).

Landau, P. S. *The Realm of the Word: Language, Gender and Christianity in a Southern African Kingdom* (London: James Currey, 1995).

Tlou, T., N. Parsons and W. Henderson. *Seretse Khama, 1921–80* (Braamfontein: Macmillan, 1995).

Vaughan, O. *Chiefs, Power and Social Change: Chiefship and Modern Politics in Botswana, 1880s–1980s* (Trenton, NJ: Africa World Press, 2003).

Williams, S. *Colour Bar: The Triumph of Seretse Khama and his Nation* (London, Allen Lane, 2006).

Select Bibliography

This bibliography is intended as a brief guide to the general historiography of religion and the British Empire. More specialized reading on particular topics may be found at the end of each chapter; complete references to archival and other sources printed prior to 1900 are contained in endnotes. For more, see A. N. Porter's, *Bibliography of Imperial, Colonial, and Commonwealth History since 1600* (Oxford: Oxford University Press, 2002), pp. 479–533, and *Religion versus Empire? British Protestant Missionaries and Overseas Expansion, 1700–1914* (Manchester: Manchester University Press, 2004), pp. 331–61.

Ballantyne, T. 'Introduction: Debating Empire', *Journal of Colonialism and Colonial History*, III, no. 1 (2002).

Ballantyne, T. 'Religion, Difference, and the Limits of British Imperial History', *Victorian Studies*, XLVII, no. 3 (2005) 427–55.

Bataldon, S., K. Cann and J. Dean, eds. *Sowing the Word: the Cultural Impact of the British and Foreign Bible Society* (London: Sheffield Phoenix Press, 2006).

Bebbington, D. 'Atonement, Sin, and Empire, 1880–1914'. In *The Imperial Horizons of British Protestant Missions, 1880–1914*, ed. A. N. Porter (Grand Rapids, MI: William B. Eerdmans, 2003), pp. 14–31.

Bell, D. *The Idea of Greater Britain: Empire and the Future of World Order, 1860–1900* (Princeton, N.J.: Princeton University Press, 2007).

Bollen, J. D. 'English Christianity and the Australian Colonies, 1788–1860', *Journal of Ecclesiastical History*, XXVIII (1977) 361–85.

Breitenbach, E. 'Empire, Religion and National Identity: Scottish Christian Imperialism in the 19th and early 20th centuries', Ph D thesis (Edinburgh: University of Edinburgh, 2005).

Brock, P. 'Mission Encounters in the Colonial World: British Columbia and South-west Australia', *Journal of Religious History*, XXIV, no. 2 (2000) 159–79.

Brock, P. 'Nakedness and Clothing in Early Encounters between Aboriginal People of Central Australia, Missionaries and Anthropologists', *Journal of Colonialism and Colonial History*, VIII, no. 1 (2007).

Brown, C. L. *Moral Capital. Foundations of British Abolitionism* (Chapel Hill, NC: University of North Carolina Press, 2006).

Brown, S. J. *The National Churches of England, Ireland, and Scotland 1801–1846* (Oxford: Oxford University Press, 2001).

Brown, S. J. and T. Tackett, eds. *Enlightenment, Reawakening and Revolution 1660–1815. The Cambridge History of Christianity, Vol. 7* (Cambridge: Cambridge University Press, 2006).

Brown, S. J. *Providence and Empire: Religion, Politics and Society in the United Kingdom* (Harlow: Pearson Longman, 2008).

Buckner, P., ed. *Canada and the British Empire*, Oxford History of the British Empire, Companion Series (Oxford: Oxford University Press, 2008).

Carey, H. M. 'Religion and Society'. In *Australia's Empire*, Oxford History of the British Empire Companion Series, ed. D. Schreuder and S. Ward (Oxford: Oxford University Press, 2008), pp. 186–210.

Carey, H. M. 'Religion and the "Evil Empire"', *Journal of Religious History,* XXXII (2008) 179–92.

Carpenter, M. W. *Imperial Bibles, Domestic Bodies: Women, Sexuality and Religion in the Victorian Market* (Athens, OH: Ohio University Press, 2003).

Chambers, D. 'The Kirk and the Colonies in the Early Nineteenth Century', *Australian Historical Studies,* XVI, no. 64 (1975) 381–401.

Coleman, D. *Romantic Colonization and British Anti-Slavery* (Cambridge: Cambridge University Press, 2005).

Colley, L. *Britons: Forging the Nation, 1770–1837* (New Haven, CT: Yale University Press, 1992).

Comaroff, J. and J. L. Comaroff, *Of Revelation and Revolution: Christianity, Colonialism and Consciousness in South Africa* (Chicago, IL: University of Chicago Press, 1991).

Cooper, F. and A. L. Stoler, *Tensions of Empire: Colonial Cultures in a Bourgeois World* (Berkeley, CA: University of California Press, 1997).

Coplan, I. 'Christianity as an Arm of Empire: The Ambiguous Case of India under the Company, c. 1813–1858', *The Historical Journal*, XLIX, no. 4 (2006) 1025–54.

Cox, J. *Imperial Fault Lines: Christianity and Colonial Power in India, 1818–1940* (Standford: Standford University Press, 2002).

Cuthbertson, G. 'The English-speaking Churches and Colonialism'. In *Theology and Violence: The South African Debate,* ed. C. Villa-Vicencio (Grand Rapids, MI: William B. Eerdmans, 1987).

Daunton, M. and R. Halpern, eds. *Empire and Others: British Encounters with Indigenous Peoples, 1600–1850* (Philadelphia, PA: University of Pennsylvania Press, 1999).

Davidson, A. K. 'Colonial Christianity: The Contribution of the Society for the Propagation of the Gospel to the Anglican Church in New Zealand, 1840–80', *Journal of Religious History,* XVI, no. 2 (1996) 144–66.

Daw, E. D. *Church and State in the Empire: The Evolution of Imperial Policy, 1846–1856* (Canberra: Dept. of Government, Faculty of Military Studies, University of New South Wales, 1977).

Delavignette, R. *Christianity and Colonialism,* tr. J. R. Foster (London: Burns & Oates, 1964).

Elbourne, E. *Blood Ground: Colonialism, Mission and the Contest for Christianity in the Cape Colony and Britain, 1799–1853* (Montreal and Kingston: McGill-Queens University Press, 2002).

Elbourne, E. 'The Sin of the Settler: The 1835–36 Select Committee on Aborigines and Debates over Virtue and Conquest in the Early Nineteenth-Century British White Settler Empire', *Journal of Colonialism and Colonial History,* IV, no. 3 (2003). <http://muse.jhu.edu/journals/journal_of_colonialism_and_colonial_history/>

Elbourne, E. 'Indigenous Peoples and Imperial Networks in the Early Nineteenth Century: The Politics Of Knowledge'. In *Rediscovering the British World,* ed. P. A. Buckner and R. D. Francis (Calgary, Alta: University of Calgary Press, 2005), pp. 59–85.

Elbourne, E. 'Religion in the British Empire'. In *The British Empire: Themes and Perspectives,* ed. Sarah Stockwell (Oxford: Blackwell, 2008).

Etherington, N. ed. *Missions and Empire,* Oxford History of the British Empire, Companion Series (Oxford: Oxford University Press, 2005).

Fabian, J. 'Religious and Secular Colonization: Common Ground', *History and Anthropology*, IV (1990), 339–55.

Gascoigne, J. *The Enlightenment and the Origins of European Australia* (Cambridge: Cambridge University Press, 2002).

Gascoigne, J. 'The Expanding Historiography of British Imperialism', *The Historical Journal*, XLIX, no. 2 (2006) 577–92.

Gascoigne, J. 'Introduction: Religion and Empire', *Journal of Religious History*, XXXII (2008) 159–178.

Gavreau, M. 'The Empire of Evangelicalism: Varieties of Common Sense in Scotland, Canada, and the United States'. In *Evangelicalism: Comparative Studies of Popular Protestantism in North America, the British Isles and Beyond, 1700–1990*, ed. M. Noll and D.W. Bebbington, George A. Rawlyk (New York and Oxford: Oxford University Press, 1994), pp. 219–52.

Gilley, S. 'The Roman Catholic Church and the Nineteenth Century Irish Diaspora', *Journal of Ecclesiastical History*, XXXV, no. 2 (1984) 188–207.

Gilley, S. and B. Stanley, eds. *World Christianities, c.1815–c.1914. The Cambridge History of Christianity, Vol. VIII* (Cambridge: Cambridge University Press, 2006).

Hall, C., ed. *Cultures of Empire* (Manchester: Manchester University Press, 2000).

Hall, C. *Civilising Subjects: Metropole and Colony in the English Imagination 1830–1867* (Oxford: Polity Press, 2002).

Hall, C. and S. O. Rose, eds. *At Home with the Empire: Metropolitan Culture and the Imperial World* (Cambridge: Cambridge University Press, 2006).

Harper, M. 'Making Christian Colonists: An Evaluation of the Scottish Churches and Christian Organisations between the Wars', *Records of the Scottish Church History Society*, XXVIII (1996) 173–216.

Hart, W. D. *Edward Said and the Religious Effects of Empire* (Cambridge: Cambridge University Press, 2000).

Hastings, A. *The Construction of Nationhood: Ethnicity, Religion and Nationalism* (Cambridge: Cambridge University Press, 1997).

Hastings, A. 'Christianity and Nationhood: Congruity or Antipathy?' *Journal of Religious History*, XXV, no. 3 (2001) 247–60.

Hilton, B. *The Age of Atonement: The Influence of Evangelicalism on Social and Economic Thought, 1785–1865* (Oxford: Clarendon Press, 1988).

Ileto, R. 'Religion and Anti-Colonial Movements'. In *The Cambridge History of South East Asia, Vol. 2: The Nineteenth and Twentieth Centuries*, ed. N. Tarling (Cambridge: Cambridge University Press, 1992), pp. 197–248.

Ion, A. H. 'The Empire that Prays Together Stays Together: Imperial Defence and Religion, 1857–1956'. In *Imperial Defence: the Old World Order 1856–1956*, ed. G. C. Kennedy (London: Routledge, 2008), pp. 197–217.

Jaffary, N. E. *Gender, Race and Religion in the Colonization of the Americas* (Aldershot: Ashgate, 2007).

Jeffery, K., ed. *'An Irish Empire'? Aspects of Ireland and the British Empire* (Manchester: Mancherster University Press, 1996).

Johnston, A. *Missionary Writing and Empire, 1800–1860* (Cambridge: Cambridge University Press, 2003).

Kenny, K, ed. *Ireland and the British Empire*. Oxford History of the British Empire, Companion Series (Oxford: Oxford University Press, 2004).

Koss, S. E. 'Wesleyanism and Empire', *Historical Journal*, XVIII, no. 1 (1975) 105–18.

Latourette, K. S. *A History of the Expansion of Christianity*, 7 vols (New York and Evanston, IL: Harper & Brothers, 1937–1945).

MacKenzie, J. M. *The Scots in South Africa: Ethnicity, Identity, Gender and Race* (Manchester: Manchester University Press, 2007).

McLeod, H. 'Protestantism and British National Identity 1815–1945'. In *Nation and Religion*, ed. P. van der Veer P. and H. Lehman (Princeton, NJ: Princeton University Press, 1999).

McLeod, H., ed. *World Christianities c.1914–c.2000. The Cambridge History of Christianity, Vol. IX* (Cambridge: Cambridge University Press, 2006).

Midgley, C. *Gender and Imperialism* (Manchester: Manchester University Press, 1998).

Miller, P. N. *Defining the Common Good: Empire, Religion, and Philosophy in Eighteenth-Century Britain* (Cambridge: Cambridge University Press, 1994).

Mills, W. G. 'Victorian Imperialism as Religion'. In *The Man on the Spot: Essays on British Empire History*, ed. R. D. Long (Westport, CT: Greenwood, 1995).

Neill, S. *A History of Christian Missions*, 2nd edn (Harmondsworth: Penguin, 1986).

Neill, S. *Colonialism and Christian Missions* (London: Lutterworth Press, 1966).

O'Brien, A. *God's Willing Workers: Women and Religion in Australia* (Sydney: University of New South Wales Press, 2005).

Pocock, J. G. A. *Barbarism and Religion, Vol. 2. Narratives of Civil Government* (Cambridge: Cambridge University Press, 1999).

Porter, A. N. '"Commerce and Christianity": The Rise and Fall of a Nineteenth-Century Missionary Slogan'. *Historical Journal*, XXVIII, no. 3 (1985) 597–621.

Porter, A. N. 'Religion and Empire: British Expansion in the Long Nineteenth Century, 1780–1914', *Journal of Imperial and Commonwealth History*, XX, no. 3 (1992) 370–90.

Porter, A. N., ed. *The Oxford History of the British Empire, Vol. 3. The Nineteenth Century* (Oxford: Oxford University Press, 1999).

Porter, A. N. *Bibliography of Imperial, Colonial, and Commonwealth History since 1600* (Oxford: Oxford University Press, 2002).

Porter, A. N., ed. *The Imperial Horizons of British Protestant Missions, 1880–1914* (Grand Rapids, MI: William B Eerdmans Publishing Co, 2003).

Porter, A. N. *Religion versus Empire? British Protestant Missionaries and Overseas Expansion, 1700–1914* (Manchester: Manchester University Press, 2004).

Porter, A. N. 'Missions and Empire: An Overview, 1700–1914'. In *Missions and Empire*, ed. N. Etherington (Oxford: Oxford University Press, 2005).

Porter, A. N. 'Missions and Empire, c. 1873–1914'. In *World Christianities, c.1815–c.1914, The Cambridge History of Christianity, Vol. 8*, ed. S. Gilley and B. Stanley (Cambridge: Cambridge University Press, 2006), pp. 560–75.

Proctor, J. H. 'Scottish Missionaries and Jamaican Slaveholders', *Slavery & Abolition*, XXV, no. 1 (2004) 51–70.

Samson, J., ed. *The British Empire* (Oxford: Oxford University Press, 2001).

Samson, J. 'Are You What You Believe? Some Thoughts on Ornamentalism and Religion', *Journal of Colonialism and Colonial History*, III, no. 1 (2002).

Schreuder, D. and S. Ward, eds. *Australia's Empire*, Oxford History of the British Empire, Companion Series (Oxford: Oxford University Press, 2008).

Scott, J. S. and G. Griffiths, eds. *Mixed Messages: Materiality, Textuality, Missions* (New York: Palgrave Macmillan, 2005).

Semple, R. A. *Missionary Women: Gender, Professionalism and the Victorian Idea of Christian Mission* (Rochester, NY: Boydell, 2003).

Southerwood, W. T. *Catholics in British Colonies: Planting a Faith Where No Sun Sets* (London: Minerva Press, 1998).

Stanley, B. 'Commerce and Christianity: Providence Theory, the Missionary Movement, and the Imperialism of Free Trade, 1842–1860', *The Historical Journal*, XXVI (1983) 71–94.

Stanley, B. *The Bible and the Flag: Protestant Missions and the Imperialism of Free Trade, 1842–1860* (Leicester: Inter-Varsity Press, 1990).

Stanley, B., ed. *Christian Missions and the Enlightenment* (Richmond, Surrey: Curzon Press, 2001).

Stanley, B. and A. M. Low, eds. *Missions, Nationalism and the End of Empire* (Grand Rapids, MI: William B. Eerdmans, 2003).

Strong, R. *Anglicanism and Empire* (Oxford: Oxford University Press, 2007).

Thorne, S. *Congregational Missions and the Making of an Imperial Culture in Nineteenth-Century England* (Stanford, CA: Stanford University Press, 1999).

Thorne, S. 'Religion and Empire'. In *At Home with the Empire: Metropolitan Culture and the Imperial World*, ed. C. Hall and S. Rose (Cambridge: Cambridge University Press, 2006).

Thorne, S. 'Imperial Pieties', *History Workshop Journal*, LXIII, no. 1 (2007) 319–27.

Van der Veer, P., ed. *Religion and Nationalism in Europe and Asia* (Princeton, NJ: Princeton University Press, 1999).

Van der Veer, P. *Imperial Encounters: Religion and Modernity in India and Britain* (Oxford and Princeton, NJ: Princeton University Press, 2001).

Walker, G. 'Empire, Religion and Nationality in Scotland and Ulster before the First World War'. In *Scotland and Ulster*, ed. I. S. Wood (Edinburgh: Mercat, 1994), pp. 97–115.

Williams, C. P. 'British Religion and the Wider World: Mission and Empire, 1800–1940'. In *A History of Religion in Britain*, ed. S. Gilley and W. J. Sheils (Oxford and Cambridge, MA: Blackwell, 1994), pp. 381–405.

Wilson, K. *Island Race: Englishness, Empire and Gender in the Eighteenth century* (London: Routledge, 2003).

Wilson, K., ed. *A New Imperial History: Culture, Identity, and Modernity in Britain and the Empire, 1660–1840* (Cambridge: Cambridge University Press, 2004).

Withycombe, R. S. M. 'Mother Church and Colonial Daughters: New Scope for Tensions in Anglican Unity and Diversity', *Studies in Church History*, XXXII (1996) 427–40.

Wolffe, J. *God and Greater Britain: Religion and National Life in Britain and Ireland, 1843–1945* (London: Routledge, 1994).

Wolffe, J. *Religion in Victorian Britain, Vol. 5. Culture and Empire* (Manchester: Manchester University Press, 1997).

Wright, L. B. *Religion and Empire: The Alliance between Piety and Commerce in English Expansion, 1558–1625*, repr. edn (New York: Octagon Books, 1965).

Yeo, G. 'A Case Without Parallel: The Bishops of London and the Anglican Church Overseas, 1660–1748', *Journal of Ecclesiastical History*, XLIV (1993) 450–75.

[illegible] Missionary [illegible] [illegible], 200[illegible])
[illegible] [illegible] Press, 199[illegible]
[illegible] [illegible] *History* [illegible]
[illegible] *Imperialism* [illegible] Press, 1990) [illegible]
[illegible] *Anglican Missions and the* [illegible] [illegible]
[illegible] [illegible] Paternalism [illegible]
[illegible] University Press, 200[illegible]
[illegible] Missions and the [illegible] Press, 199[illegible]
[illegible] Cambridge [illegible] Cambridge University Press [illegible]
[illegible] *Journal*, XIII, no. 1 [illegible]
[illegible] Press [illegible]
[illegible] and Nationality in Scotland and Ulster before the First World War [illegible] (Edinburgh [illegible])
[illegible] Cambridge [illegible] 199[illegible]
[illegible] 200[illegible]
Wilson, K. (ed.) *A New Imperial History: Culture, Identity and Modernity in Britain and the Empire* [illegible] (Cambridge: Cambridge University Press, 200[illegible])
[illegible] *Mothers* [illegible] Daughters [illegible]
[illegible] America [illegible]
Wolffe, J. *God and Greater Britain: Religion and National Life in Britain and Ireland* [illegible] (London: Routledge, 199[illegible])
Wolffe, J. *The Protestant Crusade in Great Britain* [illegible] (Oxford: Clarendon Press, 19[illegible])
[illegible] (New York [illegible])
[illegible] and the Anglican [illegible] *Journal of Ecclesiastical History* [illegible]

Index